# Frontiers in Structural Biology

*(Volume 1)*

***Role of Molecular Chaperones in Structural Folding, Biological Functions, and Drug Interactions of Client Proteins***

**Edited by**

**Mario D. Galigniana**

*Laboratory of Nuclear Receptors, Institute of Biology & Experimental Medicine, Buenos Aires, Argentina*

# Frontiers in Structural Biology

*Volume # 1*

*Role of Molecular Chaperones in Structural Folding, Biological Functions, and Drug Interactions of Client Proteins*

Editor: Mario D. Galigniana

ISSN (Online): 2589-4374

ISSN (Print): 2589-4366

ISBN (Online): 978-1-68108-615-6

ISBN (Print): 978-1-68108-616-3

## General:

1. Any dispute or claim arising out of or in connection with this License Agreement or the Work (including non-contractual disputes or claims) will be governed by and construed in accordance with the laws of the U.A.E. as applied in the Emirate of Dubai. Each party agrees that the courts of the Emirate of Dubai shall have exclusive jurisdiction to settle any dispute or claim arising out of or in connection with this License Agreement or the Work (including non-contractual disputes or claims).
2. Your rights under this License Agreement will automatically terminate without notice and without the need for a court order if at any point you breach any terms of this License Agreement. In no event will any delay or failure by Bentham Science Publishers in enforcing your compliance with this License Agreement constitute a waiver of any of its rights.
3. You acknowledge that you have read this License Agreement, and agree to be bound by its terms and conditions. To the extent that any other terms and conditions presented on any website of Bentham Science Publishers conflict with, or are inconsistent with, the terms and conditions set out in this License Agreement, you acknowledge that the terms and conditions set out in this License Agreement shall prevail.

**Bentham Science Publishers Ltd.**
Executive Suite Y - 2
PO Box 7917, Saif Zone
Sharjah, U.A.E.
Email: subscriptions@benthamscience.org

BENTHAM
SCIENCE

# CONTENTS


**PREFACE** ..... i

**LIST OF CONTRIBUTORS** ..... iii

**CHAPTER 1 REGULATORY ROLES FOR HSP70 IN CANCER INCIDENCE AND TUMOR PROGRESSION** ..... 1
*Taka Eguchi, Benjamin J. Lang, Ayesha Murshid, Thomas Prince, Jianlin Gong* and *Stuart K Calderwood*
**1. INTRODUCTION** ..... 1
**2. HSP70 PROTEINS IN THE CYTOPLASM AND NUCLEUS** ..... 2
**3. MUTATION AND OVEREXPRESSION OF HSP72 IN CANCER** ..... 7
**4. HSP72 AND THE HALLMARKS OF CANCER** ..... 9
4A. HSP72 Suppresses Apoptotic Cell Death in Cancer ..... 9
4B. HSP72 and Senescence ..... 10
4C. HSP72 in Tumor Initiation and Metastasis ..... 10
4D. HsSP72 in Sustained Angiogenesis ..... 11
**5. DRUGGING HSP70 IN CANCER: ISOFORMS AND DRUGGABLE DOMAINS** ..... 11
5A. Targeting the HSP70 Substrate-Binding Domain (SBD) ..... 12
5B. Targeting the HSP70 Nucleotide-Binding Domain ..... 13
5C. Perturbation of HSP70-Protein Interactions ..... 14
**CONCLUSIONS** ..... 15
**CONSENT FOR PUBLICATION** ..... 15
**CONFLICT OF INTEREST** ..... 15
**ACKNOWLEDGEMENTS** ..... 15
**REFERENCES** ..... 15

**CHAPTER 2 USE OF COARSE-GRAINED AND ALL-ATOM MOLECULAR DYNAMICS TO STUDY HSP70 AND HSP40 CHAPERONE ACTION** ..... 23
*Ewa I. Gołaś, Magdalena A. Mozolewska, Paweł Krupa, Cezary Czaplewski, Harold A. Scheraga* and *Adam Liwo*
**INTRODUCTION** ..... 23
**METHODS** ..... 24
**RESULTS** ..... 25
Mechanism of Chaperone Cycle ..... 25
Modeling Iron-sulfur Cluster Biogenesis ..... 32
Modeling the Structure of Isu1 from Yeast ..... 33
Modeling the Structure of the Binary Isu1-Jac1 Complex and Assessing the Stability of its Interactions ..... 35
Preliminary Molecular-modeling Study of the Structure of the Isu1-Jac1-Ssq1 Ternary Complex ..... 38
**CONCLUSIONS AND OUTLOOK** ..... 41
**CONSENT FOR PUBLICATION** ..... 42
**CONFLICT OF INTEREST** ..... 42
**ACKNOWLEDGEMENTS** ..... 42
**REFERENCES** ..... 43

**CHAPTER 3 QUATERNARY STRUCTURE OF CHAPERONES FROM THE HSP70 SYSTEM DETERMINED BY SMALL ANGLE X-RAY SCATTERING (SAXS) AND ANALYTICAL ULTRA-CENTRIFUGATION** ..... 47
*Júlio C. Borges* and *Carlos H.I. Ramos*
**INTRODUCTION** ..... 47

Protein Folding and Molecular Chaperones ..... 47
Small Angle X-ray Scattering (SAXS) ..... 50
Analytical Ultracentrifugation ..... 51
The Hsp70-folding System ..... 53
Human Mitochondrial GrpE, Conformational Modification upon Hsp70 Binding ..... 55
Eukaryotic Hsp40s Types I and II, on the Position of the J Domain ..... 57
**FINAL REMARKS** ..... 64
**CONSENT FOR PUBLICATION** ..... 64
**CONFLICT OF INTEREST** ..... 64
**ACKNOWLEDGMENTS** ..... 64
**REFERENCES** ..... 64

**CHAPTER 4 STRUCTURAL CHARACTERISTICS OF THE TPR PROTEIN-HSP90 INTERACTION: A NEW TARGET IN BIOTECHNOLOGY** ..... 73
*Ana Cauerhff* and *Mario D. Galigniana*
**INTRODUCTION** ..... 74
**TPR PROTEIN CHARACTERISTICS** ..... 75
Definition and Prediction of the Sequence and Basic Structure of TPR Motifs ..... 75
Three Dimensional Structure ..... 78
*Curvature and Shape of the TPR Domain* ..... 78
*Examples of TPR Protein Structures* ..... 79
Ligand Binding ..... 80
Folding and Stability of TPR Proteins ..... 81
*Oligomerization, Stability and Biological Functions* ..... 83
*Novel TPR Protein Design* ..... 85
**SEQUENCE, FUNCTION, AND BASIC STRUCTURE OF HSP90 PROTEINS: HSP90 ALPHA AND BETA** ..... 86
Introduction ..... 86
Hsp90 Isoforms ..... 88
Sequence and Basic Structure of Hsp90 Proteins: Hsp90 α and β ..... 89
*Difference in Structure of Hsp90 α−and β−isoforms* ..... 93
Conformational Changes in Hsp90 ..... 94
*Sequence of Conformational Changes Induced by ATP Binding to Hsp90* ..... 97
**HSP90-TPR PROTEIN INTERACTIONS** ..... 98
Introduction ..... 98
Chaperone Cycle in SHRs ..... 99
Hop/Sti1 ..... 100
*Overall Structure* ..... 100
*Hop-Hsp90 Interaction: Structural and Biophysical Aspects* ..... 102
CyP40, FKBP51 and FKBP52 ..... 106
*Function* ..... 106
*CyP40 Overall Structure* ..... 107
*Cyp40-Hsp90 Interaction: Structural Studies* ..... 110
*FKBP51: Overall Structure and Implications* ..... 111
*FKBP-like Domains* ..... 112
*TPR Domain* ..... 113
*FKBP51-Hsp90 Interaction: Structural Studies* ..... 115
*FKBP52 Structure* ..... 116
*FKBP-like Domains* ..... 116
*TPR Domain* ..... 119
*Comparison Between FKBP51 and FKBP52* ..... 120

*FKBP52-Hsp90 Complex* ........ 122
*FKBP52-FK506 Complex* ........ 122
*Binding Studies of Immunophilins* ........ 123
*CyP40, FKBP51 and FKBP52 Interaction with Hsp90: Physico-chemical Binding Studies* ........ 124
*Secondary Structure and Stability of CyP40, FKBP51 and FKBP52* ........ 125
*Stability of the CyP40, FKBP51 and FKBP52* ........ 126
PP5 ........ 126
*Introduction* ........ 126
*PP5 Overall Structure* ........ 127
*TPR Domain* ........ 128
*Catalytic Phosphatase Domain* ........ 128
*Relationship Between Phosphatase Activity and Hsp90 Binding* ........ 129
*Structural Aspects of PP5-Hsp90 Interaction* ........ 130
*PP5-Hsp90 Binding Studies* ........ 132
*Structural Considerations* ........ 133
*PP5 Folding Biophysical Studies* ........ 134
CHIP ........ 135
*Introduction* ........ 135
*Structure* ........ 136
*Hsp90 C-Terminal Binding to the CHIP TPR Domain* ........ 139
*Folding and Degradation Balance Mechanism* ........ 140
**TARGETING HSP90 INTERACTIONS FOR BIOTECHNOLOGICAL APPLICATIONS** 142
Introduction ........ 142
Hsp90 and Cancer ........ 144
Inhibitors of the N-terminal Domain of Hsp90 ........ 145
Inhibitors of the C-terminal Domain of Hsp90 ........ 150
*Inhibitors of the Hsp90-TPR Interactions* ........ 151
**FINAL CONSIDERATIONS** ........ 155
**CONSENT FOR PUBLICATION** ........ 156
**CONFLICT OF INTEREST** ........ 156
**ACKNOWLEDGEMENTS** ........ 156
**REFERENCES** ........ 156

**CHAPTER 5 GROEL CHAPERONIN: INTERACTION WITH POLYPEPTIDES LACKING A RIGID TERTIARY STRUCTURE** ........ 174
*Victor V. Marchenkov, Natalia Yu Marchenko* and *Gennady V. Semisotnov*
**INTRODUCTION** ........ 174
Driving Forces of GroEL Interaction with Substrate Polypeptides ........ 175
Location of Substrate Polypeptides within a GroEL Particle and Stoichiometry of this Complex ........ 177
The Role of Ligands in GroEL Functioning ........ 178
Biotechnological Applications of GroEL Interaction with Substrate Polypeptides ........ 181
**CONCLUSION** ........ 182
**CONSENT FOR PUBLICATION** ........ 182
**CONFLICT OF INTEREST** ........ 182
**ACKNOWLEDGEMENTS** ........ 182
**REFERENCES** ........ 182

**CHAPTER 6 MECHANISMS OF PROTEIN FOLDING BY TYPE II CHAPERONINS** ........ 190
*Rebecca L. Plimpton, José M. Valpuesta* and *Barry M. Willardson*
**INTRODUCTION** ........ 191

Molecular Chaperones and Proteostasis ..... 191
**THE CHAPERONINS** ..... 193
**CCT STRUCTURE** ..... 195
General Features ..... 195
Subunit Arrangement ..... 196
**SUBSTRATE RECOGNITION** ..... 197
Substrate Binding Sites on CCT ..... 197
CCT Binding Sites on Substrates ..... 200
**MECHANISM OF FOLDING** ..... 201
**CCT CO-CHAPERONES** ..... 204
Hsc70 ..... 204
PFD ..... 204
PhLP1 ..... 205
Pdcd5 ..... 206
**CONCLUDING REMARKS** ..... 206
**CONSENT FOR PUBLICATION** ..... 207
**CONFLICT OF INTEREST** ..... 207
**ACKNOWLEDGEMENTS** ..... 207
**REFERENCES** ..... 207

**CHAPTER 7 MECHANISMS AND FUNCTIONS OF THE CYTOSOLIC DNAJ-HSP70 CHAPERONE SYSTEM** ..... 214
*Imad Baaklini* and *Jason C. Young*
**INTRODUCTION** ..... 214
HSP70 ..... 215
Hsp70 Structure ..... 216
ATPase Cycle ..... 217
**HSP40/DNAJ CO-CHAPERONES** ..... 218
Classification ..... 218
Domains of DNAJs ..... 219
*J Domain* ..... 219
*G/F-rich Linker* ..... 220
*Substrate Binding and Cys(Zn) Regions* ..... 221
*Quaternary Structure* ..... 222
**NUCLEOTIDE EXCHANGE FACTORS (NEF)** ..... 223
**CHAPERONE FUNCTION** ..... 225
**FUNCTIONS IN CELLS AND TISSUES** ..... 229
Neurological Diseases ..... 230
Cancer ..... 232
Mitochondrial Import ..... 233
Ion Channels ..... 234
Androgen Receptor ..... 235
Activation-Induced Cytidine Deaminase ..... 236
Viruses ..... 236
**OUTLOOK** ..... 237
**CONSENT FOR PUBLICATION** ..... 237
**CONFLICT OF INTEREST** ..... 237
**ACKNOWLEDGEMENTS** ..... 237
**REFERENCES** ..... 238
**SUBJECT INDEX** ..... 252

# PREFACE

From a social perspective, the word chaperone refers to a matron who used to accompany young people in public, especially ladies, and supervise them at a social gathering to ensure proper behavior. Similarly, those proteins that assist others in their proper folding and biological functions are also referred to as chaperones. During the late '70s, it was coined the term 'molecular chaperones' to make reference to the ability of nucleoplasmin to prevent the aggregation of histones with DNA during the assembly of nucleosomes. As a consequence, this nomenclature was extended to all proteins able to mediate the post-translational assembly of protein complexes. Although the primary concept of molecular chaperone was related to its ability to ensure the correct folding of newly synthesized peptides and refolding of stress-denatured proteins, it should be noted that chaperones are also involved in essential and more sophisticated functions such as promoting the correct assembly of oligomeric complexes. One of the most remarkable examples for this special feature is the ability of a particular subfamily of molecular chaperones, the heat-shock proteins, to assist the proper assembly of steroid receptors with chaperones and co-chaperones. This important feature permits the binding of steroids to activate isoform of the receptor, which functions as a ligand-dependent transcription factor.

The term 'heat-shock protein' stems from the original observation that heat-stress greatly enhances the production of this particular class of molecular chaperones. This means that all heat-shock proteins are molecular chaperones, but not all molecular chaperones are necessarily heat-shock proteins. Temperature is not the only stimulus able to induce heat-shock proteins. Upon the onset of several environmental types of stress or due to the exposure to damaging and extreme insults, the cells increase dramatically the production of molecular chaperones, which play prominent roles in many of the most basic cellular processes by stabilizing unfolded or misfolded peptides, giving the cell time to repair or re-synthesize damaged proteins. In addition to commanding the proper folding of a factor exposed to an environmental injury, many chaperones are also related to other key functions such as enzyme activity, cytoskeletal architecture, nuclear organization, protein trafficking, transcriptional regulation, epigenetic alterations of gene expression and, even more intriguingly, heritable alterations in chromatin state.

The biological relevance of molecular chaperones during the modern times was discovered during the early 1960s when the Italian scientist Ferruccio Ritossa was studying nucleic acid synthesis in puffs of *Drosophila* salivary glands. A colleague accidentally changed the temperature of the cell incubator and an incredible transcriptional activity of new chromosomal puffs was evidenced. The induction of these proteins is one of the most important manifestations of environmentally induced changes in gene expression.

The whole proteome of the cell is successfully maintained thanks to the assistance of molecular chaperones. In addition, the subcellular localization, local concentration, and biological activity of each protein must be strictly regulated in response to both intrinsic and environmental stimuli. The recently coined portmanteau word *proteostasis* describes this equilibrated state of the healthy proteome balance, whereas the term *proteostasis network* refers to the group of cellular events and factors involved in proteostasis maintenance. Failures of proteostasis regulation are responsible for a number of diseases as well as for the deleterious consequences of physiologic processes such as ageing. Since molecular chaperones play a key role in the maintenance of this proteostasis network, they became potential pharmacological targets to preserve that proteostatic function and to improve the biology of the cells by enhancing certain activities (or preventing others). In this regard, several endeavors are currently focused in targeting Hsp90 and some of its cochaperones such as high molecular weight immunophilins and p23. Currently, this is being tested as an exciting alternative for molecular-based therapies, particularly in both malignant and neurodegenerative diseases.

In this book entitled *Role of molecular chaperones in structural folding, biological functions and drug interactions of client proteins,* several aspects of the biology of these key proteins have been addressed with the purpose of providing an updated overview of the field. The major aim is to present a broad spectrum of the molecular mechanisms of action of several molecular chaperones. Understanding these mechanisms will permit to focus on the design of small molecules able to regulate such functions in the complex cellular milieu affecting the proteostasis network in diseases characterized by aberrant protein folding. Above all, I wish to acknowledge the valuable viewpoint of all contributing authors and hope that this assemblage of perspectives will be a valuable resource for researchers in this and other related fields. Also, I hope that the high enthusiasm showed by all our contributors to make this endeavor possible will be appreciated by the readers. Finally, I must express my greatest thanks to the editorial for the encouraging support to face this endeavor.

**Mario D. Galigniana**
Laboratory of Nuclear Receptors
Institute of Biology & Experimental Medicine
Buenos Aires
Argentina

# List of Contributors


| | |
|---|---|
| **Adam Liwo** | Faculty of Chemistry, University of Gdansk, Wita Stwosza 63, 80-308 Gdańsk, Poland |
| **Ana Cauerhff** | Laboratorio de Biología Molecular y Celular, Departamento de Química Biológica, Facultad de Ciencias Exactas y Naturales, Universidad de Buenos Aires, Consejo Nacional de Investigaciones Científicas y Técnicas, Buenos Aires, Argentina |
| **Ayesha Murshid** | Department of Radiation Oncology, Beth Israel Deaconess Medical Center, Harvard Medical School, Boston, MA 02115, USA |
| **Barry M. Willardson** | Department of Chemistry and Biochemistry, Brigham Young University, Provo, UT, USA |
| **Ben Lang** | Department of Radiation Oncology, Beth Israel Deaconess Medical Center, Harvard Medical School, Boston, MA 02115, USA |
| **Carlos H.I. Ramos** | Institute of Chemistry, University of Campinas UNICAMP, Campinas SP, 13083-970, Brazil |
| **Cezary Czaplewski** | Faculty of Chemistry, University of Gdansk, Wita Stwosza 63, 80-308 Gdańsk, Poland |
| **Ewa I. Gołaś** | Faculty of Chemistry, University of Gdansk, Wita Stwosza 63, 80-308 Gdańsk, Poland<br>Baker Laboratory of Chemistry and Chemical Biology, Cornell University, Ithaca, NY, 14853-1301, USA |
| **Gennady V. Semisotnov** | Institute of Protein Research, Russian Academy of Sciences, 4 Institutskaya St., 142290 Pushchino, Moscow Region, Russia |
| **Harold A. Scheraga** | Baker Laboratory of Chemistry and Chemical Biology, Cornell University, Ithaca, NY, 14853-1301, USA |
| **Hideaki Itoh** | Department of Life Science, Graduate School and Faculty of Engineering Science, Akita University, Akita 010-8502, Japan |
| **Hiroshi Yamamoto** | Department of Life Science, Graduate School and Faculty of Engineering Science, Akita University, Akita 010-8502, Japan |
| **Imad Baaklini** | Department of Biochemistry, Groupe de Recherche Axé sur la Structure des Protéines, McGill University, Montreal, Canada |
| **Jason C. Young** | Department of Biochemistry, Groupe de Recherche Axé sur la Structure des Protéines, McGill University, Montreal, Canada |
| **Jianling Gong** | Urologic Oncology Branch, Center for Cancer Research, National Cancer Institute, Bethesda, MD 20892, USA |
| **José M. Valpuesta** | Centro Nacional de Biotecnología, Campus de la Universidad Autónoma de Madrid, Madrid, Spain |
| **Julio C. Borges** | Institute of Chemistry of Sao Carlos, University of Sao Paulo USP, São Carlos, SP, 13566-590, Brazil |

| | |
|---|---|
| **Magdalena A. Mozolewska** | Faculty of Chemistry, University of Gdansk, Wita Stwosza 63, 80-308 Gdańsk, Poland<br>Baker Laboratory of Chemistry and Chemical Biology, Cornell University, Ithaca, NY, 14853-1301, USA |
| **Mario D. Galigniana** | Laboratory of Nuclear Receptors, Institute of Biology & Experimental Medicine, Buenos Aires, Argentina |
| **Natalia Yu Marchenko** | Institute of Protein Research, Russian Academy of Sciences, 4 Institutskaya St., 142290 Pushchino, Moscow Region, Russia |
| **Paweł Krupa** | Faculty of Chemistry, University of Gdansk, Wita Stwosza 63, 80-308 Gdańsk, Poland<br>Baker Laboratory of Chemistry and Chemical Biology, Cornell University, Ithaca, NY, 14853-1301, USA |
| **Rebecca L. Plimpton** | Department of Chemistry and Biochemistry, Brigham Young University, Provo, UT, USA |
| **Ryuich Ishida** | Department of Life Science, Graduate School and Faculty of Engineering Science, Akita University, Akita 010-8502, Japan |
| **Stuart K. Calderwood** | Department of Radiation Oncology, Beth Israel Deaconess Medical Center, Harvard Medical School, Boston, MA 02115, USA |
| **Taka Eguchi** | Department of Radiation Oncology, Beth Israel Deaconess Medical Center, Harvard Medical School, Boston, MA 02115, USA |
| **Thomas Prince** | Department of Medicine, Boston University School of Medicine, Boston, MA 02114, USA |
| **Tomoya Okamoto** | Department of Life Science, Graduate School and Faculty of Engineering Science, Akita University, Akita 010-8502, Japan |
| **Victor V. Marchenkov** | Institute of Protein Research, Russian Academy of Sciences, 4 Institutskaya St., 142290 Pushchino, Moscow Region, Russia |

CHAPTER 1

# Regulatory Roles for Hsp70 in Cancer Incidence and Tumor Progression

**Taka Eguchi**[1,¶], **Benjamin J. Lang**[1,¶], **Ayesha Murshid**[1,¶], **Thomas Prince**[2], **Jianlin Gong**[3] and **Stuart K Calderwood**[1,*]

[1] *Department of Radiation Oncology, Beth Israel Deaconess Medical Center, Harvard Medical School, Boston, MA02115, USA*

[2] *Urology Department, Geisinger Clinic, Danville, PA 17822, USA*

[3] *Department of Medicine, Boston University Medical Center, Boston, MA 02118, USA*

**Abstract:** The HSP70 family of molecular chaperones plays a significant role in cancer. Notably, the inducible protein Hsp72 becomes expressed in many cancers, often to high levels and underlies escape of tumor cells from senescence and increased tumor initiation and metastasis. Examination of database suggests that mutation within the ORFs of *HSP70* genes in cancer is relatively rare suggesting a requirement for intact function. At the molecular level, Hsp72 is thought to chaperone key proteins in tumorigenesis and permit their accumulation in the malignant cell. In addition, an important role for Hsp72 in RNA metabolism is emerging, indicating mechanisms potentially involving the RNA binding protein HuR. The existence of multiple *HSP70* pseudogenes may also be important for future studies of long non-coding RNA (lncRNA) regulation through this family of chaperones. As the significance of this family of chaperones in cancer emerges, small molecule inhibitors have been developed as future potential cancer pharmaceuticals. We discuss the targeting of individual HSP70 families at key functional domains in the proteins.

**Keywords:** Heat shock protein seventy, Structure, Function, Cancer, Growth, Metastasis, Chemical inhibitor, Substrate, Domain, ATPase, Drug.

## 1. INTRODUCTION

The seventy kilodalton heat shock protein (HSP70) family is found in all cellular organisms [1]. From the beginning of cellular life, these proteins have permitted protein folding in the crowded intracellular environment and have come to the aid

* **Corresponding author Stuart K. Calderwood:** Beth Israel Deaconess Medical Center, 330 Brookline avenue, Room CLS 0610, Boston MA 02115, USA; Tel/Fax: 617 735 2497; Email: scalderw@bidmc.harvard.edu

¶ Equal Contributors

of stressed cells dealing with proteotoxic insults [1]. Over 20 years ago, Ciocca and co-workers studying clinical biopsy material noticed that human cancers contained elevated levels of HSPs and this finding has been confirmed in many subsequent studies [2 - 4]. The rationale behind these increases in HSP70 levels was not immediately apparent, although it was suspected that this change in expression might be related to alterations in the folding status of the malignant cell. Indeed, investigators studying the potential role in cancer of another member of the HSP family, Hsp90 had noticed a large increase in this chaperone in many cancer types and concluded that malignant cells were "addicted to chaperones" [5]. Withdrawal of the folding power of Hsp90 by exposure to specific drugs led to widespread denaturation and degradation of oncogenic proteins. Indeed targeting Hsp90 has become a major endeavor in cancer therapeutics [6]. Not unreasonably it could be suggested that HSP70 might be the next candidate for drug therapy based on its elevated levels and tumor dependence.

## 2. HSP70 PROTEINS IN THE CYTOPLASM AND NUCLEUS

Most cellular organisms contain a relatively large (over 10 member) HSP70 family [7]. In this review, we will discuss HSP70 family members shown to be located in the cytoplasm and nucleus, including the essential protein Hsc70, HspA1L and the inducible proteins Hsp72 and Hsp70B'.

The common structure of HSP70 family proteins includes an N-terminal ATPase domain or nucleotide binding domain (AD / NBD, 45k), a protease active short linker (8 aa), substrate binding domain/ peptide binding domain (SBD / PBD, 15k) and C-terminal helical lid region (10k) [8], [9]. The C-terminal lid region of HSP70 also contains a tetrapeptide EEVD motif at the extreme C-terminus [9] also found in HSP90, that recognizes the tetratricopeptide repeat (TPR) motif found in a range of TPR domain proteins including the Hop co-chaperone /adaptor protein [10, 11].

The human *HSP70* gene family is characterized by considerable diversity, apparently resulting from multiple duplications and retrotranspositions of the highly expressed gene *HSPA8/HSC70/HSP73* [[12]]. Indeed, Brocchieri et al identified forty-seven *Hsp70* sequences including seventeen genes and thirty pseudogenes in the human genome. The N-terminal AD was conserved at least partially in the majority of the genes while the SBD was more commonly lost homologies. Structurally well-conserved and functional *HSP70* family members include *HSC70/HSPA8, Hsp72/HSPA1A, Hsp72/HSPA1B, HSPA1L/HSP70T/ HSP70-hom, HSP70B'/HSPA6, HSP70B/HSPA7* (possibly a pseudogene) and *HSPA2*, shown in phylogenetic tree analyses [12] (Table **1**).

**Table 1. The human *HSP70* family.**

| NAME | RNA | PROTEIN | LOCATION | INDUCIBILITY | REF. |
|---|---|---|---|---|---|
| *HSPA8* | Coding | HSC70, HSP73 | Cyt/Nucl | Constitutive, mildly inducible. | 12-14 |
| *HSPA1A* | Coding | Hsp72 | Cyt/Nucl | Stress inducible. | 12 |
| *HSPA1B* | Coding | Hsp72 | Cyt/Nucl | Stress inducible | 12 |
| *HSPA1L, (HSP70L)* | Coding | Hsp70.1-like, Hsp70-HOM | Cyt/Nucl | Constitutive Spermatides. | 12 |
| *HSPA6, HSP70B'* | Coding | Hsp70B' | Cyt?Nucl, Secreted | Stress inducible in primates. | 15, 16, 17 |
| *HSPA7, HSP70B* | Coding or pseudogene | Hsp70B? | ? | Only in primates. | 12 |
| *HSPA2 (HSP70-2, HSP70-3)* | Coding | HspA2, Hsp70-2 | Cyt/Nucl | Nucleoli, Centrosome | 18-21 |

Clearly there are numerous pseudogenes of *HSP70* so that interpretation of experiments in this field should likely take this into account. For example, qRT-PCR may detect both coding *HSP70* and *HSP70* pseudogene RNAs. In addition, one antibody might detect a member of the HSP70 family but might also associate with smaller peptides translated from the same mRNA, or larger proteins such as Hsp110 or Grp170. Therefore, interpretation of western blotting data should be carefully considered.

Hsc70 (heat shock cognate protein 70) is expressed most abundantly among the HSP70 family proteins. This protein is encoded by the *HSPA8* gene and is moderately inducible by stress. Numerous pseudogenes of *HSPA8/HSC70* have been found [12] and some of these potentially might be expressed and function as lncRNA (Fig. **1**) [13, 14]. For example, the *HSC70* pseudogene ncRNA could possibly play role in Hsc70 expression by assuming a decoy / sponge function or antagonizing miRNA that target *HSC70/HSPA8*.

*HSPA1A* and *HSPA1B* are the major inducible-type *HSP70* genes and together encode Hsp72. The homology between these two genes is high and it is difficult to distinguish them at the mRNA and protein levels. In addition, *HSPA1L/HSP70T/HSP70-hom* is also homologous to *HSPA1A* and to *HSPA1B* and all three of these genes are located close together in both human and mouse genomes [12]. *HSPA1A* and *HSPA1L* are located on human chromosome 6p21.33 in plus and minus directions, respectively. *HSPA1B* is also located close by in chromosome 6p21.32, in the plus strand. Four pseudogenes of *HSP70* were also found on chromosome 6 [12]. The physical location of these genes in the chromosome suggests the possibility of functional interactions.

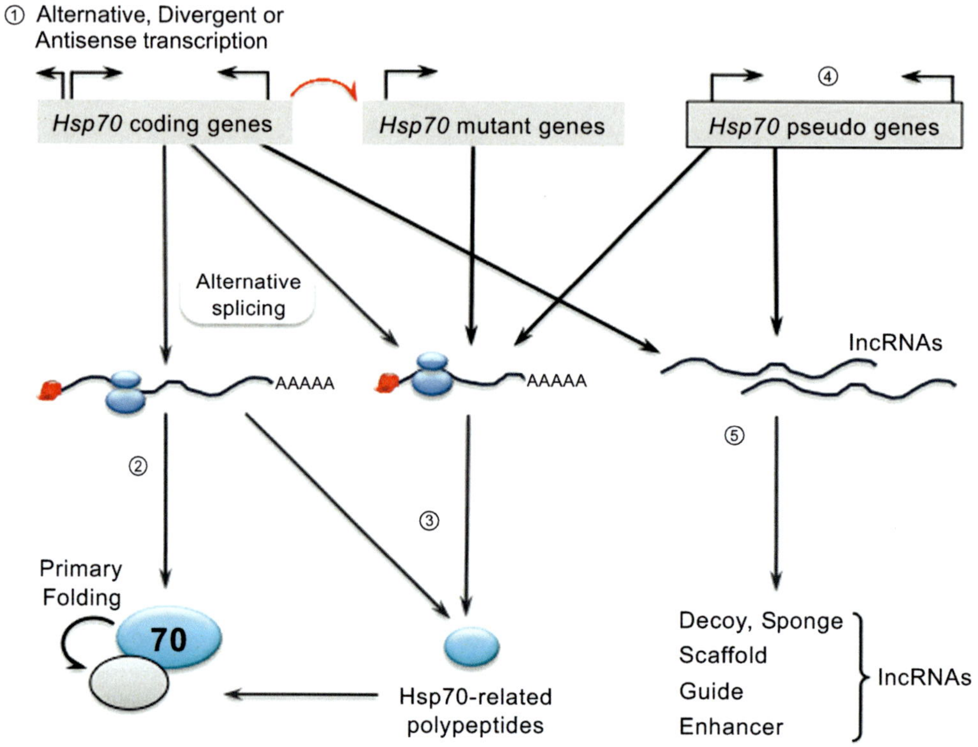


**Fig. (1).** HSP70 *transcripts and proteins derived from HSP70 genes in co-translational folding, as non-coding RNA (ncRNA) and as origins of smaller HSP70-related polypeptides.*

Two more *HSP70* genes, *HSPA6* and *HSPA7* encode the Hsp70B' and Hsp70B proteins, respectively, (although *HSPA7/HSP70B* could be a pseudogene [12]). These two *HSP70* genes exhibit strong homology and are found only in primates, suggesting the possibility that their function could be compensated for by *HspA1A/HspA1B/HspA1L* in rodents, or that *Hsp70B* and *Hsp70B'* have primate-specific roles. The *HSPA6 and HSPA7* mRNAs are minimally expressed in normal tissues but are strongly induced by stress and in cancer cells [15]. High-level expression of *HSP70B'* was correlated with elevated histone H3 lysine 4 trimethyl (H3K4me3), a marker of active transcription in the malignant prostate cancer cell line PC-3 (27% of input *HSP70B'* gene was coprecipitated with H3K4me3 in PC-3 cells, T. Eguchi & SK Calderwood, in preparation). Interestingly, it was shown that the *HSPA6/HSP70B'* product could be secreted from macrophages upon stimulation by LDL [16]. This property may indeed not be restricted to macrophages, as we recently found that Hsp70B' to be secreted from breast cancer cell line MDA-MB-231 after heat shock (T. Eguchi & SK Calderwood, unpublished). Thus extracellular Hsp70B' might play a role in human cancer [17].

Finally, the minor *HSP70* family member, HspA2 was reported to localized in the nuclei, nucleoli and centrosomes of heat shocked cancer cells [18]. HspA2 appears to be a unique HSP70 family protein and is encoded by neither of the Hsp72 genes. HSPA2 may be involved in the etiology of cancers including hepatocellular carcinoma, NSCLC and pancreatic cancer [19 - 21].

We have shown that, in the mouse mammary tumor virus (MMTV) model, in which breast cancer develops spontaneously, *hsp72* gene inactivation delayed tumor initiation and inhibited metastasis through down-regulation of the *met* oncogene [22]. We have confirmed in RNA-seq experiments that *hspa1a* and *hspa1b* are minimally expressed in the *hsp72* KO MMTV mice, and we are attempting to determine if *hsp70L1/hspa1L* is deleted at physical and/or functional levels in these mice.

HSP70 proteins are thought to be the primary chaperone for many polypeptides, encountering elongating proteins on ribosomes and then handing on such proteins to Hsp90 to complete folding [23]. HSP70 protein interaction with ribosomes and with ribonucleoprotein (RNP) has also recently emerged as an important theme. Hundley et al reported that two HSP70 proteins (Ssb and Ssz1) and one of the DnaJ/Hsp40 class (Zuo1/zuotin) of co-chaperones reside on the yeast ribosome, forming a stable heterodimer named the *ribosome-associated complex* (RAC), and that Ssb could be crosslinked to nascent polypeptide chains on ribosomes [24]. Otto et al subsequently reported that the human homolog of the J-domain protein MPP11/DNJC1/Hsp40 and its mouse homolog MIDA1/DnaJC1 formed a stable complex with Hsp70L1 and became associated with ribosomes [25]. Alternatively, it was suggested that HSP70 Ssb function might be limited to the protection of nascent polypeptide from aggregation until downstream chaperones take over and actively fold their substrates. It was recently demonstrated that deletion of Ssb leads to widespread aggregation of nascent polypeptides [26]. Ubiquitin E3 Ligases can, alternatively, associate with ribosomes and become involved in poly-ubiquitination and degradation of the nascent polypeptides in budding yeast [27, 28] and this system is also found in mammalian cells. We recently analyzed the RNA-seq transcriptome in *hsp72* KO MMTV and control MMTV mice. The quantity of ribosomal RNA was largely reduced in the *hsp72* KO mammary tumor compared to the control mammary tumor (T Eguchi, B Lang & SK Calderwood, unpublished data). Concurrently expression patterns of miRNA were also altered in *hsp72* KO mammary tumors. These data indicated that *hsp72/hspa1a* and the closely related gene *hspa1L* could be involved in ribosomal RNA stability, co-translational folding, the formation and function of processing bodies (P-bodies / GW-bodies) and stress granules (Fig. **2**) [24, 29 - 35]. These are key structures involved in miRNA and messenger ribonucleoprotein (mRNP) processing. Moreover, Hsp72 appears to be important in cancer stem cell (CSC) function. Over the last decade, the notion that tumors are maintained by their own stem cells has gained ground [36]. Such CSC are characterized by self-renewal ability, reversible differentiation-dedifferentiation or EMT, stress resistances (radio- or chemo- resistances) and supported through interactions with niche cells through ligand-receptor binding or extracellular matrices (ECM) [37, 38].

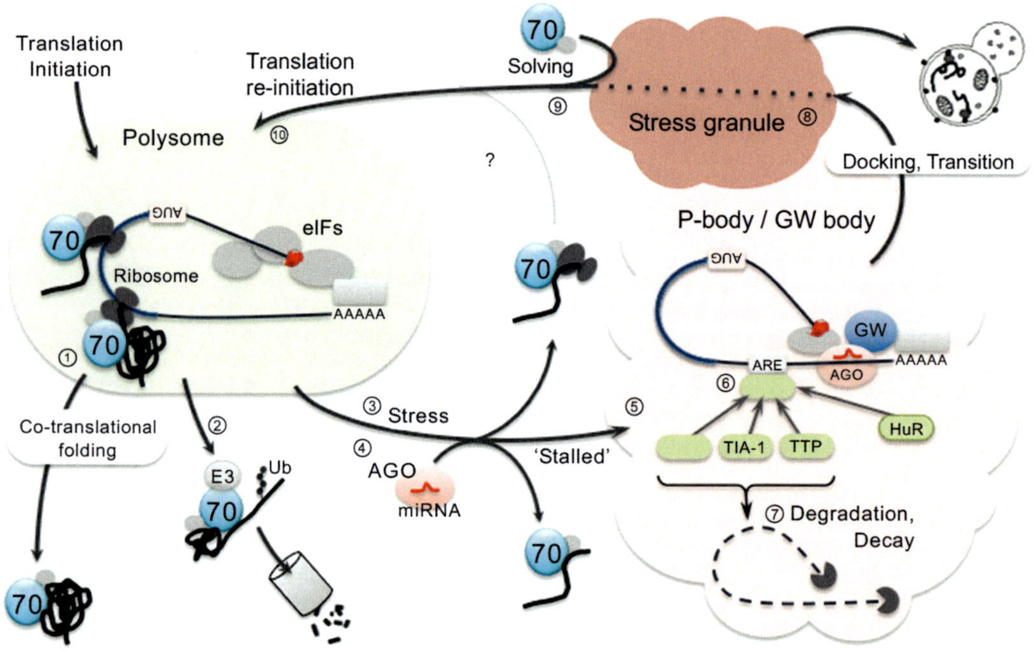


**Fig. (2).** A model for HSP70 control of messenger ribonucleoproteins (mRNP).

To determine how cell stress might be involved in the properties of CSCs, we recently stimulated CSC-abundant breast cancer cell line MDA-MB-231 with heat shock, and found that shorter polypeptides were detected after two min heat shock (Eguchi T and Calderwood SK, unpublished data). These findings regarding rapid induction of shorter polypeptides is reminiscent of the results of a HSP70 deficiency study [26]. It is known that cell stress induces processing bodies (P-bodies) and stress granules that incorporate miRNA and mRNP aggregates through which the mRNA is either degraded or recruited to ribosomes for translational reinitiation (Fig. **2**). Adenyl uridylated (AU)-rich element (ARE) binding proteins recognize AUUUA penta-ribonucleotide within U-rich sequences in the P-bodies. In addition, the Hu protein family including HuR, HuA, HuB, HuC, and HuD bind to the ARE and stabilize the mRNA, while other AU-rich element binding proteins including AUF1, TTP, BRF1, TIA-1, TIAR and KSRP destabilize the mRNA oppositely.

Indeed, one of the mRNA stabilizing protein HuR appeared to stabilize β-catenin (*CTNNB1*) mRNA under the control of activated HSF1 (pS326) and mTOR in breast cancer stem cell-rich MDA-MB-231 cells [39]. This β-catenin-producing system under the control of mTOR-HSF1 axis could be a key pathway in CSC oncology [39]. Activating mutations in the human *CTNNB1* gene and mutations in other genes in the Wnt-β-catenin-TCF signal have been frequently reported. Importantly β-catenin has been recognized a stem cell maintenance factor [36]. Therefore, under the control of the HSF1-HuR axis, active production of β-catenin and HSPs might be a key system in cancer initiation and maintenance.

## 3. MUTATION AND OVEREXPRESSION OF HSP72 IN CANCER

The Cancer Genome Atlas (TCGA) is a national effort to molecularly characterize every form of cancer and tumor type. Initiated in 2005 by the National Cancer Institute (NCI) and the National Human Genome Research Institute (NHGRI), TCGA aims to quantitatively characterize a representative number of tumor types at the molecular level through genome sequencing, promoter methylation analysis, relative mRNA expression levels and eventual proteomic profiling [40]. This information is made publicly available as it is processed, published and released to the cancer research community. Several sites and institutes host and distribute TCGA data such as cBioPortal at Memorial Sloan-Kettering Cancer Center that was used here to profile the HSP70 family of molecular chaperones [41 - 44] (Fig. **3**). Cross-cancer analysis of the genomic alterations of the 13 HSP70 family members in humans shows that bladder cancer (BLCA) at 32.3%, cutaneous melanoma (SKCM) at 30.9%, lung adenocarcinoma (LUAD) at 29.7%, stomach adenocarcinoma (STAD) at 28.6% and liver hepatocellular carcinoma (LIHC) at 25.4% have the highest rates of HSP70 family genomic alteration (Fig. **3**).

Analysis of mRNA levels in 23 tumor types indicates that the *HSP70* family is altered in a considerable number of tumor samples. Levels of mRNA expression may be influenced by a number of variables at the epigenetic, transcriptional and post-transcriptional phases of gene expression. In 22 of the 23 tumor types profiled, *HSP70* family mRNA levels were over-expressed in 25% of all tumors sampled at a standard deviation greater than 2 (Table **2**, blue columns). Tumor types that least over-expressed any of the *HSP70* homologs were colon and rectal carcinoma (COAD) at 22% and ovarian serous cystadenocarcinoma (OV) at 26%. Overall this means almost one in four tumor samples exhibited an over-expression of at least one *HSP70* homolog. Moreover, in 8 of the 23 tumor types analyzed, an *HSP70* homolog was over-expressed in 40% of tumors sampled. These included BLCA at 43%, LUAD at 43%, LIHC at 46%, kidney clear cell carcinoma (KIRC) at 45%, and chromophobe renal cell carcinoma (KICH) at 42%. Currently, three tumor types over-express an *HSP70* homolog in 50% of tumor samples. These include diffuse large B-cell lymphoma (DLBC), uterine carcinoma, and adrenocortical carcinoma (ACC). The increased percentage of mRNA over-expression in these tumor types suggests that the *HSP70* family may be especially essential towards maintaining cellular proteostasis within the cancer cell.

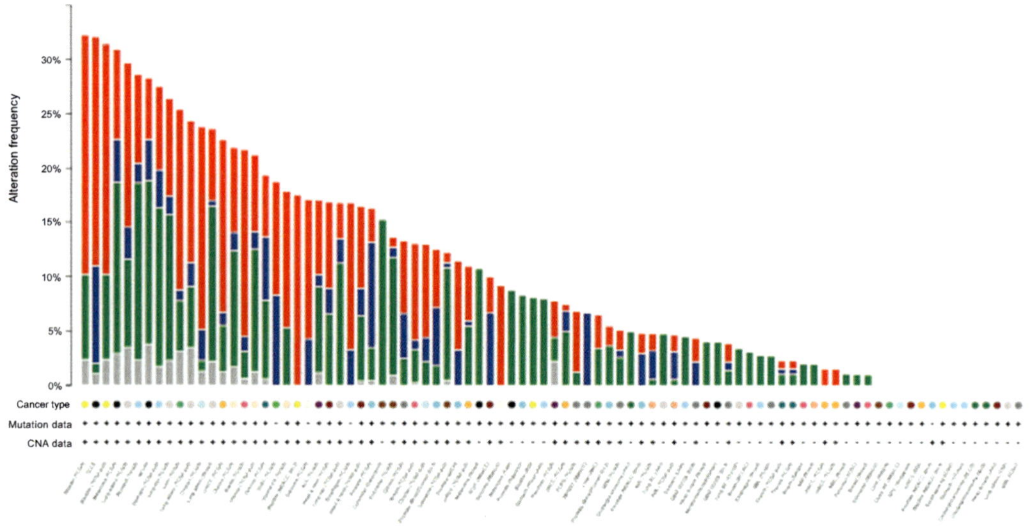


**Fig. (3).** *Frequency Of Hsp70 Homolog Alterations Across Tumor Types And Profiled By The Cancer Genome Atlas (TCGA)* [40]. Each tumor type or cell line collection listed along the x-axis were analyzed for Hsp70 family gene copy number alterations (CNA, red bars for amplification and blue bars for deletion) and open reading frame mutations (green bars). Multiple alterations are indicated by gray bars.

Analysis of the open reading frame mutations in each tumor type confirms that the *HSP70* family is highly conserved and essential. Only 6 of 22 tumor types have greater than 10% mutational alterations of any *HSP70* homolog (Table **2** green columns). DLBC currently does not have mutational data. Moreover, acute myeloid leukemia (AML), papillary thyroid carcinoma (THCA) and KICH have less than 1% mutational alterations. Stomach adenocarcinoma (STAD) and cutaneous melanoma (SKCM) have the highest *HSP70* homolog alterations at 17% of all tumor samples. Finally, throughout all tumor types analyzed, the

*HSP70* homolog genes *HSPA1L*, *HSP9A/mortalin* and *HSPA14/Hsp70L1* are found to be the most altered by either mRNA over-expression or mutation while the putative pseudogene *HSPA7/HSP70B* was found to be the least altered. These observations may indicate the significance of these *HSP70* homologs in maintaining tumor malignancy.

Table 1: HSP70 family mRNA over-expression and mutation rates across human cancer types within TCGA database

| TCGA | AML | | ACC | | BRCA | | CESC | | KICH | | KIRC | | COAD | | SKCM | |
|---|---|---|---|---|---|---|---|---|---|---|---|---|---|---|---|---|
| Sample Size | 173 | 200 | 79 | 90 | 526 | 507 | 253 | 194 | 66 | 65 | 469 | 424 | 244 | 224 | 470 | 345 |
| *HSP70s* | mRNA | Muta | mRNA | Muta | mRNA | Muta | mRNA | Muta | mRNA | Muta | mRNA | Muta | mRNA | Muta | mRNA | Muta |
| *HSPA1A* | 5% | 0% | 5% | 0% | 2% | 0% | 1% | 0% | 8% | 0% | 2% | 0% | 5% | 0% | 2% | 0% |
| *HSPA1B* | 4% | 0% | 4% | 0% | 0% | 0% | 2% | 1% | 15% | 0% | 2% | 0% | 5% | 0% | 1% | 0% |
| *HSPA1L* | 2% | 0% | 6% | 0% | 2% | 0% | 6% | 1% | 21% | 0% | 3% | 1% | 4% | 4% | 9% | 5% |
| *HSPA2* | 2% | 0% | 6% | 2% | 1% | 0% | 7% | 2% | 5% | 0% | 5% | 0% | 3% | 1% | 6% | 1% |
| *HSPA4* | 4% | 1% | 24% | 1% | 2% | 1% | 4% | 1% | 5% | 0% | 12% | 0% | 2% | 2% | 3% | 1% |
| *HSPA4L* | 5% | 0% | 14% | 0% | 3% | 0% | 3% | 1% | 6% | 0% | 4% | 0% | 4% | 2% | 5% | 2% |
| *HSPA5* | 2% | 0% | 6% | 1% | 3% | 0% | 6% | 0% | 3% | 0% | 6% | 0% | 6% | 0% | 4% | 2% |
| *HSPA6* | 6% | 0% | 1% | 1% | 11% | 0% | 1% | 1% | 3% | 0% | 0% | 0% | 1% | 1% | 3% | 3% |
| *HSPA7* | 0% | 0% | 0% | 0% | 0% | 0% | 0% | 0% | 2% | 0% | 4% | 0% | 1% | 0% | 0% | 2% |
| *HSPA8* | 3% | 0% | 6% | 1% | 0% | 1% | 1% | 3% | 6% | 0% | 7% | 1% | 2% | 1% | 2% | 2% |
| *HSPA9* | 2% | 0% | 12% | 1% | 4% | 0% | 6% | 1% | 6% | 0% | 22% | 1% | 2% | 2% | 4% | 0% |
| *HSPA12A* | 2% | 0% | 13% | 1% | 0% | 1% | 4% | 1% | 0% | 0% | 3% | 0% | 3% | 1% | 3% | 1% |
| *HSPA14* | 3% | 0% | 14% | 1% | 12% | 0% | 6% | 0% | 3% | 0% | 4% | 0% | 3% | 0% | 3% | 1% |
| Total | 28% | <1% | 61% | 9% | 31% | 3% | 37% | 8% | 42% | 0% | 46% | 5% | 22% | 11% | 32% | 17% |

| TCGA | DLBC | | GBM | | HNSC | | LIHC | | LGG | | LUAD | | LUSC | | OV | |
|---|---|---|---|---|---|---|---|---|---|---|---|---|---|---|---|---|
| Sample Size | 28 | N/A | 154 | 291 | 279 | 279 | 269 | 198 | 527 | 289 | 230 | 230 | 178 | 178 | 489 | 316 |
| *HSP70s* | mRNA | Muta | mRNA | Muta | mRNA | Muta | mRNA | Muta | mRNA | Muta | mRNA | Muta | mRNA | Muta | mRNA | Muta |
| *HSPA1A* | 4% | N/A | 5% | 0% | 7% | 0% | 6% | 0% | 2% | 0% | 3% | 0% | 7% | 0% | 0% | 0% |
| *HSPA1B* | 7% | N/A | 6% | 0% | 5% | 0% | 5% | 0% | 2% | 0% | 3% | 0% | 6% | 0% | 0% | 0% |
| *HSPA1L* | 14% | N/A | 3% | 0% | 6% | 1% | 13% | 3% | 3% | 0% | 4% | 3% | 5% | 1% | 7% | 1% |
| *HSPA2* | 7% | N/A | 4% | 0% | 8% | 0% | 3% | 0% | 6% | 0% | 7% | 2% | 8% | 1% | 2% | 0% |
| *HSPA4* | 4% | N/A | 3% | 0% | 4% | 1% | 11% | 1% | 2% | 0% | 3% | 1% | 3% | 2% | 2% | 0% |
| *HSPA4L* | 4% | N/A | 3% | 0% | 4% | 1% | 5% | 1% | 4% | 1% | 5% | 0% | 6% | 2% | 2% | 0% |
| *HSPA5* | 7% | N/A | 5% | 0% | 6% | 1% | 4% | 2% | 5% | 0% | 3% | 1% | 4% | 2% | 1% | 1% |
| *HSPA6* | 4% | N/A | 8% | 0% | 3% | 1% | 5% | 1% | 1% | 0% | 10% | 1% | 10% | 1% | 5% | 0% |
| *HSPA7* | 0% | N/A | 7% | 0% | 3% | 0% | 0% | 0% | 0% | 0% | 3% | 0% | 9% | 0% | 0% | 0% |
| *HSPA8* | 0% | N/A | 2% | 0% | 2% | 1% | 4% | 1% | 5% | 1% | 5% | 1% | 6% | 1% | 1% | 0% |
| *HSPA9* | 4% | N/A | 5% | 0% | 6% | 0% | 10% | 1% | 4% | 0% | 5% | 1% | 4% | 1% | 3% | 0% |
| *HSPA12A* | 7% | N/A | 1% | 0% | 5% | 0% | 4% | 1% | 5% | 0% | 4% | 0% | 4% | 2% | 2% | 0% |
| *HSPA14* | 7% | N/A | 1% | 0% | 2% | 0% | 10% | 0% | 7% | 0% | 8% | 0% | 7% | 0% | 6% | 0% |
| Total | 50% | N/A | 34% | 1% | 37% | 6% | 46% | 8% | 32% | 2% | 41% | 0% | 39% | 11% | 26% | 2% |

| TCGA | KIRP | | THCA | | STAD | | PRAD | | BLCA | | UCS | | UCEC | |
|---|---|---|---|---|---|---|---|---|---|---|---|---|---|---|
| Sample Size | 290 | 161 | 486 | 401 | 265 | 289 | 487 | 261 | 129 | 130 | 57 | 57 | 333 | 248 |
| *HSP70s* | mRNA | Muta | mRNA | Muta | mRNA | Muta | mRNA | Muta | mRNA | Muta | mRNA | Muta | mRNA | Muta |
| *HSPA1A* | 4% | 0% | 3% | 0% | 6% | 1% | 3% | 0% | 3% | 2% | 5% | 0% | 5% | 0% |
| *HSPA1B* | 4% | 0% | 4% | 0% | 5% | 0% | 3% | 0% | 3% | 0% | 5% | 0% | 5% | 0% |
| *HSPA1L* | 2% | 0% | 3% | 0% | 6% | 1% | 4% | 0% | 2% | 1% | 30% | 0% | 5% | 2% |
| *HSPA2* | 4% | 1% | 3% | 0% | 5% | 1% | 3% | 0% | 5% | 1% | 2% | 0% | 5% | 2% |
| *HSPA4* | 6% | 0% | 5% | 0% | 3% | 1% | 5% | 1% | 5% | 0% | 5% | 2% | 4% | 6% |
| *HSPA4L* | 3% | 1% | 4% | 0% | 4% | 3% | 4% | 0% | 2% | 2% | 2% | 2% | 5% | 5% |
| *HSPA5* | 4% | 0% | 4% | 0% | 4% | 1% | 6% | 0% | 3% | 1% | 5% | 0% | 3% | 2% |
| *HSPA6* | 2% | 1% | 2% | 0% | 5% | 3% | 2% | 0% | 3% | 1% | 5% | 2% | 5% | 4% |
| *HSPA7* | 0% | 0% | 3% | 0% | 3% | 0% | 0% | 0% | 4% | 0% | 0% | 0% | 6% | 0% |
| *HSPA8* | 5% | 1% | 3% | 0% | 4% | 1% | 3% | 1% | 3% | 2% | 2% | 0% | 3% | 3% |
| *HSPA9* | 6% | 1% | 6% | 0% | 5% | 1% | 4% | 1% | 3% | 1% | 9% | 0% | 6% | 2% |
| *HSPA12A* | 5% | 0% | 5% | 0% | 6% | 6% | 2% | 0% | 2% | 0% | 7% | 0% | 4% | 2% |
| *HSPA14* | 4% | 1% | 3% | 0% | 7% | 1% | 5% | 0% | 19% | 2% | 11% | 0% | 8% | 2% |
| Total | 31% | 5% | 28% | <1% | 35% | 17% | 22% | 3% | 41% | 10% | 53% | 5% | 36% | 12% |

## 4. HSP72 AND THE HALLMARKS OF CANCER

Hsp72 was not observed in unstressed normal tissues, but was seen in histological studies of human tumors including breast, endometrial, lung, and prostate [2, 3, 45]. Often the expression of this protein was correlated with increased tumor cell proliferation and metastasis to lymph nodes as well as weakened responses to chemotherapeutic agents. Transgenic mice expressing human Hsp72 at high levels developed multiple myeloma [46]. In addition, forced (transgenic) expression of Hsp72 caused development of tumors in nude mice [47]. This HSP70 family member was also found on the plasma membrane of some tumor cells and could also be secreted to the extracellular space from cancerous cells [3]. Abundant expression of Hsp72 in cancer cells was found to correlate with histological grade in such malignancies, suggesting its significance as a cancer marker or target for therapy [48 - 50].

In their landmark review, Hanahan and Weinberg suggested a number of hallmarks, key phenotypic characteristics that are associated with a wide spectrum of cancers [51]. We will examine the potential role of Hsp72 in some of these hallmarks.

### 4A. HSP72 Suppresses Apoptotic Cell Death in Cancer

Ability to evade programmed cell death is a characteristic of many cancers [51]. Indeed, it was thought initially that Hsp72's anti-apoptotic role played an important role in tumor development [52]. The anti-apoptotic role of Hsp72 might involve suppression of the pro-apoptotic c-Jun kinase pathway, which in turn might be important for tumor development. In addition, Hsp72 was thought to allow cancer cells to escape apoptotic cell death mediated by hypoxia, serum starvation, TNF, or FAS [53]. Overall, it was thought that depletion of this protein might cause rapid death of cancer cells whereas its overexpression prevented death. However, it was demonstrated subsequently that Hsp72 depletion was not associated with apoptotic death in prostate cancer cells such as PC-3 and Du-145 [54]. In addition, increased Hsp72 expression in cancer cells could not prevent intracellular pro-apoptotic signaling [54].

### 4B. HSP72 and Senescence

In contrast to apoptosis, Sherman et al. showed that the knockdown of Hsp72 could instead lead to senescence in a variety of cancerous cell lines such as MCF10A cells transformed by the *Her2* oncogene, but not in untransformed epithelial cells [55]. The same group described mechanisms by which Hsp72 controlled senescence [56, 57]. They deduced that senescence in Her2 positive cancer cells was regulated by Hsp72 through the CDK inhibitory protein, p21 from studies using WT and Hsp72 knockdown cells and p21 KO cells [55]. Later it was shown that Hsp72-controlled senescence was regulated by the protein survivin and that forced expression of this protein almost completely reversed Hsp72 depletion-mediated senescence [55]. However, expression of survivin only could partially reverse Hsp72 depletion-mediated senescence in non-malignant MCF10A cells that have a fully functional p21 pathway, suggesting an important role for p21 in this effect. These effects of p21 and survivin on Hsp72 depletion-mediated senescence depended on the presence of functional *p53* [55]. In Her2 positive, *p53* mutant cells, Hsp72 depletion did not lead to upregulation of p21 but downregulated survivin, suggesting that survivin function is not regulated by *p53* expression in these cells [58]. Thus, targeting Hsp72 in Her2 positive cancer cells would be an interesting approach in the anticancer drug development field as this chaperone plays an essential role in *Her2*-induced tumorigenesis in mice [55]. In addition, Hsp72 can also suppress senescence by modulating activity of the oncogenic co-chaperone Bag3 [59 - 61].

### 4C. HSP72 in Tumor Initiation and Metastasis

It was recently shown that aggressive triple negative breast cancer cells showed an increased expression of Hsp72. This protein was shown to promote survival in these cells through the mediation of the kinases Akt/PKB and PKC as well as the suppression of apoptotic signaling mentioned earlier [62, 63]. A role for Hsp72 in tumor metastasis was also confirmed earlier by many researchers. The metastatic properties of Hsp72 were thought to be regulated by 14-3-3 proteins as well as the activities of protein kinases and phosphatases in triple negative breast, cancer cells [64]. The implication of *hsp72* genes in the initiation of cancer and metastasis was confirmed recently in a spontaneous mammary cancer model.

Mice expressing the polyomavirus middle T oncogene under control of the mouse mammary tumor virus (MMTV) long terminal repeat developed spontaneous mammary carcinoma with high rate of metastasis to lung [65, 66]. Using this model, it was shown that *hsp70* was required for robust initiation of tumors, maintenance of cancer stem cell populations, dissemination of cancer cells and metastasis. Teng et al, had demonstrated a mechanism through which Hsp72 might be involved in actin polymerization and that in turn might be responsible for the increased invasive and metastatic properties [67]. They showed WASF3, an actin polymerizing protein to be bound and stabilized by Hsp72, a process required for cellular movement and metastatic properties [67]. We were able to show that inhibiting Hsp72 activity is associated with loss of WASF3 (A. Murshid & SK Calderwood, unpublished) further suggesting its role as a chaperone in metastasis. Hsp72 was also shown to regulate the *MET* oncogene, a cell surface receptor which is involved in tumorigenesis and metastasis; inactivation of *hsp72* genes led to loss of cells with stem cell markers and depleted the levels of active phospho-c-MET [22, 68].

### 4D. HsSP72 in Sustained Angiogenesis

Hsp72 also appears to play role in yet another of the hallmarks- ability to induce angiogenesis, through its influence on the primary sensor of tumor cell hypoxia- the transcription factor HIF1α [69]. HIF1 is regulated at the level of protein stability and increased concentrations of both Hsp72 and Hsp90 were required for its stabilization and accumulation in cancer cells [69]. In addition to its role in tumorigenesis, HIF1α is involved in the regulation of expression of other genes required for survival during stressful conditions such as hypoxia. Recently it has been reported that the transcription factor that mediates Hsp72 expression, HSF1 regulates a subset of HIF1α-regulated genes involved in tumor progression and also a set of cancer related miRNAs (Let-7, MiR-1991 or Mir-125B) [70].This effect might involve the RNA binding protein, HuR (a major regulator of translation, promote HIF-1α translation) which is also overexpressed in cancer and has strong correlation with cancer progression through HIF1α [71, 72]. Thus, a complex mechanism involving HSF1, Hsp72, HuR and HIF-1 regulation in tumor angiogenesis might be suggested.

## 5. DRUGGING HSP70 IN CANCER: ISOFORMS AND DRUGGABLE DOMAINS

The tumorigenic functions of HSP70 family members have therefore established them as attractive therapeutic targets for many human cancers. While research aimed at identifying HSP70 inhibitory molecules has accelerated in recent years, the development of such inhibitors is considered to be at an early stage and is a highly attractive prospect for future therapeutic developments.

The expression of *HSP70* family members and their relationship with patient outcome varies between different *HSP70* paralogs and within specific human cancer types [73 - 76]. Both pre-clinical and clinical studies have demonstrated the Hsp72 protein to be a promising target for numerous cancer types [22, 55, 76]. Targeting Hsp72 in cancer treatment fulfills important fundamental features of a promising therapeutic strategy. Firstly, Hsp72 exhibits divergent expression between cancer and normal tissue, with Hsp72 being expressed at low levels in normal tissue while being amplified in many human cancers (Fig. **3**) [73]. Furthermore, reduced Hsp72 activity is well tolerated by non-transformed cells as demonstrated by numerous *in vitro* studies and loss of Hsp72 is also well tolerated by *hsp70*$^{-/-}$ mice [77, 78]. An additional feature of Hsp72 inhibition is that tumor toxicity is not dampened by activation of a counter-survival response, as is the case for HSR activation upon Hsp90 inhibition [79]

Ablation of Hsp72 expression has been shown to be toxic in many cancer models studied to date, while other certain cancer models have required dual inhibition of Hsp72 and Hsp73 for anti-cancer effects to be observed [78 - 80]. These studies highlighted a differential importance of HSP70 family members to the tumorigenicity of specific cancers and dually, the importance of identifying which HSP70 paralogs are subject to inhibition by a given HSP70 inhibitory molecule. Addressing these considerations may ultimately enable beneficial application of HSP70 inhibitors for cancer treatment i.e. using an inhibitor that targets the HSP70 paralogs known to be important for the cancer type being treated. Inhibitors of HSP70 family members identified to date that exhibit anti-cancer activity are listed in Fig. (**4**). Each of these molecules were shown to function by perturbing HSP70: (1) substrate binding (2) ATPase activity, or (3) interactions with regulatory proteins [81].

## 5A. Targeting the HSP70 Substrate-Binding Domain (SBD)

For cancer types where inhibition of individual HSP70 paralogs is desirable, targeting the SBD may prove a conducive strategy due to the relatively low degree of conservation shared within this region between HSP70 family members [81, 82]. As discussed above, HSP70 stabilizes numerous oncogenic client substrates through interaction with its SBD. HSP70 inhibitory molecules that target the SBD such as 2-phenylethynesulfonamide (PES) perturb substrate interaction with the SBD and lead to substrate de-stabilization and/or degradation [83, 84]. PES was suggested to target both Hsp72 and Hsp73 in a non-specific 'detergent-like' manner [80, 85]. Targeting a peptide-binding cleft within the SBD is a strategy that has previously been utilized with promising chemosensitization activity [86]. It was demonstrated that ectopic expression of short peptide sequences of the HSP70-interacting region of apoptosis-inducing factor (AIF) to compete with AIF sequestration by HSP70 and thereby permit AIF nuclear entry and promote apoptosis [86, 87]. While exogenous delivery of small peptide inhibitors to tumors remains a formidable challenge, the approach taken by Schmitt *et al.,* may provide a foundation for the design and/or identification of HSP70 inhibitors through peptide mimicry.

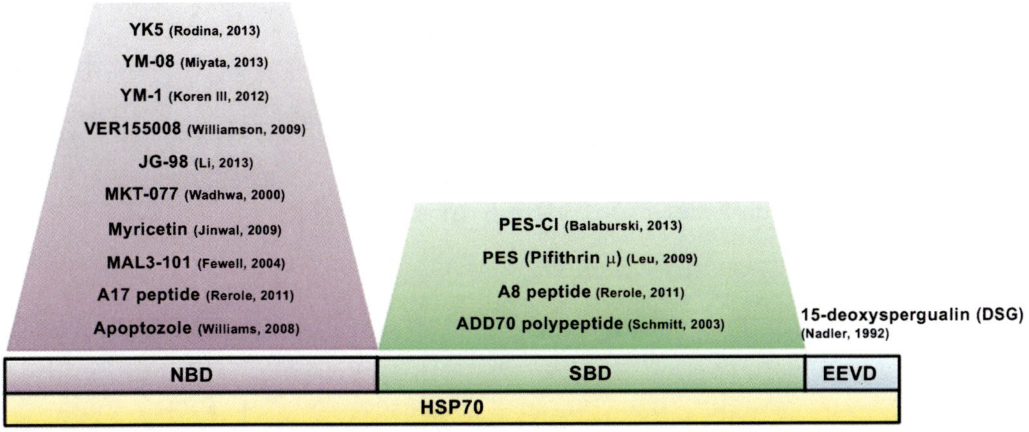


**Fig. (4). Inhibitors of Hsp 70 with anti-cancer activity.** Molecules with both anti-cancer and HSP 70-inhibitory activity are shown and grouped by the functional region of HSP70 primarily targeted by the given molecule.

## 5B. Targeting the HSP70 Nucleotide-Binding Domain

Inhibition of the ATPase activity of HSP70 proteins prevents substrate protein release and this promotes the holding of substrates by the chaperone and/or degradation via the CHIP-proteasome pathway [80, 83]. Consistent with the tumorigenic role of HSP70 in cancer, inhibitors of ATPase activity showed anti-tumor effects in various cancer models both *in vitro* and *in vivo* [88 - 90]. Inhibitors of HSP70 ATPase activity such as VER155008 and MKT-077 act upon more than one family member. VER155008 has anti-ATPase activity towards Hsp72, Hsp73 and Grp78 and MKT-077 interacts with the mitochondrial Hsp70 protein encoded by *HSPA9* as well as Hsp72 [89 - 91]. Thus, it is possible to inhibit multiple HSP70 species through one molecule that targets ATPase activity, and this approach might prove beneficial in contexts where multiple HSP70 paralogs have pro-tumorigenic functions in the same cancer type. In addition, Macias et al., demonstrated that inhibitors of the HSP70 ATPase domain could also possess specific affinity for single members of the family, further highlighting that the spectrum of HSP70 targets of a given ATPase inhibitor might be molecule specific [91].

Challenges exist, however, in targeting the ATPase domain of HSP70 proteins as a therapeutic approach. For example, the extreme affinity of the HSP70 NBD for ATP sets a high task for achieving competitive inhibition of this binding pocket [81]. In addition, the ATPase domain of HSP70 shares homology with similar structures in unrelated proteins and thus specificity may also prove problematic. For example MKT-077 was also reported to interact with F-actin in NIH-3T3 fibroblasts [81, 92]. In a recent study that utilized a structure-based modeling approach, five 'druggable' sites across HSP70 were identified [93]. Among these, the HSP70 inhibitor, YK5, was designed to form a covalent bond with the Cys267 residue of human Hsp72 and Hsp73 [93]. YK5 inhibited HSP70 refolding and ATPase activity while not directly competing for ATP binding. Targeting the Cys267 residue to form a covalent inhibitory interaction was a fine example of how intelligent molecule design could overcome some of the inherent challenges of drugging HSP70 such as its strong affinity to ATP [93].

### 5C. Perturbation of HSP70-Protein Interactions

The pleiotropic functions of HSP70 are facilitated by a large number of accessory proteins including nuclear exchange factors (NEFs), the DnaJ (HSP40) family of co-chaperones and proteins that contain a tetratricopeptide (TPR) domain that interact with the EEVD region at the C-terminal tail of Hsp70 [81]. Perturbing the interaction of HSP70 with accessory proteins is an anti-cancer strategy with promising potential, as indicated by a number of recent pre-clinical studies. For example, the inhibitor, YM-1, was shown to disrupt HSP70 complex formation with the NEF, Bag3. Inhibition of this interaction was associated with selective toxicity to transformed breast cancer cell lines over non-transformed cell lines. YM-1 was also shown to inhibit mammary and melanoma xenograft tumor growth *in vivo* [60]. A number of HSP70 inhibitors, including myricetin and MAL3-101 targeted the interaction of HSP70 with HSP40 members and had toxic properties in human cancer models including human multiple myeloma and pancreatic cancer cell lines [94, 96].

Inhibition of TPR protein interactions with the EEVD domain of HSP70 members is an avenue yet to be extensively investigated as an anti-cancer strategy [81]. The inhibitor, 15-deoxyspergualin, was shown to interact with the EEVD domain of HSP70. However limited efforts have been made to further investigate anti-cancer properties of 15-deoxyspergualin, since a phase II clinical trial of its use for metastatic breast cancer reported neuromuscular side effects with no benefit for disease [97, 98]. Whilst the outcome of disrupting interactions between HSP70 and accessory proteins may be highly specific to the given interaction, these studies have indicated that at least a selection of these HSP70-accessory protein relationships can indeed be targeted to achieve anti-cancer effects. These studies highlighted the prospect for more targets to emerge as the roles of HSP70-accessory protein interactions are characterized within different cancer contexts.

## CONCLUSIONS

Although Hsp72 is important in the etiology of cancer, the mechanisms behind its involvement in the disease are still in flux. Indeed, it is possible that the chaperone may play distinct roles according to the dominant driver oncogenes in individual malignancies. At the molecular level, the canonical function of Hsp72 is to chaperone key proteins in tumorigenesis and permit their accumulation in the malignant cell. However, in addition, important roles for Hsp72 in RNA metabolism are emerging and may be significant. The existence of multiple *HSP70* pseudogenes may also be important for future studies of potential lncRNA regulation of this family of chaperones. As the significance of this family of chaperones in cancer emerges, small molecule inhibitors are undergoing development as future potential cancer pharmaceuticals and this approach may emerge as an important "second front" in the war on cancer chaperones.

## CONSENT FOR PUBLICATION

Not applicable.

## CONFLICT OF INTEREST

The author declares no conflict of interest, financial or otherwise.

## ACKNOWLEDGEMENTS

Declared none.

## REFERENCES

[1] Lindquist S, Craig EA. The heat-shock proteins. Annu Rev Genet 1988; 22: 631-77. [http://dx.doi.org/10.1146/annurev.ge.22.120188.003215] [PMID: 2853609]

[2] Ciocca DR, Oesterreich S, Chamness GC, McGuire WL, Fuqua SA. Biological and clinical implications of heat shock protein 27,000 (Hsp27): a review. J Natl Cancer Inst 1993; 85(19): 1558-70. [http://dx.doi.org/10.1093/jnci/85.19.1558] [PMID: 8411230]

[3] Ciocca DR, Calderwood SK. Heat shock proteins in cancer: diagnostic, prognostic, predictive, and treatment implications. Cell Stress Chaperones 2005; 10(2): 86-103. [http://dx.doi.org/10.1379/CSC-99r.1] [PMID: 16038406]

[4] Ciocca DR, Arrigo AP, Calderwood SK. Heat shock proteins and heat shock factor 1 in carcinogenesis and tumor development: an update. Arch Toxicol 2013; 87(1): 19-48. [http://dx.doi.org/10.1007/s00204-012-0918-z] [PMID: 22885793]

[5] Neckers L, Ivy SP. Heat shock protein 90. Curr Opin Oncol 2003; 15(6): 419-24. [http://dx.doi.org/10.1097/00001622-200311000-00003] [PMID: 14624223]

[6] Trepel J, Mollapour M, Giaccone G, Neckers L. Targeting the dynamic HSP90 complex in cancer. Nat Rev Cancer 2010; 10(8): 537-49.
[http://dx.doi.org/10.1038/nrc2887] [PMID: 20651736]

[7] Lindquist S, Craig EA. The heat-shock proteins. Annu Rev Genet 1988; 22: 631-77.
[http://dx.doi.org/10.1146/annurev.ge.22.120188.003215] [PMID: 2853609]

[8] Kiang JG, Gist ID, Tsokos GC. Regulation of heat shock protein 72 kDa and 90 kDa in human breast cancer MDA-MB-231 cells. Mol Cell Biochem 2000; 204(1-2): 169-78.
[http://dx.doi.org/10.1023/A:1007016822939] [PMID: 10718636]

[9] Zorzi E, Bonvini P. Inducible hsp70 in the regulation of cancer cell survival: analysis of chaperone induction, expression and activity. Cancers (Basel) 2011; 3(4): 3921-56.
[http://dx.doi.org/10.3390/cancers3043921] [PMID: 24213118]

[10] Watanabe K, Tachibana M, Kim S, Watarai M. EEVD motif of heat shock cognate protein 70 contributes to bacterial uptake by trophoblast giant cells. J Biomed Sci 2009; 16: 113.
[http://dx.doi.org/10.1186/1423-0127-16-113] [PMID: 20003465]

[11] Scheufler C, Brinker A, Bourenkov G, *et al.* Structure of TPR domain-peptide complexes: critical elements in the assembly of the Hsp70-Hsp90 multichaperone machine. Cell 2000; 101(2): 199-210.
[http://dx.doi.org/10.1016/S0092-8674(00)80830-2] [PMID: 10786835]

[12] Brocchieri L, Conway de Macario E, Macario AJ. hsp70 genes in the human genome: Conservation and differentiation patterns predict a wide array of overlapping and specialized functions. BMC Evol Biol 2008; 8: 19.
[http://dx.doi.org/10.1186/1471-2148-8-19] [PMID: 18215318]

[13] Wickner S, Maurizi MR, Gottesman S. Posttranslational quality control: folding, refolding, and degrading proteins. Science 1999; 286(5446): 1888-93.
[http://dx.doi.org/10.1126/science.286.5446.1888] [PMID: 10583944]

[14] Rinn JL, Chang HY. Genome regulation by long noncoding RNAs. Annu Rev Biochem 2012; 81: 145-66.
[http://dx.doi.org/10.1146/annurev-biochem-051410-092902] [PMID: 22663078]

[15] Tang D, Khaleque MA, Jones EL, *et al.* Expression of heat shock proteins and heat shock protein messenger ribonucleic acid in human prostate carcinoma in vitro and in tumors in vivo. Cell Stress Chaperones 2005; 10(1): 46-58.
[http://dx.doi.org/10.1379/CSC-44R.1] [PMID: 15832947]

[16] Noonan EJ, Place RF, Giardina C, Hightower LE. Hsp70B′ regulation and function. Cell Stress Chaperones 2007; 12(4): 393-402.
[http://dx.doi.org/10.1379/CSC-278e.1] [PMID: 18229458]

[17] Calderwood SK, Mambula SS, Gray PJ Jr, Theriault JR. Extracellular heat shock proteins in cell signaling. FEBS Lett 2007; 581(19): 3689-94.
[http://dx.doi.org/10.1016/j.febslet.2007.04.044] [PMID: 17499247]

[18] Scieglińska D, Pigłowski W, Mazurek A, *et al.* The HspA2 protein localizes in nucleoli and centrosomes of heat shocked cancer cells. J Cell Biochem 2008; 104(6): 2193-206.
[http://dx.doi.org/10.1002/jcb.21778] [PMID: 18452162]

[19] Fu Y, Zhao H, Li XS, *et al.* Expression of HSPA2 in human hepatocellular carcinoma and its clinical significance. Tumour Biol 2014; 35(11): 11283-7.
[http://dx.doi.org/10.1007/s13277-014-2430-y] [PMID: 25117073]

[20] Zhang H, Gao H, Liu C, Kong Y, Wang C, Zhang H. Expression and clinical significance of HSPA2 in pancreatic ductal adenocarcinoma. Diagn Pathol 2015; 10: 13.
[http://dx.doi.org/10.1186/s13000-015-0253-9] [PMID: 25890028]

[21] Scieglinska D, Krawczyk Z. Expression, function, and regulation of the testis-enriched heat shock

HSPA2 gene in rodents and humans. Cell Stress Chaperones 2015; 20(2): 221-35. [http://dx.doi.org/10.1007/s12192-014-0548-x] [PMID: 25344376]

[22] Gong J, Weng D, Eguchi T, Murshid A, Sherman MY, Song B, *et al.* Targeting the hsp70 gene delays mammary tumor initiation and inhibits tumor cell metastasis. Oncogene 2015 Oct; 34(43): 5460-71. [http://dx.doi.org/10.1038/onc.2015.1] [PMID: 25659585]

[23] Kirschke E, Goswami D, Southworth D, Griffin PR, Agard DA. Glucocorticoid receptor function regulated by coordinated action of the Hsp90 and Hsp70 chaperone cycles. Cell 2014; 157(7): 1685-97. [http://dx.doi.org/10.1016/j.cell.2014.04.038] [PMID: 24949977]

[24] Hundley H, Eisenman H, Walter W, *et al.* The in vivo function of the ribosome-associated Hsp70, Ssz1, does not require its putative peptide-binding domain. Proc Natl Acad Sci USA 2002; 99(7): 4203-8. [http://dx.doi.org/10.1073/pnas.062048399] [PMID: 11929993]

[25] Otto H, Conz C, Maier P, *et al.* The chaperones MPP11 and Hsp70L1 form the mammalian ribosome-associated complex. Proc Natl Acad Sci USA 2005; 102(29): 10064-9. [http://dx.doi.org/10.1073/pnas.0504400102] [PMID: 16002468]

[26] Willmund F, del Alamo M, Pechmann S, *et al.* The cotranslational function of ribosome-associated Hsp70 in eukaryotic protein homeostasis. Cell 2013; 152(1-2): 196-209. [http://dx.doi.org/10.1016/j.cell.2012.12.001] [PMID: 23332755]

[27] Duttler S, Pechmann S, Frydman J. Principles of cotranslational ubiquitination and quality control at the ribosome. Mol Cell 2013; 50(3): 379-93. [http://dx.doi.org/10.1016/j.molcel.2013.03.010] [PMID: 23583075]

[28] Pechmann S, Willmund F, Frydman J. The ribosome as a hub for protein quality control. Mol Cell 2013; 49(3): 411-21. [http://dx.doi.org/10.1016/j.molcel.2013.01.020] [PMID: 23395271]

[29] Löffek S, Wöll S, Höhfeld J, *et al.* The ubiquitin ligase CHIP/STUB1 targets mutant keratins for degradation. Hum Mutat 2010; 31(4): 466-76. [http://dx.doi.org/10.1002/humu.21222] [PMID: 20151404]

[30] Ballinger CA, Connell P, Wu Y, *et al.* Identification of CHIP, a novel tetratricopeptide repeat-containing protein that interacts with heat shock proteins and negatively regulates chaperone functions. Mol Cell Biol 1999; 19(6): 4535-45. [http://dx.doi.org/10.1128/MCB.19.6.4535] [PMID: 10330192]

[31] Chan SP, Slack FJ. microRNA-mediated silencing inside P-bodies. RNA Biol 2006; 3(3): 97-100. [http://dx.doi.org/10.4161/rna.3.3.3499] [PMID: 17179742]

[32] Horton LE, James P, Craig EA, Hensold JO. The yeast hsp70 homologue Ssa is required for translation and interacts with Sis1 and Pab1 on translating ribosomes. J Biol Chem 2001; 276(17): 14426-33. [http://dx.doi.org/10.1074/jbc.M100266200] [PMID: 11279042]

[33] Buchan JR, Parker R. Eukaryotic stress granules: the ins and outs of translation. Mol Cell 2009; 36(6): 932-41. [http://dx.doi.org/10.1016/j.molcel.2009.11.020] [PMID: 20064460]

[34] Buchan JR, Kolaitis RM, Taylor JP, Parker R. Eukaryotic stress granules are cleared by autophagy and Cdc48/VCP function. Cell 2013; 153(7): 1461-74. [http://dx.doi.org/10.1016/j.cell.2013.05.037] [PMID: 23791177]

[35] Carpenter S, Ricci EP, Mercier BC, Moore MJ, Fitzgerald KA. Post-transcriptional regulation of gene expression in innate immunity. Nat Rev Immunol 2014; 14(6): 361-76. [http://dx.doi.org/10.1038/nri3682] [PMID: 24854588]

[36] Clevers H. The cancer stem cell: premises, promises and challenges. Nat Med 2011; 17(3): 313-9.

[http://dx.doi.org/10.1038/nm.2304] [PMID: 21386835]

[37] Borovski T, De Sousa E Melo F, Vermeulen L, Medema JP. Cancer stem cell niche: the place to be. Cancer Res 2011; 71(3): 634-9.
[http://dx.doi.org/10.1158/0008-5472.CAN-10-3220] [PMID: 21266356]

[38] Dean M. Cancer stem cells: Implications for cancer causation and therapy resistance. Discov Med 2005; 5(27): 278-82.
[PMID: 20704888]

[39] Chou SD, Murshid A, Eguchi T, Gong J, Calderwood SK. HSF1 regulation of beta-catenin in mammary cancer cells through control of HuR/elavL1 expression. Oncogene 2014 Apr; 23;34(17): 2178-88.

[40] TCGA. The Cancer Genome Atlas 2015.http://cancergenome.nih.gov/

[41] Tavaria M, Gabriele T, Kola I, Anderson RL. A hitchhiker's guide to the human Hsp70 family. Cell Stress Chaperones 1996; 1(1): 23-8.
[http://dx.doi.org/10.1379/1466-1268(1996)001<0023:AHSGTT>2.3.CO;2] [PMID: 9222585]

[42] Cerami E, Gao J, Dogrusoz U, *et al.* The cBio cancer genomics portal: an open platform for exploring multidimensional cancer genomics data. Cancer Discov 2012; 2(5): 401-4.
[http://dx.doi.org/10.1158/2159-8290.CD-12-0095] [PMID: 22588877]

[43] Gao J, Aksoy BA, Dogrusoz U, *et al.* Integrative analysis of complex cancer genomics and clinical profiles using the cBioPortal. Sci Signal 2013; 6(269): pl1.
[http://dx.doi.org/10.1126/scisignal.2004088] [PMID: 23550210]

[44] cBioPortal. cBioPortal for Cancer Genomics 2015 Available from: http://www.cbioportal.org/

[45] Vargas-Roig LM, Gago FE, Tello O, Aznar JC, Ciocca DR. Heat shock protein expression and drug resistance in breast cancer patients treated with induction chemotherapy. Int J Cancer 1998; 79(5): 468-75.
[http://dx.doi.org/10.1002/(SICI)1097-0215(19981023)79:5<468::AID-IJC4>3.0.CO;2-Z] [PMID: 9761114]

[46] Zhang L, Fok JJ, Mirabella F, *et al.* Hsp70 inhibition induces myeloma cell death via the intracellular accumulation of immunoglobulin and the generation of proteotoxic stress. Cancer Lett 2013; 339(1): 49-59.
[http://dx.doi.org/10.1016/j.canlet.2013.07.023] [PMID: 23887058]

[47] Volloch VZ, Sherman MY. Oncogenic potential of Hsp72. Oncogene 1999; 18(24): 3648-51.
[http://dx.doi.org/10.1038/sj.onc.1202525] [PMID: 10380887]

[48] Kang Y, Jung WY, Lee H, *et al.* Prognostic significance of heat shock protein 70 expression in early gastric carcinoma. Korean J Pathol 2013; 47(3): 219-26.
[http://dx.doi.org/10.4132/KoreanJPathol.2013.47.3.219] [PMID: 23837014]

[49] Kluger HM, Chelouche Lev D, Kluger Y, *et al.* Using a xenograft model of human breast cancer metastasis to find genes associated with clinically aggressive disease. Cancer Res 2005; 65(13): 5578-87.
[http://dx.doi.org/10.1158/0008-5472.CAN-05-0108] [PMID: 15994930]

[50] Kalogeraki A, Giannikaki E, Tzardi M, Kafousi M, Ieromonachou P, Dariviannaki K, *et al.* Correlation of heat shock protein (HSP70) expression with cell proliferation (MIB1), estrogen receptors (ER) and clinicopathological variables in invasive ductal breast carcinomas. Journal of experimental & clinical cancer research : CR 2007 Sep; 26(3): 367-8.
[PMID: 17987797]

[51] Hanahan D, Weinberg RA. Hallmarks of cancer: the next generation. Cell 2011; 144(5): 646-74.
[http://dx.doi.org/10.1016/j.cell.2011.02.013] [PMID: 21376230]

[52] Clemons NJ, Buzzard K, Steel R, Anderson RL. Hsp72 inhibits Fas-mediated apoptosis upstream of

the mitochondria in type II cells. J Biol Chem 2005; 280(10): 9005-12.
[http://dx.doi.org/10.1074/jbc.M414165200] [PMID: 15632129]

[53] Igney FH, Krammer PH. Death and anti-death: tumour resistance to apoptosis. Nat Rev Cancer 2002; 2(4): 277-88.
[http://dx.doi.org/10.1038/nrc776] [PMID: 12001989]

[54] Gabai VL, Budagova KR, Sherman MY. Increased expression of the major heat shock protein Hsp72 in human prostate carcinoma cells is dispensable for their viability but confers resistance to a variety of anticancer agents. Oncogene 2005; 24(20): 3328-38.
[http://dx.doi.org/10.1038/sj.onc.1208495] [PMID: 15735699]

[55] Meng L, Hunt C, Yaglom JA, Gabai VL, Sherman MY. Heat shock protein Hsp72 plays an essential role in Her2-induced mammary tumorigenesis. Oncogene 2011; 30(25): 2836-45.
[http://dx.doi.org/10.1038/onc.2011.5] [PMID: 21297664]

[56] Sherman MY, Meng L, Stampfer M, Gabai VL, Yaglom JA. Oncogenes induce senescence with incomplete growth arrest and suppress the DNA damage response in immortalized cells. Aging Cell 2011; 10(6): 949-61.
[http://dx.doi.org/10.1111/j.1474-9726.2011.00736.x] [PMID: 21824272]

[57] Yaglom JA, Gabai VL, Sherman MY. High levels of heat shock protein Hsp72 in cancer cells suppress default senescence pathways. Cancer Res 2007; 67(5): 2373-81.
[http://dx.doi.org/10.1158/0008-5472.CAN-06-3796] [PMID: 17332370]

[58] O'Callaghan-Sunol C, Gabai VL, Sherman MY. Hsp27 modulates p53 signaling and suppresses cellular senescence. Cancer Res 2007; 67(24): 11779-88.
[http://dx.doi.org/10.1158/0008-5472.CAN-07-2441] [PMID: 18089808]

[59] Calderwood SK. Molecular cochaperones: tumor growth and cancer treatment. Scientifica 2013; 2013: 217513.
[http://dx.doi.org/10.1155/2013/217513] [PMID: 24278769]

[60] Colvin TA, Gabai VL, Gong J, *et al.* Hsp70-Bag3 interactions regulate cancer-related signaling networks. Cancer Res 2014; 74(17): 4731-40.
[http://dx.doi.org/10.1158/0008-5472.CAN-14-0747] [PMID: 24994713]

[61] Rauch JN, Gestwicki JE. Binding of human nucleotide exchange factors to heat shock protein 70 (Hsp70) generates functionally distinct complexes in vitro. J Biol Chem 2014; 289(3): 1402-14.
[http://dx.doi.org/10.1074/jbc.M113.521997] [PMID: 24318877]

[62] Beere HM. Death versus survival: functional interaction between the apoptotic and stress-inducible heat shock protein pathways. J Clin Invest 2005; 115(10): 2633-9.
[http://dx.doi.org/10.1172/JCI26471] [PMID: 16200196]

[63] Gabai VL, Mabuchi K, Mosser DD, Sherman MY. Hsp72 and stress kinase c-jun N-terminal kinase regulate the bid-dependent pathway in tumor necrosis factor-induced apoptosis. Mol Cell Biol 2002; 22(10): 3415-24.
[http://dx.doi.org/10.1128/MCB.22.10.3415-3424.2002] [PMID: 11971973]

[64] Kluger HM, Chelouche Lev D, Kluger Y, *et al.* Using a xenograft model of human breast cancer metastasis to find genes associated with clinically aggressive disease. Cancer Res 2005; 65(13): 5578-87.
[http://dx.doi.org/10.1158/0008-5472.CAN-05-0108] [PMID: 15994930]

[65] Guy CT, Cardiff RD, Muller WJ. Induction of mammary tumors by expression of polyomavirus middle T oncogene: a transgenic mouse model for metastatic disease. Mol Cell Biol 1992; 12(3): 954-61.
[http://dx.doi.org/10.1128/MCB.12.3.954] [PMID: 1312220]

[66] Weng D, Penzner JH, Song B, Koido S, Calderwood SK, Gong J. Metastasis is an early event in mouse mammary carcinomas and is associated with cells bearing stem cell markers. Breast Cancer Res

2012; 14(1): R18.
[http://dx.doi.org/10.1186/bcr3102] [PMID: 22277639]

[67] Teng Y, Ngoka L, Mei Y, Lesoon L, Cowell JK. HSP90 and HSP70 proteins are essential for stabilization and activation of WASF3 metastasis-promoting protein. J Biol Chem 2012; 287(13): 10051-9.
[http://dx.doi.org/10.1074/jbc.M111.335000] [PMID: 22315230]

[68] Bladt F, Riethmacher D, Isenmann S, Aguzzi A, Birchmeier C. Essential role for the c-met receptor in the migration of myogenic precursor cells into the limb bud. Nature 1995; 376(6543): 768-71.
[http://dx.doi.org/10.1038/376768a0] [PMID: 7651534]

[69] Zhou J, Schmid T, Frank R, Brüne B. PI3K/Akt is required for heat shock proteins to protect hypoxia-inducible factor 1alpha from pVHL-independent degradation. J Biol Chem 2004; 279(14): 13506-13.
[http://dx.doi.org/10.1074/jbc.M310164200] [PMID: 14726529]

[70] Kulshreshtha R, Davuluri RV, Calin GA, Ivan M. A microRNA component of the hypoxic response. Cell Death Differ 2008; 15(4): 667-71.
[http://dx.doi.org/10.1038/sj.cdd.4402310] [PMID: 18219318]

[71] Danilin S, Sourbier C, Thomas L, *et al.* Role of the RNA-binding protein HuR in human renal cell carcinoma. Carcinogenesis 2010; 31(6): 1018-26.
[http://dx.doi.org/10.1093/carcin/bgq052] [PMID: 20219773]

[72] López de Silanes I, Fan J, Yang X, *et al.* Role of the RNA-binding protein HuR in colon carcinogenesis. Oncogene 2003; 22(46): 7146-54.
[http://dx.doi.org/10.1038/sj.onc.1206862] [PMID: 14562043]

[73] Ciocca DR, Calderwood SK. Heat shock proteins in cancer: diagnostic, prognostic, predictive, and treatment implications. Cell Stress Chaperones 2005; 10(2): 86-103.
[http://dx.doi.org/10.1379/CSC-99r.1] [PMID: 16038406]

[74] Calderwood SK, Khaleque MA, Sawyer DB, Ciocca DR. Heat shock proteins in cancer: chaperones of tumorigenesis. Trends Biochem Sci 2006; 31(3): 164-72.
[http://dx.doi.org/10.1016/j.tibs.2006.01.006] [PMID: 16483782]

[75] Tang D, Khaleque MA, Jones EL, *et al.* Expression of heat shock proteins and heat shock protein messenger ribonucleic acid in human prostate carcinoma in vitro and in tumors in vivo. Cell Stress Chaperones 2005; 10(1): 46-58.
[http://dx.doi.org/10.1379/CSC-44R.1] [PMID: 15832947]

[76] Yang Z, Zhuang L, Szatmary P, *et al.* Upregulation of heat shock proteins (HSPA12A, HSP90B1, HSPA4, HSPA5 and HSPA6) in tumour tissues is associated with poor outcomes from HBV-related early-stage hepatocellular carcinoma. Int J Med Sci 2015; 12(3): 256-63.
[http://dx.doi.org/10.7150/ijms.10735] [PMID: 25798051]

[77] Hunt CR, Dix DJ, Sharma GG, *et al.* Genomic instability and enhanced radiosensitivity in Hsp70.1- and Hsp70.3-deficient mice. Mol Cell Biol 2004; 24(2): 899-911.
[http://dx.doi.org/10.1128/MCB.24.2.899-911.2004] [PMID: 14701760]

[78] Nylandsted J, Brand K, Jäättelä M. Heat shock protein 70 is required for the survival of cancer cells. Ann N Y Acad Sci 2000; 926(122-125): 122-5.
[PMID: 11193027]

[79] Powers MV, Clarke PA, Workman P. Dual targeting of HSC70 and HSP72 inhibits HSP90 function and induces tumor-specific apoptosis. Cancer Cell 2008; 14(3): 250-62.
[http://dx.doi.org/10.1016/j.ccr.2008.08.002] [PMID: 18772114]

[80] Schlecht R, Scholz SR, Dahmen H, *et al.* Functional analysis of Hsp70 inhibitors. PLoS One 2013; 8(11): e78443.
[http://dx.doi.org/10.1371/journal.pone.0078443] [PMID: 24265689]

[81] Assimon VA, Gillies AT, Rauch JN, Gestwicki JE. Hsp70 protein complexes as drug targets. Curr

Pharm Des 2013; 19(3): 404-17.
[http://dx.doi.org/10.2174/138161213804143699] [PMID: 22920901]

[82] Murphy ME. The HSP70 family and cancer. Carcinogenesis 2013; 34(6): 1181-8. [http://dx.doi.org/10.1093/carcin/bgt111] [PMID: 23563090]

[83] Budina-Kolomets A, Balaburski GM, Bondar A, Beeharry N, Yen T, Murphy ME. Comparison of the activity of three different HSP70 inhibitors on apoptosis, cell cycle arrest, autophagy inhibition, and HSP90 inhibition. Cancer Biol Ther 2014; 15(2): 194-9. [http://dx.doi.org/10.4161/cbt.26720] [PMID: 24100579]

[84] Leu JI, Pimkina J, Frank A, Murphy ME, George DL. A small molecule inhibitor of inducible heat shock protein 70. Mol Cell 2009; 36(1): 15-27. [http://dx.doi.org/10.1016/j.molcel.2009.09.023] [PMID: 19818706]

[85] Leu JI, Pimkina J, Pandey P, Murphy ME, George DL. HSP70 inhibition by the small-molecule 2-phenylethynesulfonamide impairs protein clearance pathways in tumor cells. Mol Cancer Res 2011; 9(7): 936-47. [http://dx.doi.org/10.1158/1541-7786.MCR-11-0019] [PMID: 21636681]

[86] Schmitt E, Parcellier A, Gurbuxani S, *et al.* Chemosensitization by a non-apoptogenic heat shock protein 70-binding apoptosis-inducing factor mutant. Cancer Res 2003; 63(23): 8233-40. [PMID: 14678980]

[87] Gurbuxani S, Schmitt E, Cande C, *et al.* Heat shock protein 70 binding inhibits the nuclear import of apoptosis-inducing factor. Oncogene 2003; 22(43): 6669-78. [http://dx.doi.org/10.1038/sj.onc.1206794] [PMID: 14555980]

[88] Koren J III, Miyata Y, Kiray J, *et al.* Rhodacyanine derivative selectively targets cancer cells and overcomes tamoxifen resistance. PLoS One 2012; 7(4): e35566. [http://dx.doi.org/10.1371/journal.pone.0035566] [PMID: 22563386]

[89] Massey AJ, Williamson DS, Browne H, *et al.* A novel, small molecule inhibitor of Hsc70/Hsp70 potentiates Hsp90 inhibitor induced apoptosis in HCT116 colon carcinoma cells. Cancer Chemother Pharmacol 2010; 66(3): 535-45. [http://dx.doi.org/10.1007/s00280-009-1194-3] [PMID: 20012863]

[90] Deocaris CC, Widodo N, Shrestha BG, *et al.* Mortalin sensitizes human cancer cells to MKT-07--induced senescence. Cancer Lett 2007; 252(2): 259-69. [http://dx.doi.org/10.1016/j.canlet.2006.12.038] [PMID: 17306926]

[91] Macias AT, Williamson DS, Allen N, *et al.* Adenosine-derived inhibitors of 78 kDa glucose regulated protein (Grp78) ATPase: insights into isoform selectivity. J Med Chem 2011; 54(12): 4034-41. [http://dx.doi.org/10.1021/jm101625x] [PMID: 21526763]

[92] Maruta H, Tikoo A, Shakri R, Shishido T. The anti-RAS cancer drug MKT-077 is an F-actin cross-linker. Ann N Y Acad Sci 1999; 886: 283-4. [http://dx.doi.org/10.1111/j.1749-6632.1999.tb09437.x] [PMID: 10667240]

[93] Rodina A, Patel PD, Kang Y, *et al.* Identification of an allosteric pocket on human hsp70 reveals a mode of inhibition of this therapeutically important protein. Chem Biol 2013; 20(12): 1469-80. [http://dx.doi.org/10.1016/j.chembiol.2013.10.008] [PMID: 24239008]

[94] Phillips PA, Sangwan V, Borja-Cacho D, Dudeja V, Vickers SM, Saluja AK. Myricetin induces pancreatic cancer cell death via the induction of apoptosis and inhibition of the phosphatidylinositol 3-kinase (PI3K) signaling pathway. Cancer Lett 2011; 308(2): 181-8. [http://dx.doi.org/10.1016/j.canlet.2011.05.002] [PMID: 21676539]

[95] Chang L, Miyata Y, Ung PM, *et al.* Chemical screens against a reconstituted multiprotein complex: myricetin blocks DnaJ regulation of DnaK through an allosteric mechanism. Chem Biol 2011; 18(2): 210-21. [http://dx.doi.org/10.1016/j.chembiol.2010.12.010] [PMID: 21338918]

[96] Braunstein MJ, Scott SS, Scott CS, Behrman S, Walter P, Wipf P, *et al.* Antimyeloma Effects of the Heat Shock Protein 70 Molecular Chaperone Inhibitor MAL3-101. Journal of Oncology 2011; 2011

[97] Nadler SG, Dischino DD, Malacko AR, Cleaveland JS, Fujihara SM, Marquardt H. Identification of a binding site on Hsc70 for the immunosuppressant 15-deoxyspergualin. Biochem Biophys Res Commun 1998; 253(1): 176-80.
[http://dx.doi.org/10.1006/bbrc.1998.9775] [PMID: 9875240]

[98] Dhingra K, Valero V, Gutierrez L, *et al.* Phase II study of deoxyspergualin in metastatic breast cancer. Invest New Drugs 1994; 12(3): 235-41.
[http://dx.doi.org/10.1007/BF00873965] [PMID: 7896543]

## CHAPTER 2

# Use of Coarse-Grained and All-Atom Molecular Dynamics to Study Hsp70 and Hsp40 Chaperone Action

**Ewa I. Gołaś[1,2,¶], Magdalena A. Mozolewska[1,2,¶], Paweł Krupa[1,2], Cezary Czaplewski[1], Harold A. Scheraga[2] and Adam Liwo[1,*]**

[1] *Faculty of Chemistry, University of Gdańsk, Wita Stwosza 63, 80-308 Gdańsk, Poland*

[2] *Baker Laboratory of Chemistry and Chemical Biology, Cornell University, Ithaca, NY, 14853-1301, USA*

**Abstract:** The 70 kDalton (Hsp70) chaperones perform a variety of functions in living cells, the most crucial being assisting correct protein folding, refolding misfolded proteins, and participating in iron-sulfur cluster biogenesis. The chaperones consist of the nucleotide-binding domain which, upon transitions between the ADP- to ATP-bound state, undergoes slight conformational changes, which trigger major conformational changes in the conformation of the whole molecule, ultimately leading to substrate binding or release. This chapter summarizes our work on the simulations of the chaperone cycle by means of all-atom and coarse-grained molecular dynamics, and on modeling the structure and interactions of two complexes that are formed during the process of iron-sulfur biogenesis: the binary complex composed of the Iron-sulfur protein 1 and the Jac1 Hsp40 cochaperone from yeast, and the ternary complex composed of the Iron-sulfur protein 1, the Jac1 Hsp40 cochaperone, and the Stress-seventy subfamily Q protein 1 Hsp70 chaperone from yeast.

**Keywords:** Hsp70 chaperones, Chaperone cycle, Iron-sulfur biogenesis, Molecular modeling, Molecular dynamics.

## INTRODUCTION

The Hsp70 (70-kDalton) heat-shock proteins are key molecular chaperones that occur in all living cells and are involved in a wide range of cellular processes, whose orchestration is based on reiterative cycles of substrate-binding and release [1, 2]. The chaperones contain two domains: a substrate-binding domain (SBD) and the nucleotide-binding domain (NBD), which is an ATP-ase; the two are

* **Corresponding author Adam Liwo:** Faculty of Chemistry, University of Gdansk, Gdańsk, Poland; Tel: +48 58 523 5124; Fax: +48 58 523 5012; Email: adam@sun1.chem.univ.gda.pl

¶ These authors contributed equally to the paper

**Mario D. Galigniana (Ed.)**

connected through a short hydrophobic linker. The chaperones are implicated in nascent protein folding, membrane transport events, the mediation of protein aggregates and re-folding, and the stress response—a suspiciously diverse set of cellular functions for only two domains, and the transfer of Fe-S clusters [3]. Hsp70s rely on interactions with other cycle members such as nucleotide exchange factors (NEFs) and J-proteins, which tune the protein for its destined function [4]. Moreover, Hsp70s rely on a complex allosteric network for communication with not only other players, but also between its substrate-binding domain and nucleotide-binding domain, whose mutual interaction and response to substrate and/or nucleotide are the chaperone's central function [2].

Computational methods, together with insight from the available experimental data, are valuable tools to study biomolecular processes. In this chapter, we summarize the results of our theoretical studies of the mechanism of the Hsp70 chaperone cycle and of the docking of the iron-sulfur protein 1 (Isu1) onto the Jac1 Hsp40 cochaperone from yeast, and modeling the ternary complex between Isu1, Jac1 and Ssq1 (a yeast Hsp70). Because of the large size of the systems, both all-atom and coarse-grained molecular dynamics, as well as comparative-modeling methods were used in the research.

## METHODS

All-atom molecular dynamics simulations were run with the AMBER11 force field [5, 6] with explicit water in a periodic box. The TIP3P water model [7] was used. Because of the long computation time and large size of the system, the calculations were run on the ANTON supercomputer dedicated to all-atom molecular dynamics simulations [8].

Coarse-grained molecular dynamics was run with the physics-based UNRES model and force field developed in our laboratory [9 - 12]. In the UNRES model, a polypeptide chain is reduced to a sequence of α-carbon atoms, which serve as geometric points, with attached united side chains and united peptide groups positioned halfway between two consecutive α-carbon atoms. The solvent is implicit in the side chain – side chain interactions in the model. The effective energy function originates from the potential of mean force of polypeptide chain(s) in aqueous solution and its components were carefully derived from the energy surfaces of model systems, and the total energy function was subsequently calibrated to reproduce the structures and thermodynamics of selected training proteins [11]. Molecular dynamics with UNRES is carried out in the Langevin-dynamics mode [13, 14], in which the non-conservative forces from the solvent are represented as net friction and random forces. The reduction of the number of degrees of freedom not only lowers the cost of force evaluation by orders of magnitude compared with that of all-atom calculations but also, because of the elimination of the fast-moving degrees of freedom including the solvent degrees of freedom, the time scale is extended by approximately three orders of magnitude compared to that of all-atom simulations. As a result, simulated folding of a small protein which takes several days with the 512-core dedicated ANTON machine takes only a couple of hours with a single processor on a standard off-shelf desktop with UNRES [13, 14]. Details of the force field [9 - 12] as well as the details of molecular-dynamics implementation [13, 14] of the UNRES model are in the references cited.

The Iterative Threading ASSEmbly Refinement (I-TASSER) server [15] and YASARA (Yet Another Scientific Artificial Reality Application) [16] software were used for homology modeling of the structure of the iron-sulfur-binding protein (Isu1) and of the Hsp70 chaperone (Ssq1) from yeast. The ZDOCK server [17] and the AutoDock software were used to obtain the initial structure of the complexes between Isu1 and Jac1 and Isu1, Jac1, and Ssq1.

## RESULTS

### Mechanism of Chaperone Cycle

As stated in the Introduction, the Hsp70 chaperone molecules contain two domains: SBD and NBD, which is an ATP-ase; the two are connected through a short hydrophobic linker [2]. In the substrate-bound conformation, the two domains are separated from each other while, in the substrate-free (ATP-bound) conformation, the SBD opens up to interact tightly with the NBD. The complete structures of the substrate-bound conformation were determined for many chaperones, including the bacterial Hsp70 chaperone (PDB: 2KHO) [18], while the structure of the complete ATP-bound form was determined only recently for the bacterial Hsp70 chaperone (PDB: 4B9Q) [19]. The experimental structures of both forms are shown in Fig. (**1**).

Information regarding the binding state of the NBD is transmitted to the SBD and vice-versa. The binding of ATP results in an increase of the rate of dissociation of the substrate from the substrate-binding domain; conversely, ADP favors substrate binding [4]; the presence of bound substrate stimulated ATP hydrolysis. ATP binding and release must thus induce conformational changes in the NBD which are relayed, through allostery, to the whole protein. Experimental studies have suggested that these changes involve a relative twist motion of the two sub-domains comprising NBD: NBD-1 and NBD-2 [20]. An understanding of such details necessitates insight from protein dynamics at the per-residue and/or atomic scales. Computational methods - and in particular, molecular dynamics simulations - provide an unambiguous and facile method with which the motion of the protein may be observed and analyzed.

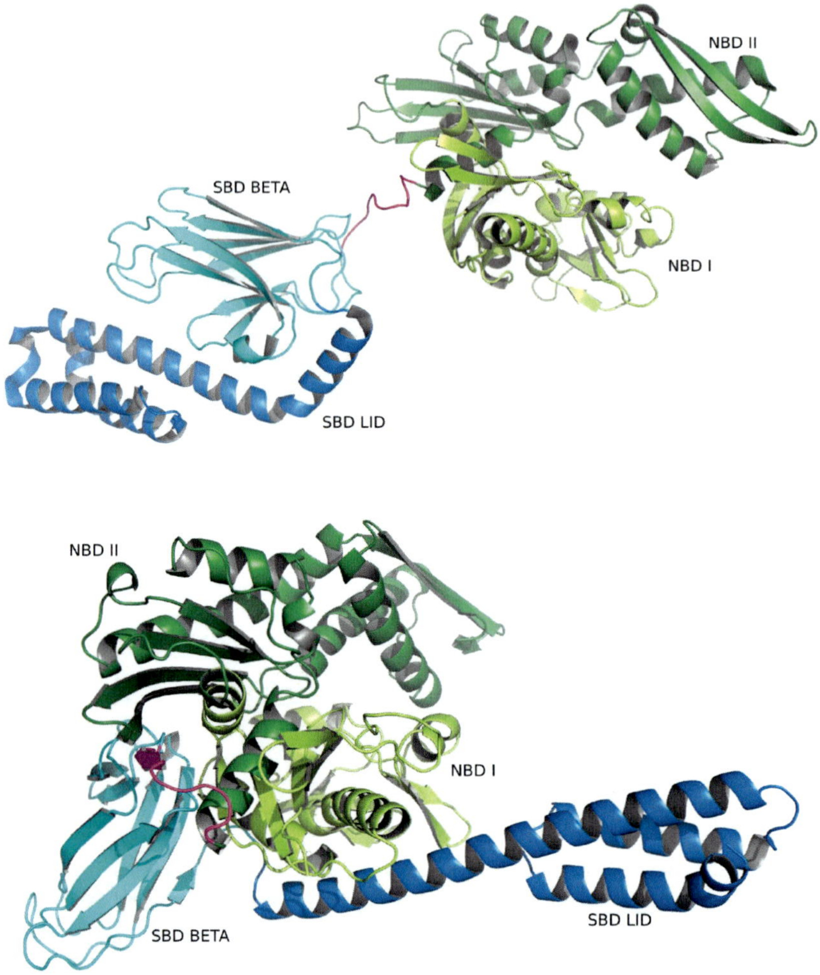


**Fig. (1).** (A) The SBD-closed; PDB: 2KHO and (B) SBD-open; PDB: 4B9Q:A structure of the Hsp70 chaperone from *E.coli.*. The respective subdomains: NBD I, NBD II. SBD I, SBD β and SBD lid (or SBD α) are indicated and colored light green, dark green, light blue, and dark blue, respectively.

To study the conformational dynamics of the NBD for different nucleotide-bound states, we used all-atom molecular dynamics [21]. The Hsp70 NBD from *Bos taurus* (PDB: 3C7N, chain B) was simulated for 1-2 μs in a series of canonical Molecular Dynamics simulations; the system—the NBD domain solvated in explicit water—was simulated in three states: with bound ADP, bound ATP, and without nucleotide. Random thermal motion is an inherent element of each trajectory—which makes the net motion of organized protein structures difficult to identify, track, and correlate. An effective filter for such motion, and furthermore, for the extraction of key trends, is Essential Dynamics [22], which is a Principal Component Analysis (PCA) technique [23] applied to the time-dependent coordinates of the protein. Eigenvectors of the coordinate co-variance matrix represent trend motions, with the possibility of reproducing a desired level of variance in the system (usually 70-80%) by considering a representative set of eigenvectors, whose individual contributions may be assessed by the relative size of each eigenvalue. Taking into account the eigenvectors across all trajectories and binding states of 3C7N, vectors may be grouped by similarity based on the absolute value of their scalar product, the criterion being $|\mathbf{v}_i \circ \mathbf{v}_j| >= 0.45$, where $\mathbf{v}_i$ and $\mathbf{v}_j$ are the eigenvectors being compared and further, by binding state, allowing sorting into classes.

Following such treatment, conclusions can be drawn regarding the protein dynamics of the NBD [21]; in particular, trajectory dynamics can be transcribed as a sum of representative vectors from specific classes of motion. Possible classes entail: the No-Nucleotide (NN)-mutual, the Nucleotide (NUC)-mutual, the All-mutual, the ADP-unique, the ATP-unique, and the Apo-unique classes, respectively (Fig. **2**). As stated, each of the above may be labeled as being either of mutual- or unique-type: mutual-type vectors are shared between several binding states, while unique-type vectors are binding-state specific. For any given trajectory, the largest vector contributions - and hence motions that dominate the behavior of the NBD - originate from vectors belonging to mutual-type classes. Projecting such vectors onto the structure of the protein reveals that their trend motions pertain to the block-like rotation of the sub-domains (Ia, IIa, Ib, and IIb) of the NBD (see Fig. **2** for the definition of these subdomains). This result is coherent with Nuclear Magnetic Resonance chemical shift perturbation studies, which suggested such motion as an instrument for signal transduction within the domain [24]. Further, a rotational order for the subdomains was determined as IIb > Ib > IIa > Ia. That the IIb subdomain is most active is consistent with residual dipolar coupling analyses that identified its significant structural (20°) rotation [20]. All-mutual vectors appear regardless of the binding state, and are inherent to the structure of the NBD. Although NN-mutual vectors originate from a likeness to the unbound NBD, they are not restricted to the Apo state; their trademark is a distinguished preference for IIb rotation over the remaining domains.

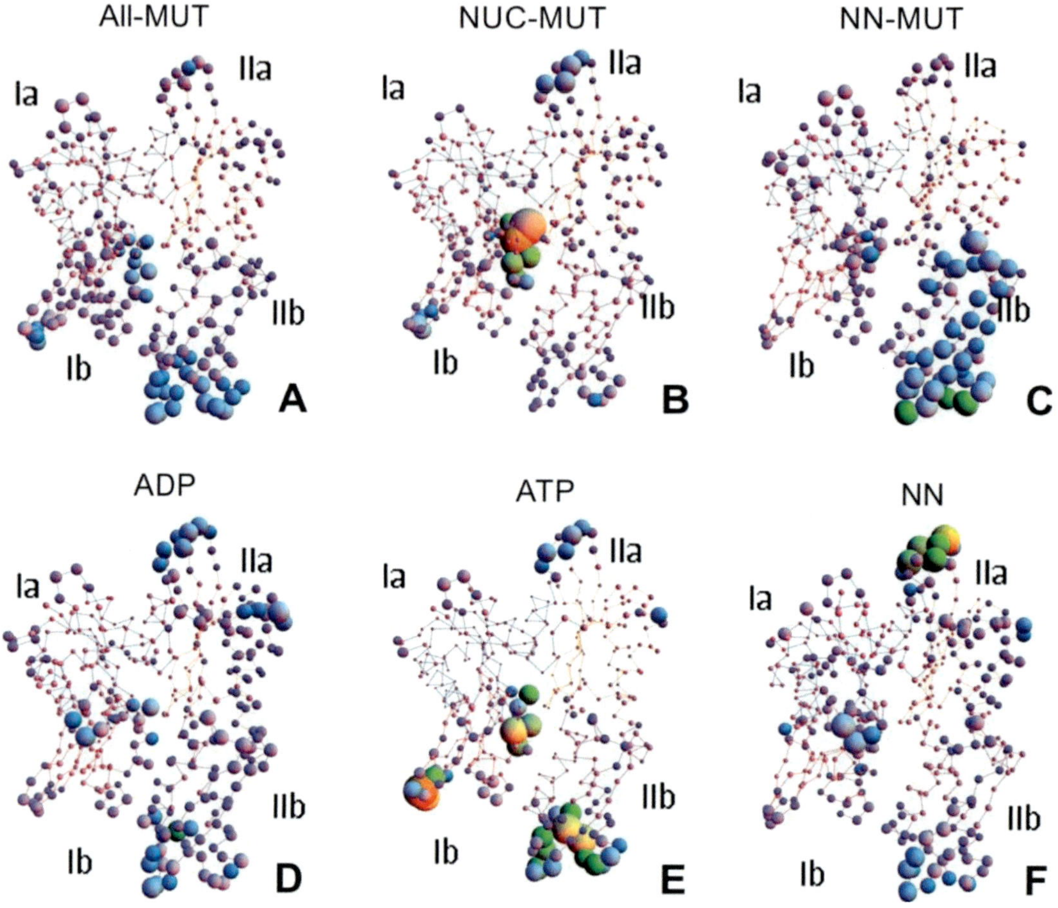


**Fig. (2).** Imaging of the $C^{\alpha}$ atoms targeted by the weighted coefficient vector for each trend motion class. Sphere size reflects the relative mobility (contribution) of an atom to the vector. A. The all-mutual vector class. B. The NUC-mutual vector class. C. The NN-mutual (apo) vector class. D. The ADP-unique vector class. E. The ATP-unique vector class. F. The NN-unique (apo) vector class. Reproduced with permission from Fig. (**7**) of Ref [21].

Such motion is in accord with experimental studies that have implicated rotation of the IIb subdomain to NEF activity and to the Apo state [25]. NUC-mutual class eigenvectors pertain only to systems bearing nucleotide - they principally do not describe subdomain rotation, but rather the motion of "hotspots" on the surface (via external loops) and at interstices of the NBD. Hotspots are localized regions (generally short fragments of chain) that exhibit heightened mobility (Fig. **3**). Unique-class vectors deal entirely with the motion of hot-spots. Several such zones are simultaneously activated and appear for each unique-type class: although some hot-spots re-appear over the different binding states, their exact combination is singular and binding-state dependent, effectively describing a characteristic network of mobile sites on the protein that selectively awakens upon nucleotide binding (such as, *e.g.,* the motions of the top and of the bottom loops of the ATP-bound NBD that are marked by representing the respective part of the $C^{\alpha}$-trace in Fig. (**3C**) as thick red sticks). Such a multi-site response is coherent with a combinatorial mechanism to allosteric control, and opens a pathway for tunable and diverse chaperone function, especially when the possibility of running the network backwards - that is, accessing hotspots by, in principal, J-proteins or other cycle members, and hence activating the NBD - is considered. The existence of hotspots has been confirmed by chemical shift perturbation NMR studies [24]; and many of the hotspots detected overlap with those identified computationally.

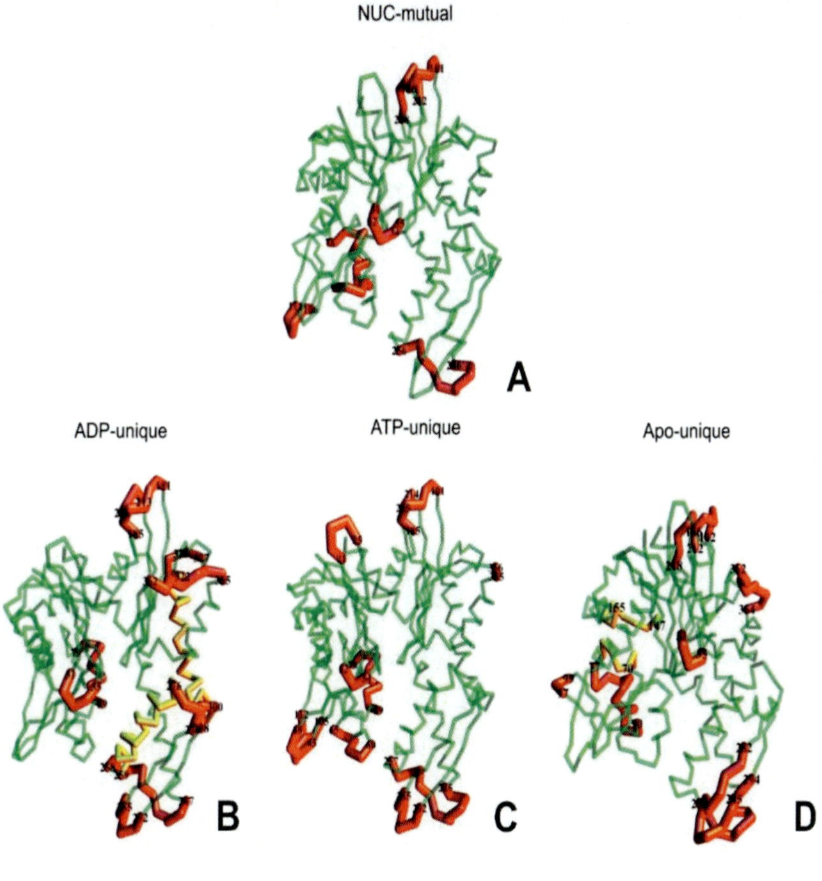


**Fig. (3).** Hotspot locations in the NBD as a function of the A. NUC-mutual, B. ADP-unique, C. ATP-unique, and D. Apo-unique (NN) motion classes. Reproduced with permission from Fig. (**8**) of Ref [21].

Stepping back to look at the NBD from a lobe-based perspective (the Ia-Ib and IIa-IIb subdomains), the higher number of eigenvectors necessary to describe the

ATP-bound state (when compared against its ADP-bound analogue) is reflected in a more restricted freedom of in-plane and out-of-plane angles that the lobes can assume - a corollary to the depressed contribution of mutual-type sub-domain vectors that orchestrate the module-like rotation of subdomains that combine to define the relative lobe-lobe interaction.

Faced with the taxing demands of a large all-atom system size, the protein dynamics of the whole chaperone can be efficiently studied using a coarse-grained model; the UNRES force-field and model was used for this purpose [26]. Canonical Langevin molecular dynamics simulations of the bacterial Hsp70 DnaK, starting from the SBD-closed experimental structure (*E. coli*, PDB code: 2KHO) were carried out for approximately 0.4 ms of real time (0.18 μs in UNRES time; the difference of the time scale occurs because of averaging out the secondary degrees of freedom, including the solvent degrees of freedom in UNRES [13]). Restraints were applied to the lobes of the NBD in a series of simulations so as to mimic the ADP-bound, the ATP-bound (with subdomains I and II of the protein in a more compact configuration, modeled after the structure of a homologous [ATP-bound] Hsp110), and the Apo states (lobes I and II free with respect to each other), respectively. The SBD was free of substrate. Three binding interaction arrangements were identified for the SBD and NBD domains, namely: Type I, Type II, and "Closed Binding" (Fig. **4**). The most frequent binding event observed was closed binding, in which the closed SBD bound to the NBD, docking for the most part at the NBD-I surface. Binding Types I and II entailed opening of the SBD and the consequent binding of its component α- and β-subdomains to locations on the surface of the NBD. The type I conformation bound α-SBD to the front of NBD-I and β-SBD to the top (and back) of the NBD I and II interstice. Type II binding reversed the orientation of the α- and β-subdomains. The type I conformation resembles the experimental structure of the open ATP-bound homologue Hsp110 [27] and the DnaK ATP-bound crystal structure [19], which was determined after the results of our studies were published [26]. The conformation of the NBD influenced the frequency with which binding events occurred. The ATP-bound state favored open-binding (Type I and II) incidents, in contrast to the ADP-bound state, which portrayed the lowest propensity for open-bound conformations. This is coherent with the ATP-state favoring substrate loss and ADP-state substrate binding [24]. Moreover, the ATP-bound state supported the existence of a flexible and open-SBD, even without binding to the NBD, while its ADP-bound counterpart always terminated SBD opening with docking to the NBD.

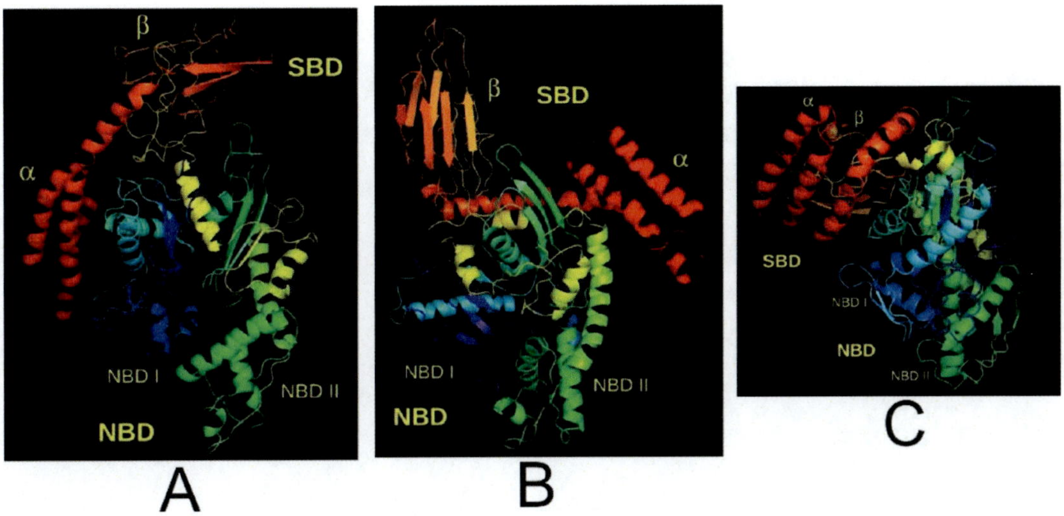


**Fig. (4).** Docking conformations of Hsp70 following A. Type I, B. Type II, and C. Closed binding, respectively. Reproduced with permission from Fig. (**5**) of Ref [26], copyright 2012 American Chemical Society.

Based on the results of coarse-grained simulations of the transition from the SBD-closed to the SBD-open conformation of the bacterial Hsp70 chaperone, we proposed a plausible mechanism of chaperone cycle, which is illustrated in Fig. (**5**). As shown in the Figure, in the SBD-closed structure, NBD I and NBD II are in a "crossed" orientation, and the three-helix-bundle "lid" (SBD-α) binds to the SBD-β, the substrate (S; thin black line) remaining trapped between the SBD-β and SBD-α. SBD-β is connected to NBD-II by a linker (black line), which is connected to the "switch" (SW) helix (extending from E369 to G380 and represented as a long yellow rectangle with ragged-line border, partially covered by the NBD-I and NBD-II lobes in Fig. (**5**)) that is held between the main body of NBD-I and its HO helix (extending from E171 to Y179 and shown as a small green rectangle with dashed border covered by the NBD-I and NBD-II lobes in Fig. (**5**)). Thus, the SW helix is quite firmly tied to NBD-I. ATP binding results in out-of-plane (scissor-like) rotation of NBD-I and NBD-II toward each other. Consequently, the SW helix rotates toward the back side of NBD-II. Following this event, the linker is moved close to the outer β-strand of the IIa subdomain and, as a result of this motion, forms an additional strand of the β-sheet of IIa; this event is depicted as locking the blue circles representing backbone amide protons in the half-rings. The binding of the linker to NBD II brings the SBD close to NBD, thus enabling interdomain communication. Simultaneously or following the above-mentioned event, helix A+B of SBD-α (part 2 of Fig. **5**) straightens, and this domain dissociates from SBD-β, and binds to NBD-I, which leads to the formation of the "open" structure (2) and substrate release. When ATP is removed by a nucleotide-exchange factor, and substrate is delivered by Hsp40 cochaperone (the J-domain protein), the NBD-I and NBD-II domains return to the "cross" orientation and the SW helix rotates, peeling the linker segment off from the β-sheet, and pulling SBD-β away from NBD-II, which initiates the transition to the closed structure.

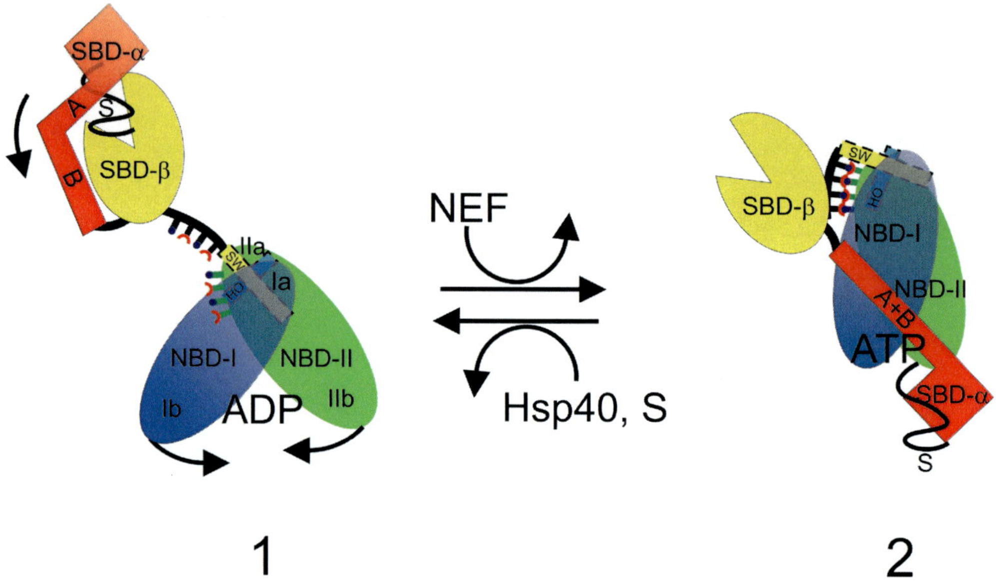


**Fig. (5).** A plausible mechanism for Hsp70 chaperone opening (transition from structure 1 (SBD-closed), in which SBD-α and SBD-β are in contact, to structure 2 (SBD-open), in which SBD-α and SBD-β are far apart from each other and bind to opposite faces of the NBD) suggested by coarse-grained simulations of DnaK (PDB ID: 2KHO). NBD-I is shown as a blue transparent ellipse, NBD-II is represented as a green ellipse, SBD-β is shown as a yellow incomplete ellipse, and SBD-α is shown as a red broken L-shape in structure 1 and as a red L-shape in structure 2. The linker segment is represented as a thick black line, and the substrate (S) is represented as a thin black line. The vertical sticks stretching out of the linker segment and out of the left upper corner of NBD-II represent backbone amide (with blue circles representing hydrogen atoms) and backbone carbonyl (with red halfrings representing oxygen atoms) groups, respectively. The "switch" (SW) and "holder" (HO) helices (see text for definition) are shown as a yellow and a blue rectangle, respectively, with ragged-line borders. The subdomains, the helices, and the substrate are labeled except for subdomains Ia and IIa in structure 2. Reproduced with permission from Fig. (**12**) of ref [26], copyright 2012 American Chemical Society.

## Modeling Iron-sulfur Cluster Biogenesis

One of many important functions of Hsp70s is Fe-S cluster biogenesis, which is usually studied experimentally for yeast chaperones. The yeast Hsp70 chaperones duplicated during evolution [28] to create the Stress-seventy subfamily Q protein 1 (Ssq1), an Hsp70 chaperone, which is highly specialized in Fe-S cluster biogenesis. The J-protein, also termed co-chaperone, is a mitochondrial chaperone which, together with Hsp70, assists in the transfer of the Fe-S clusters from the scaffold Iron Sulfur 1 (Isu1) protein to a recipient protein, such as ferredoxin and aconitase. The formation of the complex between Jac1 and Isu1 is a key element of the Fe-S cluster biogenesis process [29 - 35]. In next stage of the process, both proteins interact with Ssq1. Biochemical studies of the analogues of Jac1 and Isu1 in *Escherichia coli*, HscB and IscU [36], respectively, demonstrated that their interaction is important for maintaining Hsp70 ATPase activity and for transfer of Fe-S clusters. The disturbance of the balance of these processes can have very serious and dangerous consequences. In *Homo sapiens*, they can cause diseases such as cerebellar and Friedrich's ataxia, myopathy and microcytic anemia [37].

The iron-sulfur clusters are the most ancient co-factors of proteins involved in many processes like catalysis and electron transfer. These are essential processes in *Saccharomyces cerevisiae* and bacterial systems. The release of the Fe/S cluster from Isu1, and its transfer and incorporation into recipient apoproteins (Apo) is facilitated by late components of the Iron Sulfur Cluster (ISC) assembly machinery including the ATP-dependent Hsp70 chaperone Ssq1 and the DnaJ-like co-chaperone Jac1 [32].

In the proposed scheme of iron-sulfur cluster biogenesis (Fig. **6**) [38], the cluster is transferred from Isu1 to the respective target protein in a cascade of processes, which is initiated by sequestering the Fe-S cluster bound to Isu1 from the cysteine desulphurase complex Isd11-Nfs1 by Jac1 (an Hsp40 cochaperone). The next step involves binding the Hsp70 chaperone and transferring the Fe-S cluster to the respective target protein; thus it is one of the key stages of the whole process [32]. The mechanism of this process is unknown, although mutational studies suggest the mode of interaction Isu1 with Jac1 and Hsp70. We, therefore, studied the binding of Isu1 to Jac1 with Hsp70 by using large-scale coarse-grained molecular dynamics (MD) simulations with the UNRES force field.

## Modeling the Structure of Isu1 from Yeast

The Isu1 and Jac1 proteins are highly conserved during evolution and their equivalents can be found in every eukaryote organisms. Isu1 contains 165 amino-acid residues of which the first 27 residues are a fragment responsible for transport of the protein into the mitochondrion and it serves as a "transit" or "signal" peptide, which has no relevance to its function, and the mature protein does not have this fragment [38 - 41]. The structure of Isu1 is unknown and we, therefore, applied the comparative-modeling techniques [15, 16] to predict the 3D structure of this protein.

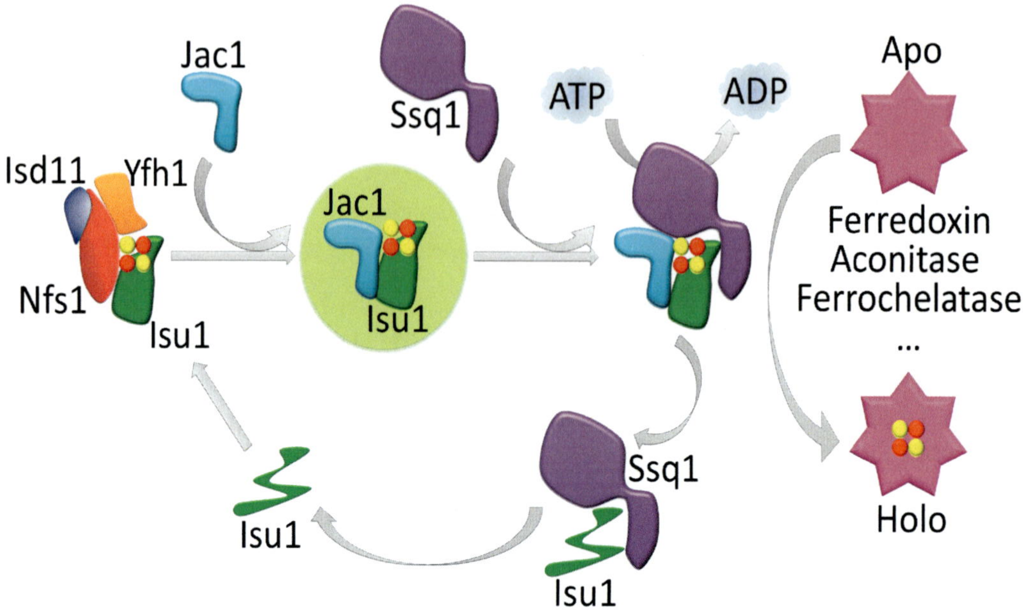


**Fig. (6).** Schematic representation of Fe-S cluster biogenesis cycle. The green circle marks the main stage of our investigation – Isu1-Jac1 complex. Small yellow and red circles represent the Fe-S cluster.

Four models were obtained from the I-TASSER server [15] and a hybrid model resulted from application of YASARA [16]. To construct the homology models, various Isu1 homologue proteins from different organisms were used (e.g. *Mus musculus*, *Escherichia coli*, *Haemophilus influenzae*, and *Aquifex aeolicus*) (Fig. **7**). The models from the I-TASSER server and from the YASARA program were similar, but the hybrid model from the YASARA program contained the most compact and regular structure with the best agreement to a PSIPRED prediction [42]. To refine the structure of the chosen homology model and to check the stability of the model we performed the molecular dynamics simulation with the use of the AMBER11 all-atom force field [5, 6]. This simulation showed that the structure was stable during 100 ns of the simulation and only small fluctuations (RMSD between 1.5 and 2.0 Å) could be observed, which were related to structure relaxation.

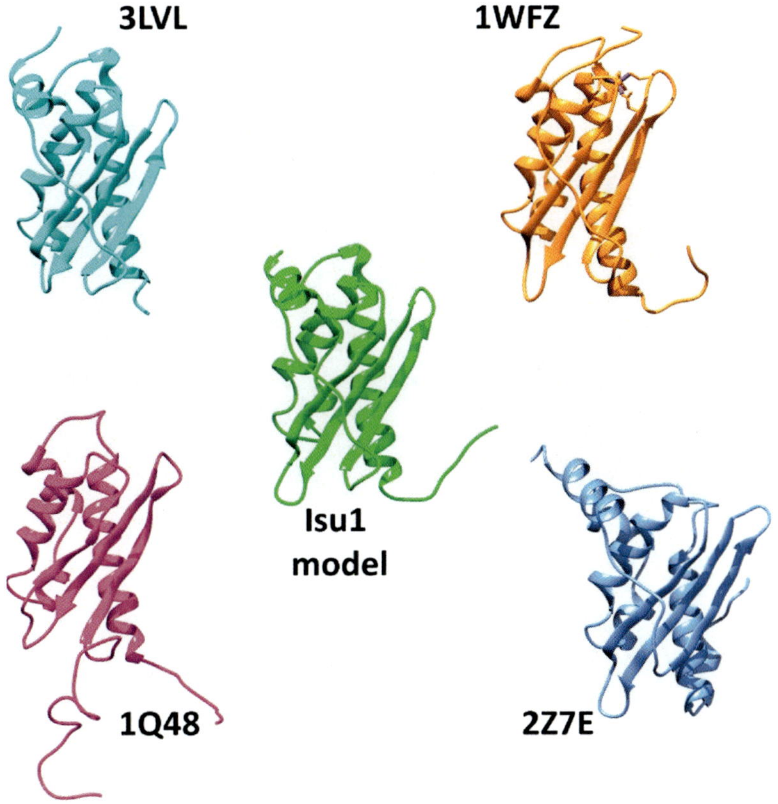


**Fig. (7).** Homologue proteins used in comparative modeling to obtain the structure of Isu1 protein.

## Modeling the Structure of the Binary Isu1-Jac1 Complex and Assessing the Stability of its Interactions

In the next step, we carried out an extensive search of the docking space of Isu1 to Jac1 (global docking) to determine the approximate binding position of these proteins. For this purpose we fed the homology model of Isu1 and the crystal structure of Jac1 (PDB: 3UO3) [34], to ZDOCK server [17]. From the prediction, we obtained three structures (Fig. **8A**, **8B** and **8C**). In the first two structures, Isu1 binds to the J-domain of Jac1. However, this is the binding site of the Ssq1 Hsp70 chaperone, whose cochaperone is Jac1. We, therefore, selected the third structure as a plausible structure of the Isu1-Jac1 complex and refined it by performing a limited search of the docking space in the neighborhood of the initial structures (local docking) by using the ZDOCK server and the modified AutoDock program [43]. As a result, three main conformations (Fig. **8D**, **8E** and **8F**) were obtained. Consequently, all these latter three structures were selected for further studies.

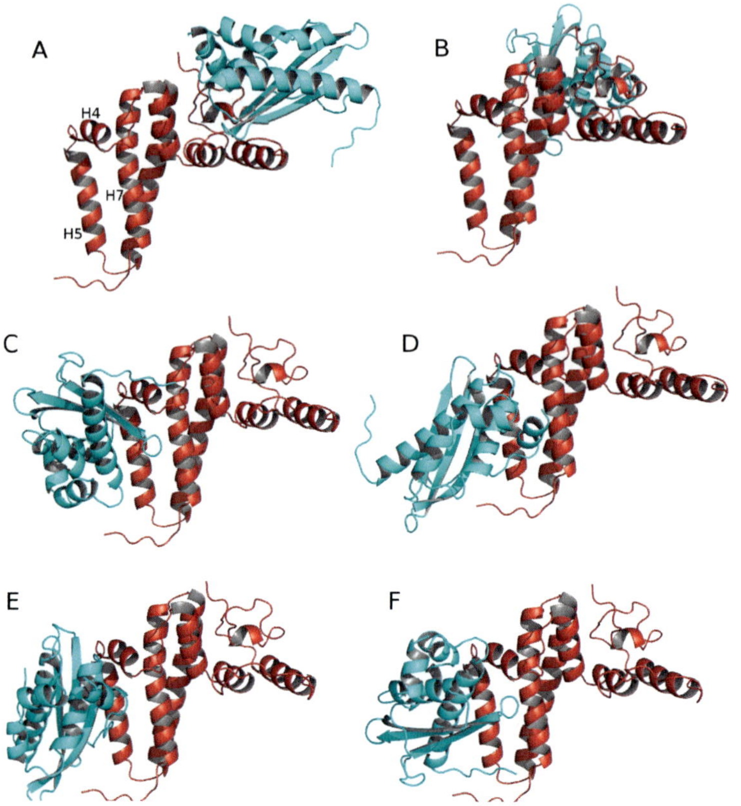


**Fig. (8).** Structures of the Isu1-Jac1 complex. **A, B, C,** structures obtained from global docking; **D, E, F,** structures obtained from local docking. The helices of Jac1 are labeled in panel A.

To check the stability of the selected three models of the Isu1-Jac1 complex, we performed molecular dynamics simulations using the coarse-grained UNRES force field, which provides a roughly 1000-times speed-up compared to all-atom approaches [9 - 12]. A total of 16 trajectories were simulated for each model, each trajectory consisting of 40 million steps with a 4.89 fs time step. The total length of each trajectory was 200 ns of UNRES time (around 200 μs of real time). All trajectories were combined together and clustered using Ward's minimum-variance method [44 - 46], with the $C^{\alpha}$ RMSD as a measure of the distances between conformations. The conformations corresponding to cluster centers were considered as representatives of the clusters. A total of 8 conformations thus obtained represent different orientations of Jac1 with respect to Isu1; they are shown in Fig. (**9**). Two of them represent the most populated clusters and are, therefore, the most probable (Fig. **9**). The structure of the Isu1-Jac1 complex, in which Isu1 binds through helix H5 (Fig. **9A**) suggests that Jac1 extracts Isu1 from the complex with protein Nfs1, Isd11 and Yfh1 (Fig. **6**). Once Isu1 is extracted and bound to Jac1, the interface undergoes moderate changes, and Isu1 then interacts with Jac1 mainly through the B1-B3 β-sheets (Fig. **9B**).

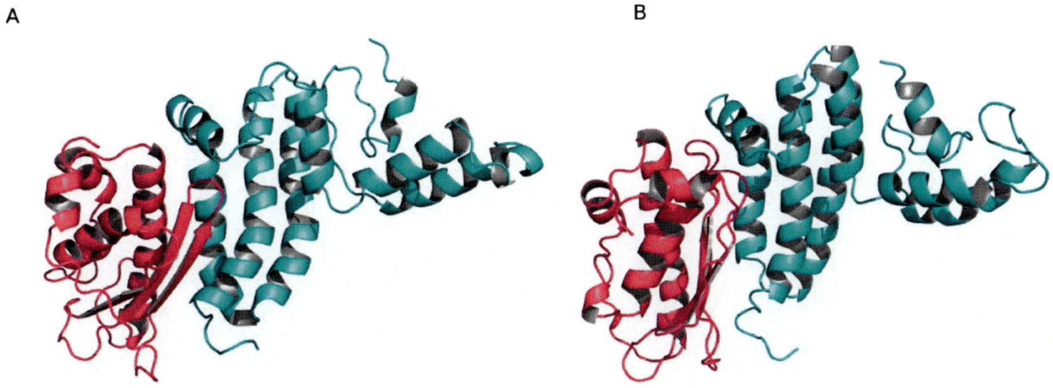

**Fig. (9).** Two most probable structures of the Isu1-Jac1 complex. **A)** Isu1 binds through helix H5; **B)** Isu1 binds through β-sheet section.

By analyzing the interactions between the components of the heterodimer, we found the residues, which could play crucial roles in binding to Isu1 (*i.e.*, which are in contact with the residues of the counterpart protein in the heterodimer). A map of the distances between the Isu1 and Jac1 residues averaged over complexes is shown in Fig. (**10**). Residues $L_{105}$, $L_{109}$, and $Y_{163}$ of Jac1, which were also found to be important for yeast-cell functioning by mutational experiments [34], form tight interactions. To verify the importance of these residues in the stabilization of the structure of the complex, we carried out canonical MD simulations with the UNRES force field on the mutated complex, in which the $L_{105}$, $L_{109}$, and $Y_{163}$ residues of Jac1 were replaced with alanine residues. After simulation performed with the coarse-grained UNRES force field, we observed that substituting alanine for tyrosine in position 163 resulted in destabilization of helices H5 and H7 (see Fig. **8A** for definition), which changed the binding pattern of Isu1 to Jac1 so that Isu1 binds by its top part and not by the β-sheet face. The destabilization of the complex also took place when the leucine residues in positions 105 and 109 were mutated to alanine residues. The residues of Isu1 found important for binding in our studies and found important for the functioning of yeast cells by mutational experiments are $L_{63}$, $V_{72}$, and $F_{94}$ [34]. The residues which were found important in our study in the formation of the binding interface but not yet tested in mutational experiments were residues $L_{104}$, $K_{107}$, $D_{110}$, $D_{113}$, $E_{114}$, and $Q_{117}$ of Jac1 and residues $N_{95}$, $T_{98}$, $P_{102}$, and $H_{112}$ of Isu1 located on the β-sheet section of this protein and residues $V_{159}$, $L_{167}$, $A_{170}$, and $W_{174}$, located on helix H5 [38]. The complete binding interface is shown on Fig. (**11**).

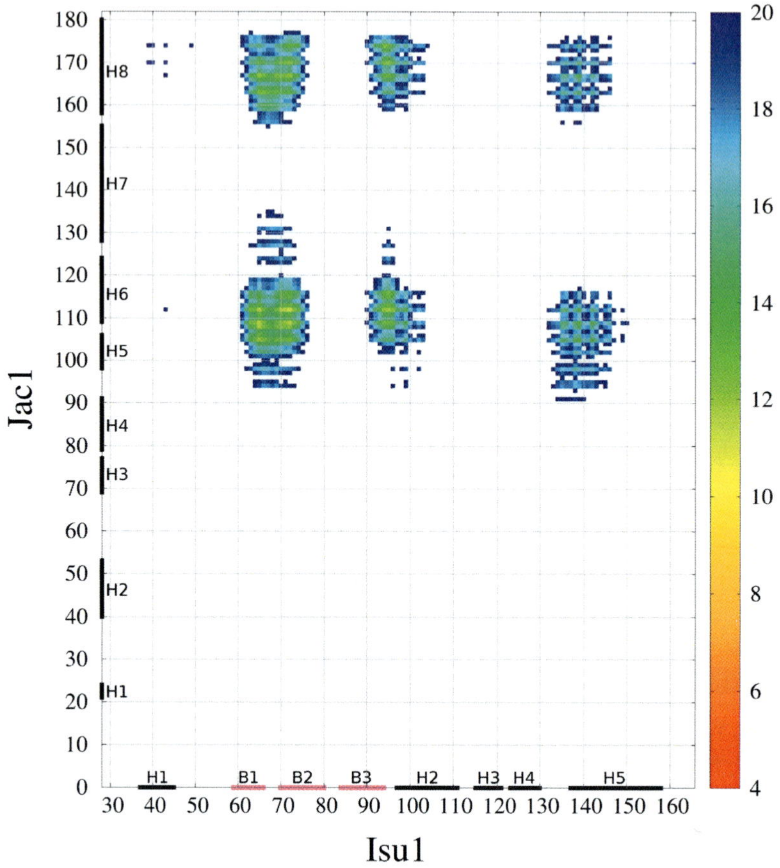


**Fig. (10).** Distance map (Cα...Cα atom distances smaller than 20 Å) averaged over all clusters of the simulated structures of the Isu1/Jac1 complex. The Cα...Cα atom distances are shown in color scale in Å from red (smallest distances to dark blue (20 Å). Distances greater than 20 Å are not mapped. Close contacts occur when residues are at a distance shorter than 7 Å. Secondary structure elements are marked by lines at the bottom with labels of the respective helix (H, black) or β-strand (B, pink) on the x and y axes. Residue numbers are marked on the horizontal axis and on the vertical axis. Reproduced with permission from Fig. (**5**) of Ref [38].

## Preliminary Molecular-modeling Study of the Structure of the Isu1-Jac--Ssq1 Ternary Complex

Having obtained the most probable structure of the Isu1-Jac1complex [38], we made a preliminary attempt at modeling the structure of the complete ternary complex between Isu1, Jac1, and Ssq1 from yeast. Because there is no experimental structure of Ssq1, we have modeled it by using I-TASSER and Robetta [47]. The structures of other homologous chaperones (*e.g.,* from *Escherichia coli* and *Saccharomyces cerevisiae*) were used as templates. Both tools produced the SBD-closed and the SBD-open conformations, respectively

(Fig. **12**). Similar to the other Hsp70 chaperones (*e.g.,* the Hsp70 from *Escherichia coli*; Fig. **1**), these two conformations differ in the position of the three-α-helix-bundle C-terminal 'lid', which is separated from the SBD-β subdomain in the SBD-open conformation, while the lid is very close to the SBD-β subdomain in the SBD-closed conformation. For further investigation, we used the open conformation of protein Ssq1 because only this form can bind the substrate.

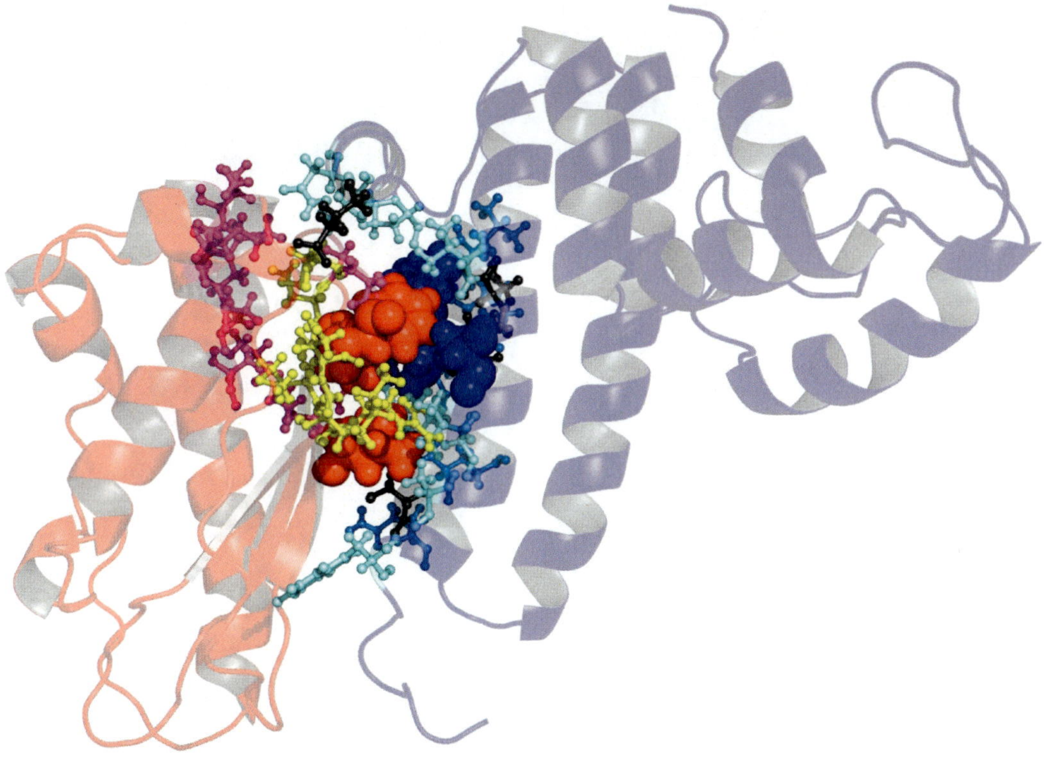

**Fig. (11).** A representative structure of the Isu1-Jac1 complex (from group C of Fig. **8**) obtained by molecular docking followed by UNRES/MD simulations. The bulk of the structure is shown in cartoon representation (blue: Jac1, red: Isu1), while the residues of the binding interface are shown in atomic-detailed representation. The Jac1 residues found to interact tightly with their Isu1 counterparts in our simulation studies [38] and found important for cell functioning in earlier experimental studies ($L_{105}$, $L_{109}$, and $Y_{163}$) [34] are colored dark-blue and shown in space-filling representation; the Jac1 residues found to interact tightly with their Isu1 counterparts and suggested by experiment to be important in cell functioning ($L_{104}$, $K_{107}$, $D_{110}$, $D_{113}$, $E_{114}$, and $Q_{117}$) are colored blue and shown in ball-and-stick representation, residues that were found to form interactions are colored cyan and shown in ball-and stick representation (residues $N_{95}$, $T_{98}$, $P_{102}$, $H_{112}$, $V_{159}$, $L_{167}$, $A_{170}$, $W_{174}$), while the residues of Jac1 found by simulations to make less tight contacts with Isu1 but also possibly important for binding (residues $E_{91}$, $V_{108}$, $S_{116}$, and $E_{160}$) are colored dark-gray and shown in ball-and-stick representation. The same hierarchy of representation and colors red ($L_{63}$, $V_{72}$, and $F_{94}$), yellow ($V_{64}$, $A_{66}$, $D_{71}$, $M_{73}$, $R_{74}$, $K_{92}$, $T_{93}$, $C_{96}$), and magenta ($G_{65}$, $G_{70}$, $G_{95}$, $V_{135}$, $K_{136}$, $H_{138}$, $C_{139}$, $L_{142}$) are used for the residues of Isu1 found to be important for binding by both simulation and experiment, by simulation and, to a lesser extent, by experiment, and by simulation only, respectively. Reproduced with permission from Fig. (**9**) of Ref [38].

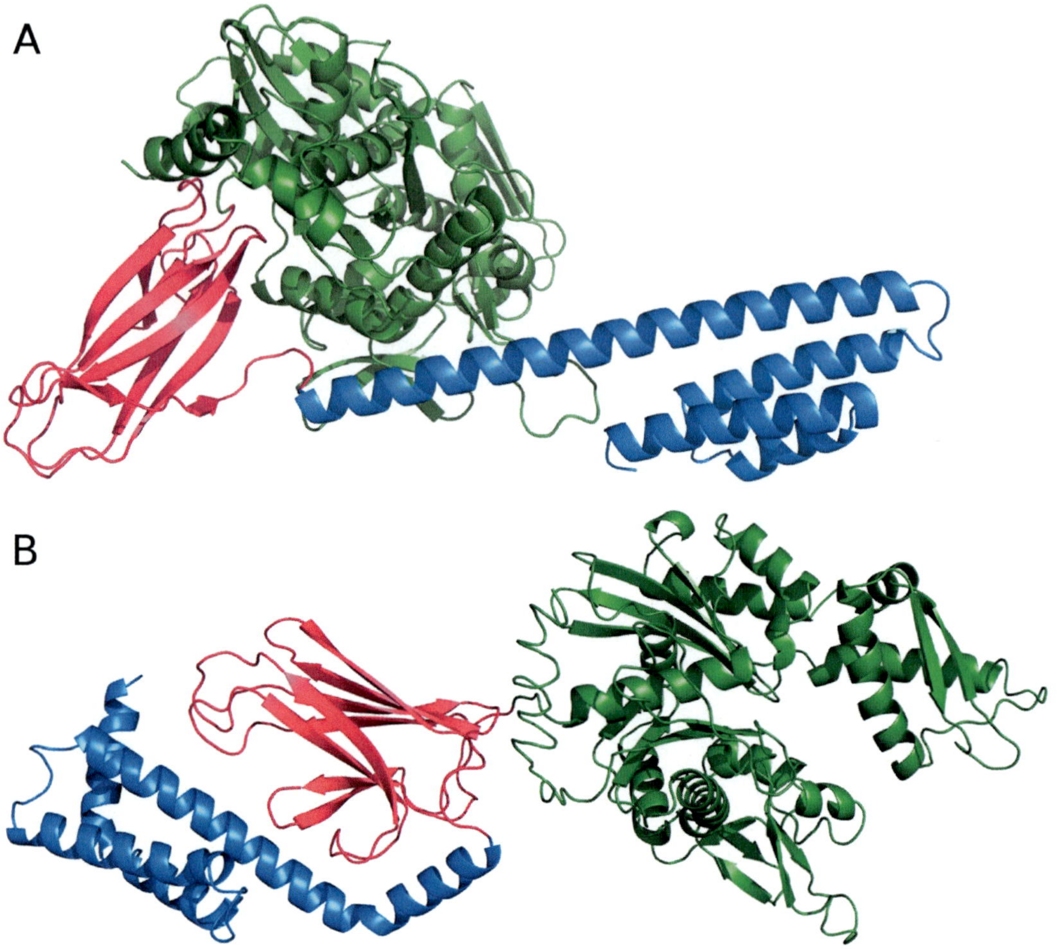


**Fig. (12).** Homology models of yeast Ssq1 chaperone colored by the characteristic fragments: NDB domain (green), SBD domain (pink and blue), and 'lid' (blue). A) open conformation; B) closed conformation.

Docking was performed by using the ZDOCK [17] and HADDOCK [48] servers. Because of the large size of the Ssq1 protein, global docking did not produce satisfactory result. To restrict the search space we, therefore, used mutation-experiment data, which show that residues $L_{97}$, $P_{98}$, $P_{99}$, $V_{100}$, and $K_{101}$ of Isu1 and residues $F_{462}$ and $V_{472}$ of Ssq1 are involved in binding Isu1 to Ssq1 [48 - 50], while the Jac1 protein binds to Ssq1 by HPD motif, and from Ssq1 residue responsible

for interaction with Jac1 are $S_{145}$, $G_{147}$, $L_{148}$, $S_{400}$, $E_{526}$ and $P_{539}$ [49, 51]. The most common structure from restricted docking is presented on Fig. (**13**).

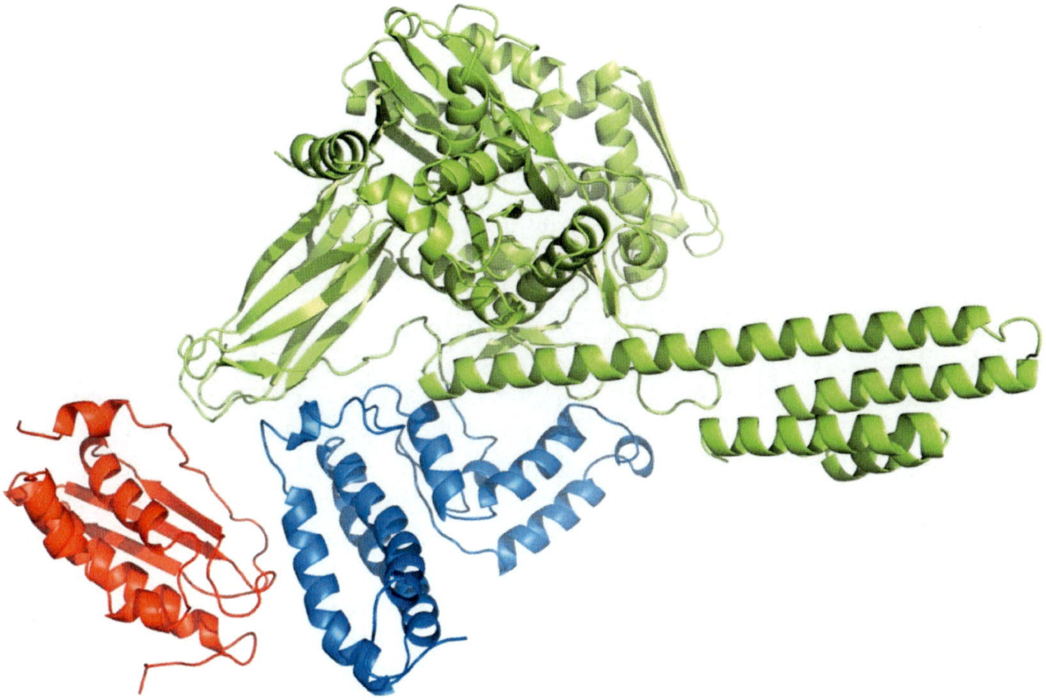

**Fig. (13).** The most common structure of the Isu1 Jac1 and Ssq1 complex, which appeared during the molecular-docking calculations

The obtained structure of the Isu1-Jac1-Ssq1 complex (Fig. **13**) was subsequently subjected to UNRES/MD simulations with newly implemented side-chain – backbone correlation potentials [52, 53] and turned out to be stable after a 200 ns (or 200 μs real time) simulations, this suggesting that the structure from docking is a very probable model of the complex. Because the size of the system is larger than that of the Isu1-Jac1 complex, longer coarse-grained simulations must be carried out to assess the stability of the ternary complex and also simulations of mutated complex need to be carried out to determine which residues influence its stability the most. These studies are being carried out in our laboratory.

## CONCLUSIONS AND OUTLOOK

In the studies summarized in this chapter, we used various molecular-modeling tools to study the mechanism of the chaperone cycle and the key part of iron-sulfur-cluster biosynthesis, which involves the binding of the Hsp70 chaperone and its Hsp40 co-chaperone. To investigate the chaperone cycle, we used all-atom molecular dynamics, while the dynamics of the complete chaperone was studied with the coarse-grained UNRES force field because of the large size of the system. The results of both studies were in full agreement with the available experimental data [21, 26]; moreover, one of the simulated SBD-open structures turned out to be very similar to the crystal structure of the ATP-analogue-bound Hsp70 chaperone from *Escherichia coli*, which was determined [19] after our theoretical results were published [26], demonstrating that the applied method is predictive. In addition, from our simulation studies we also found two other modes of the docking of the SBD to the NBD [26], one involving SBD opening but docking it to the other face of the NBD (Fig. **4B**) and one involving the docking of the closed SBD to the NBD. These docking modes are probably not productive as far as the chaperone cycle is concerned, and are not stable in the crystal state but should be considered as possible kinetic traps when modeling the conformational transition from the SBD-closed to the SBD-open conformation and vice versa. Having assessed that the results of modeling agree with the available experimental data pertaining to the structure and mobility of the system under study, an advanced analysis of the obtained molecular dynamics trajectory could then be performed [mainly through PCA (essential dynamics)] to identify the key functionally-important motions, the 'hot-spots' on the protein surface and, ultimately, to propose a plausible mechanism of the transition from the SBD-closed to the SBD-open conformation [26].

Modeling the interactions of Isu1 with Jac1 and the structure of the Isu1-Jac-Ssq1 ternary complex required the use of all molecular-modeling tools available; these included comparative modeling, coarse-grained molecular dynamics and all-atom molecular dynamics; in addition, experimental information was included to model the structure of the ternary complex. The obtained structure of the Isu1-Jac1 complex [38] was consistent with the available mutation data [34]. Use of coarse-grained molecular dynamics enabled us to assess the stability of the modeled structure and its sensitivity to mutations, the results of which turned out to be consistent with the experimental data. In summary, the applied methodology also turned out to be able to deal with the complex mechanisms of cellular processes that involve molecular chaperones.

## CONSENT FOR PUBLICATION

Not applicable.

## CONFLICT OF INTEREST

The author declares no conflict of interest, financial or otherwise.

## ACKNOWLEDGEMENTS

We thank Dr. Jarosław Marszałek (Intercollegiate Faculty of Biotechnology,

University of Gdańsk and Medical University of Gdańsk, Poland) for helpful discussions and feedback while carrying out the research described in this chapter. This work was supported by grants from the Foundation for Polish Science (grant MPD/2010/5), the U.S. National Institutes of Health (GM-14312), the U.S. National Science Foundation (MCB10-19767), and the Polish National Science Center (DEC-2012/06/A/ST4/00376). Computational resources were provided by (a) Pittsburgh Supercomputer Center through use of the 512-node ANTON supercomputer (grant PSCA10025P), which was made possible through the NIH Award (NIH RC2GM093307) awarded to CMU through the NRBSC (b) Argonne Leadership Computing Facility at Argonne National Laboratory, which is supported by the Office of Science of the U.S. Department of Energy under contract DE-AC02-06CH11357, (c) the Academic Computer Center in Gdańsk TASK and (d) Interdisciplinary Center of Mathematical and Computer Modeling (ICM) at the University of Warsaw. Our Beowulf clusters at Baker Laboratory of Chemistry, Cornell University and at the Faculty of Chemistry, University of Gdańsk, were also used to run calculations and perform trajectory analysis.

## REFERENCES

[1] Polier S, Dragovic Z, Hartl FU, Bracher A. Structural basis for the cooperation of Hsp70 and Hsp110 chaperones in protein folding. Cell 2008; 133(6): 1068-79. [http://dx.doi.org/10.1016/j.cell.2008.05.022] [PMID: 18555782]

[2] Kampinga HH, Craig EA. The HSP70 chaperone machinery: J proteins as drivers of functional specificity. Nat Rev Mol Cell Biol 2010; 11(8): 579-92. [http://dx.doi.org/10.1038/nrm2941] [PMID: 20651708]

[3] Saibil HR. Chaperone machines in action. Curr Opin Struct Biol 2008; 18(1): 35-42. [http://dx.doi.org/10.1016/j.sbi.2007.11.006] [PMID: 18242075]

[4] Clerico EM, Tilitsky JM, Meng W, Gierasch LM. How HSP70 molecular machines interact with their substrates to mediate diverse physiological functions. J Mol Biol 2015; 427(7): 1575-88. [http://dx.doi.org/10.1016/j.jmb.2015.02.004] [PMID: 25683596]

[5] Case DA, Cheatham TE III, Darden T, *et al.* The AMBER biomolecular simulation programs. J Comput Chem 2005; 26(16): 1668-88. [http://dx.doi.org/10.1002/jcc.20290] [PMID: 16200636]

[6] Roitberg A, Seabra G, Kolossváry I, *et al.* AMBER 11 Users' Manual. Univ Calif 2010; 1-300.

[7] Mark P, Nilsson L. Structure and Dynamics of the TIP3P, SPC, and SPC/E water models at 298 K. J Phys Chem A 2001; 105(43): 9954-60. [http://dx.doi.org/10.1021/jp003020w]

[8] Shaw DE, Chao JC, Eastwood MP, *et al.* Anton, a special-purpose machine for molecular dynamics simulation. Commun ACM 2008; 51(7): 91-7. [http://dx.doi.org/10.1145/1364782.1364802]

[9] Liwo A, Kaźmierkiewicz R, Czaplewski C, *et al.* United-residue force field for off lattice protein structure simulations: III. Origin of backbone hydrogen bonding cooperativity in united-residue potentials. J Comput Chem 1998; 19: 259-76. [http://dx.doi.org/10.1002/(SICI)1096-987X(199802)19:3<259::AID-JCC1>3.0.CO;2-S]

[10] Liwo A, Czaplewski C, Pillardy J, Scheraga H. Cumulant-based expressions for the multibody terms for the correlation between local and electrostatic interactions in the united-residue force field. J Chem

Phys 2001; 115: 2323-47.
[http://dx.doi.org/10.1063/1.1383989]

[11] Liwo A, Khalili M, Czaplewski C, *et al.* Modification and optimization of the united-residue (UNRES) potential energy function for canonical simulations. I. Temperature dependence of the effective energy function and tests of the optimization method with single training proteins. J Phys Chem B 2007; 111(1): 260-85.
[http://dx.doi.org/10.1021/jp065380a] [PMID: 17201450]

[12] Liwo A, Czaplewski C, Ołdziej S, *et al.* Simulation of Protein Structure and Dynamics with the Coarse-Grained UNRES Force Field. In: Coarse-Graining of Condensed Phase and Biomolecular Systems. Coarse-Graining Condens. Phase Biomol. Syst. 2008; No. 8; 107–122..

[13] Khalili M, Liwo A, Jagielska A, Scheraga HA. Molecular dynamics with the united-residue model of polypeptide chains. II. Langevin and Berendsen-bath dynamics and tests on model alpha-helical systems. J Phys Chem B 2005; 109(28): 13798-810.
[http://dx.doi.org/10.1021/jp058007w] [PMID: 16852728]

[14] Liwo A, Khalili M, Scheraga HA. Ab initio simulations of protein-folding pathways by molecular dynamics with the united-residue model of polypeptide chains. Proc Natl Acad Sci USA 2005; 102(7): 2362-7.
[http://dx.doi.org/10.1073/pnas.0408885102] [PMID: 15677316]

[15] Zhang Y. I-TASSER server for protein 3D structure prediction. BMC Bioinformatics 2008; 9(1): 40.
[http://dx.doi.org/10.1186/1471-2105-9-40] [PMID: 18215316]

[16] Venselaar H, Joosten RP, Vroling B, *et al.* Homology modelling and spectroscopy, a never-ending love story. Eur Biophys J 2010; 39(4): 551-63.
[http://dx.doi.org/10.1007/s00249-009-0531-0] [PMID: 19718498]

[17] Pierce BG, Hourai Y, Weng Z. Accelerating protein docking in ZDOCK using an advanced 3D convolution library. PLoS One 2011; 6(9): e24657.
[http://dx.doi.org/10.1371/journal.pone.0024657] [PMID: 21949741]

[18] Bertelsen EB, Chang L, Gestwicki JE, Zuiderweg ER. Solution conformation of wild-type E. coli Hsp70 (DnaK) chaperone complexed with ADP and substrate. Proc Natl Acad Sci USA 2009; 106(21): 8471-6.
[http://dx.doi.org/10.1073/pnas.0903503106] [PMID: 19439666]

[19] Kityk R, Kopp J, Sinning I, Mayer MP. Structure and dynamics of the ATP-bound open conformation of Hsp70 chaperones. Mol Cell 2012; 48(6): 863-74.
[http://dx.doi.org/10.1016/j.molcel.2012.09.023] [PMID: 23123194]

[20] Bhattacharya A, Kurochkin AV, Yip GN, Zhang Y, Bertelsen EB, Zuiderweg ER. Allostery in Hsp70 chaperones is transduced by subdomain rotations. J Mol Biol 2009; 388(3): 475-90.
[http://dx.doi.org/10.1016/j.jmb.2009.01.062] [PMID: 19361428]

[21] Gołaś EI, Czaplewski C, Scheraga HA, Liwo A. Common functionally important motions of the nucleotide-binding domain of Hsp70. Proteins 2015; 83(2): 282-99.
[http://dx.doi.org/10.1002/prot.24731] [PMID: 25412765]

[22] Amadei A, Linssen AB, Berendsen HJ. Essential dynamics of proteins. Proteins 1993; 17(4): 412-25.
[http://dx.doi.org/10.1002/prot.340170408] [PMID: 8108382]

[23] Maisuradze GG, Leitner DM. Free energy landscape of a biomolecule in dihedral principal component space: sampling convergence and correspondence between structures and minima. Proteins 2007; 67(3): 569-78.
[http://dx.doi.org/10.1002/prot.21344] [PMID: 17348026]

[24] Zhuravleva A, Gierasch LM. Allosteric signal transmission in the nucleotide-binding domain of 70-kDa heat shock protein (Hsp70) molecular chaperones. Proc Natl Acad Sci USA 2011; 108(17): 6987-92.

[http://dx.doi.org/10.1073/pnas.1014448108] [PMID: 21482798]

[25] Xu Z, Page RC, Gomes MM, *et al.* Structural basis of nucleotide exchange and client binding by the Hsp70 cochaperone Bag2. Nat Struct Mol Biol 2008; 15(12): 1309-17. [http://dx.doi.org/10.1038/nsmb.1518] [PMID: 19029896]

[26] Gołaś E, Maisuradze GG, Senet P, *et al.* Simulation of the opening and closing of Hsp70 chaperones by coarse-grained molecular dynamics. J Chem Theory Comput 2012; 8(5): 1750-64. [http://dx.doi.org/10.1021/ct200680g] [PMID: 22737044]

[27] Schuermann JP, Jiang J, Cuellar J, *et al.* Structure of the Hsp110:Hsc70 nucleotide exchange machine. Mol Cell 2008; 31(2): 232-43. [http://dx.doi.org/10.1016/j.molcel.2008.05.006] [PMID: 18550409]

[28] Knieszner H, Schilke B, Dutkiewicz R, *et al.* Compensation for a defective interaction of the hsp70 ssq1 with the mitochondrial Fe-S cluster scaffold isu. J Biol Chem 2005; 280(32): 28966-72. [http://dx.doi.org/10.1074/jbc.M503031200] [PMID: 15958384]

[29] Mühlenhoff U, Gerber J, Richhardt N, Lill R. Components involved in assembly and dislocation of iron-sulfur clusters on the scaffold protein Isu1p. EMBO J 2003; 22(18): 4815-25. [http://dx.doi.org/10.1093/emboj/cdg446] [PMID: 12970193]

[30] Lesuisse E, Lyver ER, Knight SA, Dancis A. Role of YHM1, encoding a mitochondrial carrier protein, in iron distribution of yeast. Biochem J 2004; 378(Pt 2): 599-607. [http://dx.doi.org/10.1042/bj20031387] [PMID: 14629196]

[31] Alves R, Herrero E, Sorribas A. Predictive reconstruction of the mitochondrial iron-sulfur cluster assembly metabolism: I. The role of the protein pair ferredoxin-ferredoxin reductase (Yah1-Arh1). Proteins 2004; 56(2): 354-66. [http://dx.doi.org/10.1002/prot.20110] [PMID: 15211518]

[32] Lill R. Iron-sulfur clusters : basic building blocks for life. Int J Parasitol 2007; 1(3): 17-20.

[33] Rouhier N, Couturier J, Johnson MK, Jacquot JP. Glutaredoxins: roles in iron homeostasis. Trends Biochem Sci 2010; 35(1): 43-52. [http://dx.doi.org/10.1016/j.tibs.2009.08.005] [PMID: 19811920]

[34] Ciesielski SJ, Schilke BA, Osipiuk J, *et al.* Interaction of J-protein co-chaperone Jac1 with Fe-S scaffold Isu is indispensable *in vivo* and conserved in evolution. J Mol Biol 2012; 417(1-2): 1-12. [http://dx.doi.org/10.1016/j.jmb.2012.01.022] [PMID: 22306468]

[35] Uzarska MA, Dutkiewicz R, Freibert SA, Lill R, Mühlenhoff U. The mitochondrial Hsp70 chaperone Ssq1 facilitates Fe/S cluster transfer from Isu1 to Grx5 by complex formation. Mol Biol Cell 2013; 24(12): 1830-41. [http://dx.doi.org/10.1091/mbc.E12-09-0644] [PMID: 23615440]

[36] Shi R, Proteau A, Villarroya M, *et al.* Structural basis for Fe-S cluster assembly and tRNA thiolation mediated by IscS protein-protein interactions. PLoS Biol 2010; 8(4): e1000354. [http://dx.doi.org/10.1371/journal.pbio.1000354] [PMID: 20404999]

[37] Kim JH, Füzéry AK, Tonelli M, *et al.* Structure and dynamics of the iron-sulfur cluster assembly scaffold protein IscU and its interaction with the cochaperone HscB. Biochemistry 2009; 48(26): 6062-71. [http://dx.doi.org/10.1021/bi9002277] [PMID: 19492851]

[38] Mozolewska M.A. Krupa P, Scheraga H.A., Liwo A. Molecular modeling of the binding modes of the iron-sulfur protein to the jac1 Co-chaperone from Saccharomyces cerevisiae by all-atom and coarse-grained approaches. Proteins Struct Funct Bioinforma 2015; 83(8): 1414-26. [http://dx.doi.org/10.1002/prot.24824]

[39] Rupp B. Biomolecular crystallography: principles, practice, and application to structural biology. Garland Science 2009.

[40] Rosse AH, Harrison JS. The Yeasts: Yeast Technology. Elsevier Science 2012.

[41] Teixeira PF, Glaser E. Processing peptidases in mitochondria and chloroplasts. Biochim Biophys Acta 2013; 1833(2): 360-70.
[http://dx.doi.org/10.1016/j.bbamcr.2012.03.012] [PMID: 22495024]

[42] Buchan DWA, Minneci F, Nugent TCO, Bryson K, Jones DT. Scalable web services for the PSIPRED protein analysis workbench. Nucleic Acids Res 2013; 41(Web Server issue): W349-57.
[http://dx.doi.org/10.1093/nar/gkt381]

[43] Morris G, Goodsell D, Halliday R, *et al.* Automated docking using a lamarckian genetic algorithm and an empirical binding free energy function. J Comput Chem 1998; 19(14): 1639-62.
[http://dx.doi.org/10.1002/(SICI)1096-987X(19981115)19:14<1639::AID-JCC10>3.0.CO;2-B]

[44] Murtagh F, Heck A. Multivariate Data Analysis. Kluwer Academic Publishers 1998.

[45] Ward JH. Hierarchical groupings to optimize an objective function. J Am Stat Assoc 1963; 58: 234-44.
[http://dx.doi.org/10.1080/01621459.1963.10500845]

[46] Murtagh F. Multidimensional clustering algorithms. Compstat Lect Vienna Phys Verlag 1985; 1985

[47] Kim DE, Chivian D, Baker D. Protein structure prediction and analysis using the Robetta server. Nucleic Acids Res 2004; 32(Web Server issue): W526-31.
[http://dx.doi.org/10.1093/nar/gkh468]

[48] Dominguez C, Boelens R, Bonvin AM. HADDOCK: a protein-protein docking approach based on biochemical or biophysical information. J Am Chem Soc 2003; 125(7): 1731-7.
[http://dx.doi.org/10.1021/ja026939x] [PMID: 12580598]

[49] Gässler CS, Buchberger A, Laufen T, *et al.* Mutations in the DnaK chaperone affecting interaction with the DnaJ cochaperone. Proc Natl Acad Sci USA 1998; 95(26): 15229-34.
[http://dx.doi.org/10.1073/pnas.95.26.15229] [PMID: 9860951]

[50] Smock RG, Rivoire O, Russ WP, *et al.* An interdomain sector mediating allostery in Hsp70 molecular chaperones. Mol Syst Biol 2010; 6(1): 414.
[PMID: 20865007]

[51] Suh WC, Burkholder WF, Lu CZ, Zhao X, Gottesman ME, Gross CA. Interaction of the Hsp70 molecular chaperone, DnaK, with its cochaperone DnaJ. Proc Natl Acad Sci USA 1998; 95(26): 15223-8.
[http://dx.doi.org/10.1073/pnas.95.26.15223] [PMID: 9860950]

[52] Krupa P, Sieradzan AK, Rackovsky S, *et al.* Improvement of the treatment of loop structures in the UNRES force field by inclusion of coupling between backbone- and side-chain-local conformational states. J Chem Theory Comput 2013; 9(10): 4620-32.
[http://dx.doi.org/10.1021/ct4004977] [PMID: 24273465]

[53] Sieradzan AK, Krupa P, Scheraga HA, Liwo A, Czaplewski C. Physics-based potentials for the coupling between backbone- and side-chain-local conformational states in the UNited RESidue (UNRES) force field for protein simulations. J Chem Theory Comput 2015; 11(2): 817-31.
[http://dx.doi.org/10.1021/ct500736a] [PMID: 25691834]

# CHAPTER 3

# Quaternary Structure of Chaperones from the Hsp70 System Determined by Small Angle X-Ray Scattering (SAXS) and Analytical Ultra-centrifugation

**Júlio C. Borges**[1] and **Carlos H.I. Ramos**[2,*]

[1] *São Carlos Institute of Chemistry, University of São Paulo USP, São Carlos, SP, 13566-590, Brazil*

[2] *Institute of Chemistry, University of Campinas UNICAMP, Campinas SP, 13083-970, Brazil*

**Abstract:** High-resolution techniques, such as X-ray crystallography and nuclear magnetic resonance, are not always capable of providing insight into the quaternary structure of full-length forms of eukaryotic chaperones. This has somewhat limited the field's ability to understand the mechanisms by which chaperones regulate and specify their functions. To fill this information gap, small angle X-ray scattering (SAXS) has been used to gain insight into the quaternary structure of chaperones and co-chaperones in the absence and in the presence of their ligands. This chapter will review selected structural biology publications of the Hsp70 system, in which SAXS was used to investigate the quaternary structure of these molecular chaperones, and will examine how analytical ultracentrifugation can be used as an important tool to validate SAXS data.

**Keywords:** Protein folding, Molecular chaperone, Heat shock protein, SAXS, Analytical ultracentrifugation, Hsp70, Hsp40, Nucleotide exchange factor.

## INTRODUCTION

### Protein Folding and Molecular Chaperones

Proteins are essential to the cell because they are key participants in almost all physiological functions, *e.g.*, metabolism, growth, movement, signaling, and catalysis. A protein is a polymer of amino acids, whose sequence is coded from a gene. By means of folding, this sequence dictates the structure and, thus, the function of the biomolecule [1 - 5]. Thus, protein folding is occasionally

* **Corresponding author Carlos H. I. Ramos:** Institute of Chemistry, University of Campinas UNICAMP, Campinas SP, 13083-970, Brazil; Tel: 55-19-3521-3096; Fax: 55-19-3521-3023; Email: cramos@iqm.unicamp.br

considered the second half of a so-called 'gene expression game' [6]. However, how does a polypeptide fold on a time-scale that is compatible with biologically relevant processes? The funnel fold hypothesis considers that protein folding follows multiple pathways without passing through a unique transition state [1, 7] and is based on statistical mechanics and Monte Carlo simulations of protein folding [8, 9]. In this energy landscape for folding, which is organized in a funnel, the top represents the vast number of conformations with high energy, or the unfolded state. As the polypeptide folds, the number of possible conformations and the amount of energy decreases until it reaches the lowest energy or native state [7].

Partially folded or misfolded proteins can form soluble or insoluble non-physiological associations, known as aggregates, via exposed hydrophobic regions [1, 10 - 12]. These aggregates can originate amyloid-like fibrils that are related to several neurodegenerative diseases, such as Huntington's, Parkinson's, and Alzheimer's [1, 12, 13]. Protein homeostasis or proteostasis, the regulation of the abundance and folding of the proteome, involves gene expression and protein folding and degradation. The collapse of these processes is related to the aging process [14 - 19]. Heat shock proteins and molecular chaperones are among the most important cellular systems that have evolved to facilitate the correct folding and prevent misfolding and aggregation [14, 16 - 18, 20 - 22]. These protective systems are also present in specific cell compartments, thus emphasizing the major role that these proteins play in protein homeostasis and stress situations [14, 16 - 18, 22].

A molecular chaperone is defined as "a protein that binds to and stabilizes an otherwise unstable conformer of another protein, facilitates its correct fate *in vivo*: be it in folding, oligomeric assembly, transport to a particular subcellular compartment, or controlled switching between active/inactive conformations" [23]. Chaperones assist in the folding of an unfolded protein by recognizing exposed hydrophobic surfaces that will be buried in the native state. Chaperones were initially referred to as heat shock proteins (Hsp) because of their increased abundance in cells exposed to thermal stress, though many are constitutively expressed [14, 24 - 28]. Additionally, not all chaperones are Hsps, and vice versa. There are at least five major molecular chaperone families (Table **1**), each of which assists in protein folding in a different way, and they are classified by their molecular masses [14, 29]. The Hsp70 (70 kDa heat shock protein) family plays a central role in assisting protein folding because it is involved in the folding of nascent proteins and can receive and distribute protein substrates in cooperation with other molecular chaperone families. The Hsp100 (100 kDa heat shock protein) family has the ability to recover proteins from aggregates. The Hsp90 (90 kDa heat shock protein) family is involved in regulation because numerous signal transduction pathway proteins are among its substrates. The Hsp60/Hsp10 complex forms a barrel-like structure that encapsulates partially unfolded proteins to provide the proper environment for folding and avoiding aggregation. The small Hsp (smHsp) family is composed of low molecular mass monomers compared with other chaperones, and these monomers lack ATPase activity and work by binding and protecting unfolded and partially folded proteins from aggregation.

**Table 1. General characteristics of molecular chaperones.**

| General name | Monomer molecular mass | Functional oligomer | ATP binding | Category |
|---|---|---|---|---|
| **Hsp100** | ~100 | Hexamer | *yes* | *Disaggregase, foldase* |
| **Hsp90** | ~90 | Dimer | *yes* | *Foldase* |
| **Hsp70** | ~70 | Monomer | *yes* | *Foldase, disaggregase* |
| **Hsp60/Hsp10** | ~60/~10 | Double heptamer/ heptamer | *yes* | *Foldase* |
| **Small Hsp** | 10-30 | 8-24 | *no* | *Holder* |

Molecular chaperones and Hsps are classified by the mechanism through which they interact with their substrates. This classification produces three general types of chaperones, which are referred to as disaggregases, foldases, and holders (Fig. **1**) [14, 30]. Disaggregases, as indicated by their designation, are involved in rescuing aggregated proteins. They are primarily represented by ClpB/Hsp104 chaperones, but recent evidence suggests that the Hsp70 system is also capable of disaggregase functions [31 - 36]. Foldases directly aid in folding or refolding by helping the substrate adopt its native state in an ATP-dependent manner. Classical examples of foldases are chaperonin, Hsp70s and Hsp90s [24, 37 - 39]. Holders prevent protein misfolding or aggregation by binding substrate proteins (or client proteins) in an ATP-independent manner and delivering them to other chaperones for further action, which is commonly a foldase. Classical examples of holders are small Hsps [14, 40, 41].

Due to the large molecular mass of molecular chaperones (Table **1**), it is not always possible to obtain information about their structure by using high-resolution techniques, such as X-ray crystallography and nuclear magnetic resonance. For some chaperones, high-resolution information has been acquired for protein domains, but not for an entire chaperone protein. However, other techniques, such as small angle X-ray scattering (SAXS) and analytical ultracentrifugation, have been used to gain insight into the conformation of chaperones and Hsps, in the apo and or holo forms. Here, we present the application of these techniques to investigate the structure of the Hsp70 family. A brief review of these techniques is provided in the next two sections, but the reader can skip them, and jump directly to the sections related to the conformation of the Hsp70 proteins.

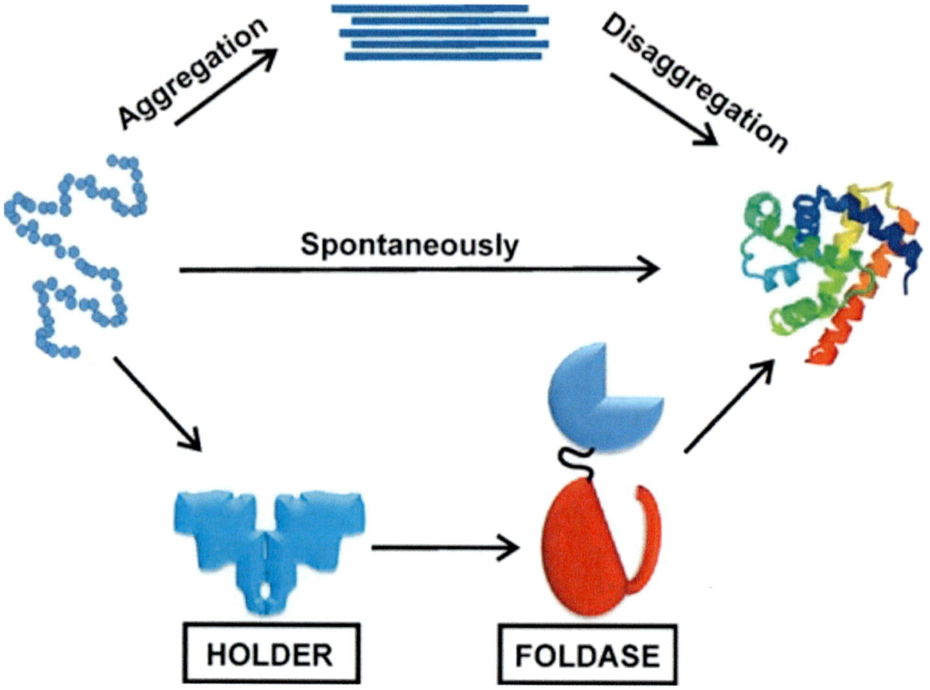


**Fig. (1).** Protein folding in the cell. Nascent proteins can fold spontaneously or with the help of chaperones. In the crowded cell environment, nascent proteins can misfold, which leads to aggregation. In this cartoon, one possible pathway is for the unfolded nascent protein to be bound by a holder chaperone, thereby precluding its aggregation. The protected protein is then delivered to a foldase chaperone that aids in folding. Even aggregation can be rescued by chaperones because disaggregases can recover protein aggregates for refolding or degradation.

## Small Angle X-ray Scattering (SAXS)

Small angle X-ray scattering is one of the best techniques to provide structural information on the shape and domain organization of proteins in solution (for recent reviews, see [42 - 44]. The strength of the SAXS technique is that it can derive a 10–15 Å molecular envelope of density for a large structure in a short period of time. In this sense, this method can provide relevant complementary information to the available high-resolution structures, or it can be the sole source for all of the structurally relevant information for those proteins that lack high-resolution structures. Either way, the molecular envelope generated has strong potential use for investigating a given protein. The derived molecular envelope is a *de novo* model that was constructed from an X-ray scattering curve generated by a highly concentrated protein in solution, which usually has a concentration of 2.5-25.0 mg.ml$^{-1}$. The scattering intensity produced by the buffer is subtracted from the protein measurement, and the difference curves are scaled to equivalent protein concentration. To generate a useful model, extensive data collection at different protein concentrations, buffers and temperatures is necessary. A reliable software package, such GNOM [45], is used to evaluate the forward scattering intensity I(0) and the radius of gyration Rg, thus producing the distance distribution function p(r) and, accordingly, the maximum dimension Dmax of the scattering intensity.

Several excellent software packages are available for modelling, such as DAMMIN [46], BUNCH [47], DAMAVER [48], GASBOR [49] and SUPCOMB [50]. To increase the confidence of the derived molecular envelope, the HydroPro software [51] is used to estimate the translational diffusion coefficient D, the Rg, the maximum distance (Dmax) and the sedimentation coefficient s from the *ab initio* models generated by SAXS and then compare them with those measured by analytical ultracentrifugation and/or dynamic light scattering. This approach provides additional support for the model structure that was predicted using SAXS, which is a good approximation of the real conformation of the protein.

## Analytical Ultracentrifugation

This topic gives a short introduction to the technique of analytical ultracentrifugation and other hydrodynamic techniques. The reader can skip this part if he is primarily interested in the structure and function of molecular chaperones; for those with more interest in the topic, excellent reviews are available (to name a few [52 - 64]. These techniques are used to investigate the molecular mass and aggregation state of proteins, and they may be used to distinguish between folded and unfolded conformations. Briefly, analytical ultracentrifugation investigates the properties of particles in solution during sedimentation, which is a property of particles that are dissolved in a solution that tend to precipitate toward the bottom of a tube due to the force of gravity or to an applied external force, such as centrifugal force. This force is generated when a tube is placed in a rotor attached to a shaft that is rotating at a high speed. Additionally, the migration of the particle through the solution, which is related to its size, shape and density, is viewed in real time, thus allowing for an accurate determination of its thermodynamic and hydrodynamic parameters.

There are two types of measurements for analysis: sedimentation velocity (SV) and sedimentation equilibrium (SE). SV uses high speed to sediment a particle, thereby forming a concentration gradient within the sample tube. Experimental analysis allows the sedimentation coefficient s, diffusion coefficient, molecular mass (M), particle shape and asymmetry to be determined for each particle of interest. M, s and D (translational diffusion coefficient, which is related to the velocity of the particle in solution) are related by the following equation:

$$M = \frac{sRT}{D(1-V_{bar}\rho)} \quad \textbf{(1)}$$

where R is the gas constant, T is the absolute temperature, $V_{bar}$ is the partial specific volume, and ρ is the density of the solvent. The Sednterp software (http://www.jphilo.mailway.com/download.htm#SEDNTERP) can be used to estimate the hydrodynamic parameters. D can also be calculated using dynamic light scattering (DLS) equipment. The measured s and D need to be extrapolated to 0 mg/mL, 20 °C in water to give $s^0_{20,w}$ and $D^0_{20,w}$, respectively. Then, one can estimate the value of these parameters for a given protein if it was a non-hydrated sphere:

$$s_{sph} = 0.012\frac{M^{2/3}(1-V_{bar}\rho)}{V_{bar}^{\ 1/3}} \quad \textbf{(2)}$$

and:

$$D_{sph} = \frac{RT}{N_{Av}6\pi\eta\left(\frac{3MV_{bar}}{4\pi N_{av}}\right)^{1/3}} \quad \textbf{(3)}$$

where $N_{AV}$ is Avogadro's number and η is the viscosity of the solution. The measured and estimated values can then be compared to determine whether the protein has a more spherical or elongated shape.

Sedimentation equilibrium (SE) provides information on M, density, homogeneous states of aggregation and the association constant. This experiment is performed at low speeds such that the sedimentation and diffusion forces oppose each other and reach equilibrium. Usually, a self-association method is used to analyze the data, and the distribution of the protein along the cell is fit using the following equation [65]:

$$C_{(r)} = C_0 e^{\left[\frac{M(1-V_{bar}\rho)\omega^2(r^2-r_0^2)}{2RT}\right]} \qquad \textbf{(4)}$$

where $C_{(r)}$ is the protein concentration at radial position $r$, $C_0$ is the protein concentration at radial position $r_0$ and $\omega$ is the angular velocity.

## The Hsp70-folding System

The 70 kDa heat shock protein (Hsp70) family is highly conserved in evolution and has a central role in the molecular chaperone system, where it is involved with the folding of nascent proteins and in the intercommunication between substrate and chaperones in the cell [14, 24, 26, 66 - 72]. This system assists in folding, transport through membranes, degradation and escape from aggregation, and its importance can be further recognized by the fact that each organelle has its own Hsp70 protein.

The Hsp70 system is formed by Hsp70, a foldase, and a holder co-chaperone, usually Hsp40 (Fig. **2**), although it can also receive client proteins from smHsps [73, 74]. Some Hsp70s also requise the assistance of a nucleotide exchange factor, such as GrpE or Hsp110 [66, 75]. Hsp70s have ATPase activity and are composed of two protein domains: one that binds the substrate and another that binds the nucleotide. The substrate binding domain (SBD) is approximately 25 kDa and binds to short and extended hydrophobic segments, whereas the nucleotide binding domain (NBD) binds to nucleotides and has low ATPase activity. An EEVD-motif is located at the C-terminus of Hsp70s and is responsible for interactions with other chaperones.

Hsp70 associates with the substrate with the help of Hsp40, a co-chaperone that works as a holder and thus binds and delivers the substrate, or client protein, to Hsp70. Substrate binding to the SBD and interaction with the Hsp40 J-domain stimulates the ATP hydrolysis by the NBD [76 - 78]. Several studies have shown that the conversion of ATP to ADP and inorganic phosphate causes a conformational change in the ATPase domain of Hsp70 that is transmitted to the substrate binding site. The dissociation of the Hsp70-substrate complex occurs during the regeneration of the Hsp70-ATP bound state, which may require the aid of nucleotide exchange factors (Fig. **2**).

Both the nucleotide and substrate binding domains have two sub-domains, named NBD1 and NBD2 and SBD1 and SBD2, respectively. In particular, the SBD2 is composed of a long α-helix that closes onto substrates when ADP is bound to the NBD, whereas an open conformation is adopted when ATP is bound to the NBD. Thus, the effect of ATP on the conformation of Hsp70 is larger than that caused by ADP, and this conformational change was confirmed by investigations of human Hsp70.1 with analytical ultracentrifugation and dynamic light scattering [79]. Sedimentation velocity AUC experiments showed that the standard sedimentation coefficient 0 mg/ml protein concentration ($s^0_{20w}$) of Hsp70.1 is 4.40 ± 0.03 S in the absence of nucleotide binding, 4.53 ± 0.05 S in the presence of ADP, and 4.72 ± 0.03 S in the presence of ATP (Table **2**). Additionally, the diffusion coefficient $D^0_{20,w}$ of Hsp70.1 obtained by dynamic light scattering is 5.7 ± 0.1 × $10^{-7}$ $cm^2s^{-1}$ in the absence of nucleotide, 5.9 ± 0.1 × $10^{-7}$ $cm^2s^{-1}$ in the presence of ADP, and 5.8 ± 0.1 × $10^{-7}$ $cm^2s^{-1}$ in the presence of ATP (Table **2**). For comparison, Schönfeld et al [80] studied the *Escherichia coli* Hsp70/DnaK complex and found an equilibrium between monomer and dimer (3 to 1). For the monomer, they measured a sedimentation coefficient s of 4.0 S and a diffusion coefficient D of 3.4 × $10^{-7}cm^2s^{-1}$.

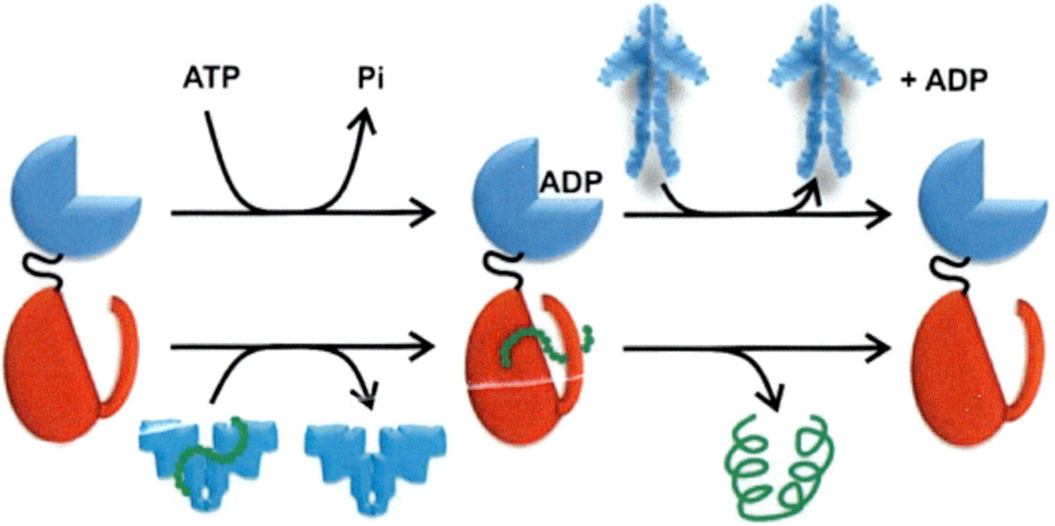


**Fig. (2).** The folding cycle of the Hsp70 system. Hsp70 associates with the substrate with the help of Hsp40, a co-chaperone that works as a holder and thus binds and delivers the substrate, or client protein, to Hsp70. Substrate binding and interaction with the Hsp40 J-domain stimulates the ATP hydrolysis. The dissociation of the Hsp70-substrate complex may require the aid of nucleotide exchange factors.

**Table 2. Hydrodynamic parameters of human Hsp70.1 from reference [79].**

| Nucleotide | $s^0_{20w}$ (S)[1] | $D^0_{20,w}$ ($10^{-7}cm^2s^{-1}$)[2] | Perrin factor[3] |
|---|---|---|---|
| **None** | 4.40 ±0.03 | 5.7 ± 0.1 | 1.40 ±0.01 |
| **ADP** | 4.53 ± 0.05 | 5.9 ± 0.1 | 1.36 ± 0.02 |
| **ATP** | 4.72 ± 0.03 | 5.8 ± 0.1 | 1.30 ± 0.01 |

1, standard sedimentation coefficient at 0 mg/ml protein concentration in Sverdbergs (S); 2, diffusion coefficient; 3, Perrin or shape factor *F* ($f/f_0$ ratio).

These parameters were also used to calculate the so-called Perrin or shape factor *F* [81], which provides information on the shape of a protein because it is the ratio of the measured frictional coefficient $f$ to the frictional coefficient $f_0$ of a hypothetical sphere for which a hypothetical radius is calculated using its molecular mass. Hsp70.1 was found to be an asymmetric protein with a $f/f_0$ ratio of approximately 1.40 ±0.01. However, in the presence of ADP, the $f/f_0$ ratio for Hsp70 was approximately 1.36 ± 0.02, and in the presence of ATP, it was approximately 1.30 ± 0.01 (Table **2**). Taken together, these results show that ATP had a larger effect on the conformation of Hsp70 than did ADP. Additionally, the $f/f_0$ ratio indicates that nucleotide binding promoted a less elongated conformation of the chaperone. Thus, these results support a model in which ATP binding to Hsp70 causes conformational changes that alter the relative position of the domains and induces a more globular shape for Hsp70.

## Human Mitochondrial GrpE, Conformational Modification upon Hsp70 Binding

GrpE increases the release of ADP and ATP bound to Hsp70, *i.e.* this co-chaperone GrpE is an Hsp70 nucleotide exchange factor that is found in prokaryotes and some organelles, such as mitochondria [80, 82, 83]. GrpE is essential for bacterial viability at all temperatures and although only one isoform is present in *E. coli* [84], two isoforms have been found in the eukaryotic mitochondria and named 1 and 2, respectively [85 - 87]. GrpE from *E. coli* is 197 residues long and is divided into four main regions [75, 88 - 92] (Fig. **3**): a) the first 34 residues form a region that enhances binding in the SBD of Hsp70/DnaK by reducing its affinity for the client protein; b) residues 40-88 form an elongated helix that works as a thermosensor; c) residues 89-137 form a middle region (four-helix bundle) that acts as a stabilization center for dimerization; and d) residues 138-197 form a beta-sheet structured domain at the C-terminus that binds to the nucleotide binding domain of Hsp70/DnaK, thus causing a mechanical opening movement that releases ADP from the nucleotide binding domain (NBD). GrpE is a dimer that is symmetric in solution [83, 93] and one of the monomers, known as proximal, changes to a more open conformation when bound to the NBD [88]. A high-resolution structure of residues 34-197 of *E. coli* GrpE complexed with the *E. coli* DnaK-NBD (EcDnaK3-383) is available (Harrison et al, 1997-) and shows that GrpE forms a dimer and that only one of the subunits binds to Hsp70.

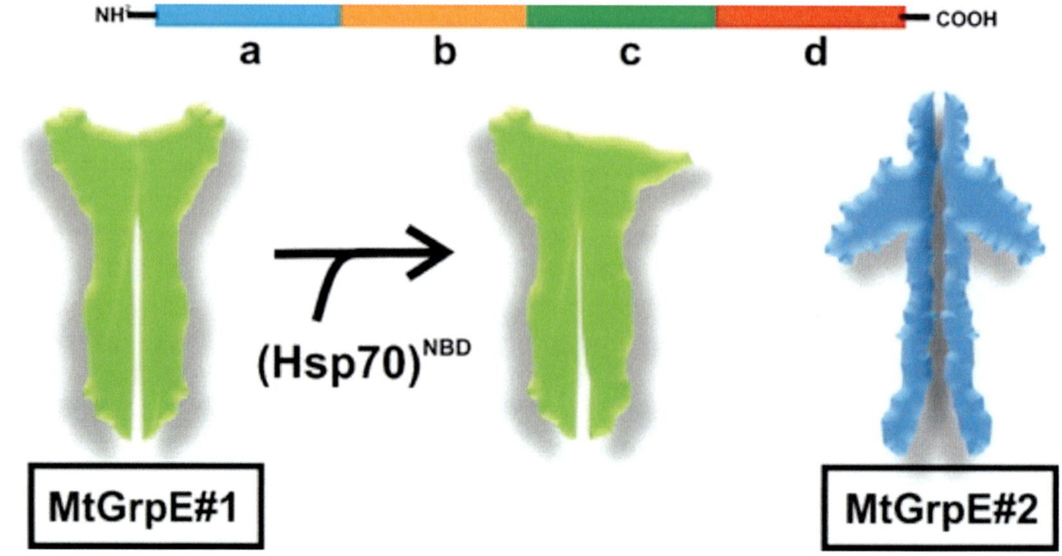


**Fig. (3).** The quaternary structure of GrpEs. Top, domains of GrpEs. a) region that enhances binding in the SBD of Hsp70/DnaK by reducing its affinity for the client protein; b) elongated helix region that works as a thermosensor; c) middle region that acts as a stabilization center for dimerization; and d) C-terminus beta-sheet structured domain that binds to the nucleotide binding domain of Hsp70/DnaK. Bottom, the quaternary structure of GrpEs and the change caused by Hsp70-NBD binding (see text). GrpEs have similar quaternary structure, *i.e.*, a symmetric dimer with a cruciform elongated shape.

Small angle X-ray scattering was used to obtain information on the human mitochondrial GrpE isoforms 1 and 2 (Mt-GrpE#1 and Mt-GrpE#2) [93, 94]. The results showed that they have similar quaternary structure, *i.e.*, a dimer with a cruciform elongated shape, and that they both have a symmetrical conformation when in solution (Fig. **3**). The molecular envelopes determined by an *ab initio* method for these isoforms are also similar to the high-resolution envelope of *E. coli* GrpE bound to DnaK obtained from single crystal x-ray diffraction [93, 94]. Additionally, the human GrpE structure generated using the bound monomer as a template does not fit well with the bead model generated from the SAXS results, whereas the structure generated using the unbound monomer as a template exhibits a much better fit. This finding suggests that in solution, when it is not bound to DnaK, GrpE is a symmetric dimer with each monomer in a "closed" conformation. Accordingly, the characterization of human GrpE in solution using analytical ultracentrifugation and dynamic light scattering clearly indicates that the protein is in a dimeric conformation with an elongated overall shape. However, the sedimentation coefficient $s^0_{20,w}$ of Mt-GrpE#2 (2.80 S) was distinct from that measured for Mt-GrpE#1 (2.34 S) (Table **3**). This difference may indicate that Mt-GrpE#2 was less asymmetric than Mt-GrpE#1, which is not supported by the SAXS data (see below), or that it was more flexible. However, it is worth noting that the sedimentation coefficient s for *E. coli* GrpE is 2.7 S [80] which indicates some similarity between this co-chaperone and Mt-GrpE#2 than to Mt-GrpE#1 [93].

**Table 3. Sedimentation coefficient *s* (in S) for GrpEs.**

| Organism | *s (S)* | REFERENCE |
|---|---|---|
| ***E. coli*** | 2.70 | [80] |
| **Human (isoform 1)** | 2.34 | [93] |
| **Human (isoform 2)** | 2.80 | [94] |

The results presented here may also add to the general understanding of the Hsp70 system mechanism of action. Unfolded proteins are presented to the Hsp70 system by the Hsp40 co-chaperone. The binding of GrpE to Hsp70 (bound to an unfolded protein) occurs by conformational changes in the GrpE C-terminal proximal 'arm' that exposes residues involved in this interaction. An interaction between the Hsp70 nucleotide-binding domain and the GrpE C-terminal proximal 'arm' may create several other changes that cause the long GrpE N-terminal helices to move. Because of their size, these long helices could interact with the Hsp70 substrate binding domain, helping release the folded protein (Fig. **3**).

## Eukaryotic Hsp40s Types I and II, on the Position of the J Domain

As introduced in the first topic of this chapter, Hsp40s are holders and co-chaperones of the Hsp70 system. They not only bind client proteins to be delivered to the Hsp70 but also stimulate the ATPase activity of Hsp70 (for reviews see [14, 95 - 105]). Hsp40s can also function as chaperones through their own activity and bind to client proteins through hydrophobic interactions and present them to DnaK/Hsp70 [76, 106 - 110].

Hsp40s are ubiquitous but increase in diversity with the complexity of the organism. There is one representative in *E. coli*, two in *Saccharomyces cerevisiae*, but more than 40 in humans and plants [86, 87, 97, 111]. The high diversity of Hsp40s in eukaryotes allow differences to exist in the architecture of their domains and in their cellular location. Despite their diversity, all Hsp40s are defined by the presence of a J-domain, which is α-helical and responsible for binding and stimulating the ATPase activity of Hsp70. The J-domain, located at the N-terminus, defines Types I and II Hsp40s, whereas Type III Hsp40s have the J-domain located in other protein region (Fig. **4**). Types I and II have a disordered Gly/Phe rich region (G/F region) that follows the J-domain that is responsible for its flexibility and functions as a linker between the J-domain and C-terminal region of DnaJ proteins [112, 113]. In Type I Hsp40s, such as human DnaJA1 and yeast Ydj1, the central portion between the G/F domain and the C-terminus contains a zinc finger-like region (ZFLR), whereas in Type II, such as yeast Sis1 and human DnaJB4, the ZFLR is replaced by a G/M-rich (Glycine/Methionine-rich) domain. In Types I and II, the region known as the C-terminal Domain (CTD) is formed by two structurally similar subdomains (named CTDI and CTDII) that are composed of beta-sheet structures that dimerize. Both CTD and ZFLR are responsible for the chaperone activity of Hsp40s [108 - 110, 112, 114 - 117].

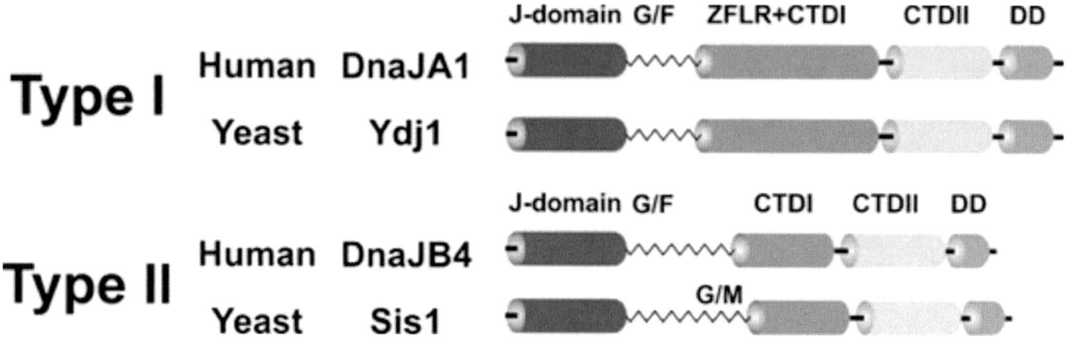


**Fig. (4).** Domains arrangement of Types I and II Hsp40s. The J-domain, located at the N-terminus, defines Types I and II Hsp40s. Types I and II have a disordered Gly/Phe rich region (G/F region) that follows the J-domain and functions as a linker between the J-domain and C-terminal region of DnaJ proteins. In Type I Hsp40s, the central portion between the G/F domain and the C-terminus contains a zinc finger-like region (ZFLR), whereas in Type II, the ZFLR is replaced by a G/M-rich (Glycine/Methionine-rich) domain. In Types I and II, the region known as the C-terminal Domain (CTD) is composed of beta-sheet structures that dimerize.

**Table 4. Differences between Types I and II Hsp40s.**

| Characteristics | Type I | Type II |
|---|---|---|
| **Examples** | yeast Ydj1 and human DnaJA1 | yeast Sis1 and human DnaJB4 |
| **Domain arrangement (from N- to C-terminus)** | J-domain; G/F rich; ZFLR; CTD | J-domain; G/F rich; G/M rich; CTD |
| **Hsp70 dependence** | no | yes |
| **Essential** | no | yes |
| **Prion assembly** | [URE3] | [RNQ+] |

Types I and II do not have equivalent functions in the cell [102, 109, 118 - 127] (Table **4**). To demonstrate these different functions, yeast Ydj1 (Type I) and Sis1 (Type II) are used as examples. Type I Hsp40s have autonomous chaperone activity and may therefore work in an Hsp70-dependent or Hsp70-independent manner, whereas Type II Hsp40s have no autonomous chaperone activity and depend on Hsp70 for full activity. Sis1 is an essential protein and high levels of Sis1 can suppress the slow-growth phenotype of *ydj1Δ*, whereas the lethal phenotype of *sis1Δ* strains cannot be complemented by Ydj1. They also play different roles in the assembly of amyloid-like yeast prions. The overexpression of Ydj1, but not Sis1, can cure the [URE3] prion phenotype, whereas Sis1, but not Ydj1, is essential for the propagation of the yeast prion [RNQ+].

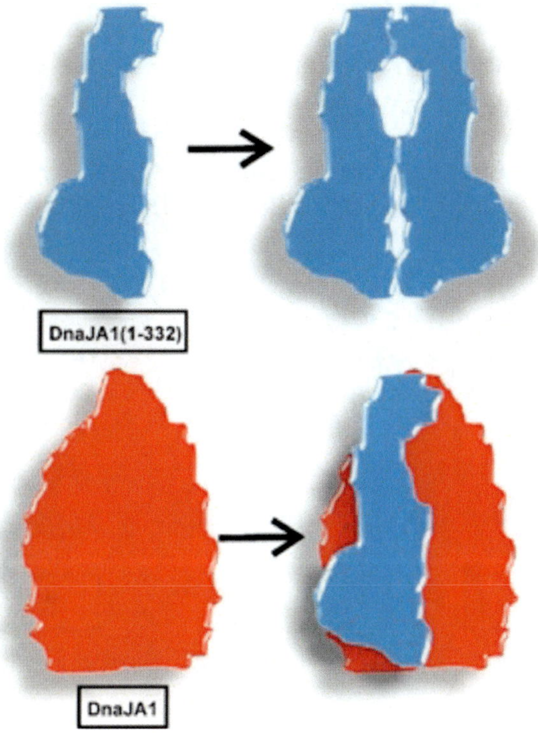


**Fig. (5).** Human Type I Hsp40 conformation construction. Figure shows DnaJA1 and DnaJA1(1-332), a C-terminal mutant lacking residues 333-394. Top, the two monomers of DnaJA1(1–332) are shown facing each other to illustrate how they could form a dimer. Bottom, the monomer envelope placed within the dimer envelope fits one half of the envelope very well.

To determine the quaternary conformations of Type I Hsp40s, pioneering studies combined SAXS and AUC using human DnaJA1 (also known as Dja1, Hdj2, Dj2, HSDJ and Rdj1) and a C-terminal mutant lacking residues 333-394 and named here DnaJA1(1-332) [95]. The deletion mutant played a key role in the evaluation of the molecular envelopes because this deletion prevent dimerization, although it had no significant effect on the global conformation of the protein. First of all, DnaJA1 and DnaJA1(1–332) had similar maximum diameters (Dmax 140 Å), which implies that they had a similar asymmetry. A cartoon representing the molecular envelope models for both DnaJA1 (left) and DnaJA1(1–332) from [95] is shown in Fig. (**5**). In this figure, the two monomers of DnaJA1(1–332) are shown facing each other to illustrate how they form a dimer. Supporting this model are the facts that the monomer envelope placed within the dimer envelope fits one half of the envelope very well and that the hydrodynamic properties of the two combined monomers of DnaJA1(1–332), as calculated by the HydroPro software, were similar to those determined for the DnaJA1 dimer model. Based on these observations, the region involved in dimer formation was located mainly in the C terminus of DnaJA1; the positions of the central domains and of the J-domains are determined based on that location (Fig. **6**). Type I human Hsp40 DnaJA1 is bullet-shaped, and the C termini monomers are located in the thinnest side of the model and are responsible for the dimerization, whereas the N-terminal monomers (J-domains) are located in the thickest part of the model (Fig. **6**).

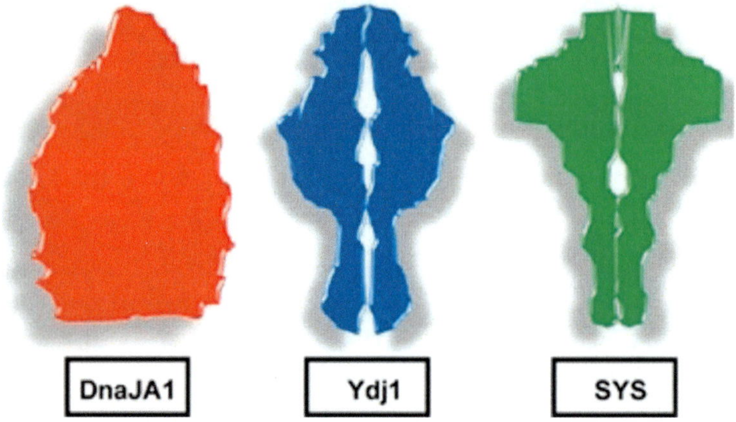


**Fig. (6).** Quaternary conformation of Type I Hsp40s. Type I Hsp40s, human DnaJA1 and yeast Ydj1, have a bullet-shaped conformation: the C termini monomers are located in the thinnest side of the model and are responsible for the dimerization, whereas the N-terminal monomers (J-domains) are located in the thickest part of the model. SYS, a chimeric protein in which the central part of Sis1 (residues 122–257) was exchanged with the central part of Ydj1 (residues 101–255), has a quaternary structure similar to that of Type I Ydj1.

The conformation of Type I Hsp40s is confirmed by studies with yeast Type I Hsp40 Ydj1, which has a fairly compact structure in which two asymmetric monomers are closely assembled [111] (Fig. **6**). To determine the quaternary conformation of Type II Hsp40s, pioneering studies combined SAXS and AUC using human DnaJB4 (also known as DjB4, Hlj1 and DnaJW). First, the human Type II Hsp40 DnaJB4 was determined to be more elongated than Type I DnaJA1, as it had a Dmax of 200 Å. Then, a molecular envelope model was constructed using the following findings. 1) Studies of C-terminal deletions of Hsp40 [95, 110, 117] indicated that this region is responsible for dimerization, which implies that the C-terminal monomers must be located near to each other and, thus, that they must be located at the center of the model (Fig. **7**). 2) The J-domains are placed at the extremities of the molecular model because they are at the N-terminus. 3) Connecting the J-domain to the C-terminal are the long and flexible region formed by the G/F-rich region (residues 68–139) and the sequence of 60 amino acids of unknown function (residues 140–199). The conformation of Type II Hsp40s is confirmed by studies with yeast Type II Hsp40 Sis1 that has N-termini that points away from the compact C-termini core responsible for dimerization [111].

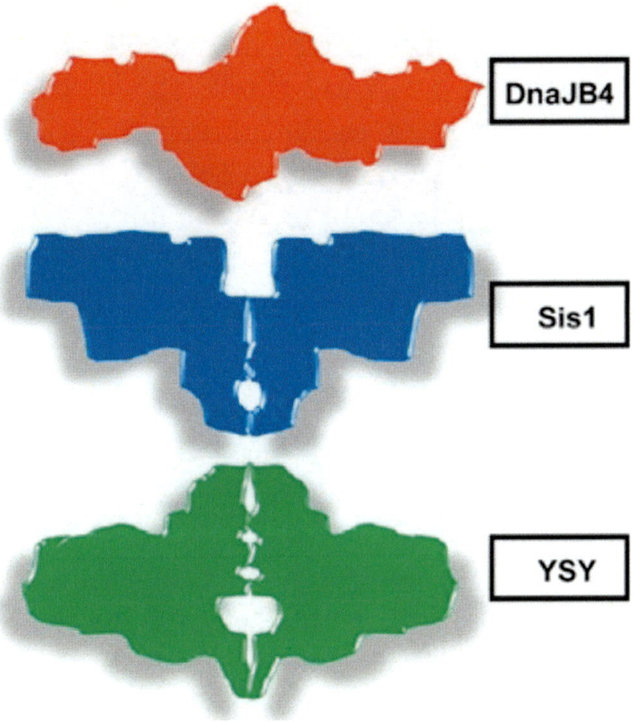


**Fig. (7).** Quaternary conformation of Type II Hsp40s. In Type II Hsp40s, human DnaJB4 and yeast Sis1, the C-terminal monomers are located near at the center of the model and J-domains are located at the extremities of the molecular at the N-terminus. YSY, a chimeric protein in which the central part of Ydj1 (residues 101–255) was exchanged with the central part of Sis1 (residues 122–257), has a quaternary structure similar to that of Type II Sis1.

It is clear that the shapes of human DnaJA1 and DnaJB4 are different and that the quaternary structures of Type I and Type II Hsp40s are conserved from yeast to humans. The arrangement of the molecular envelope models from SAXS together with information about the crystallographic and NMR structures of Hsp40 domains available in the Protein Data Bank can be found in ref [95]. The final models give more detailed insight into the structural conformation of Types I and II Hsp40s. It is therefore possible that differences in the ability of the J-domains of Ydj1 and Sis1 to interact with Hsp70 in the context of protein folding contribute to the observed differences in co-chaperone activity of Type I and Type II Hsp40s. Additionally, the hydrodynamic properties of the final models were analyzed using the Hydropro software, and they agree beautifully with those measured for the purified proteins [112] (Table **5**). Investigation into yeast Types I and II Hsp40s by SAXS and AUC supported the results found with their human counterparts [111].

**Table 5. Summary of the hydrodynamic data obtained from DLS and AUC experiments for wild-type and mutants of Types I and II Hsp40s.**

| Protein | $s^0_{20w}$ (S)[1] | $D^0_{20,w}$ ($10^{-7}cm^2s^{-1}$)[2] | Reference |
|---|---|---|---|
| **Human DnaJA1 (I)** | 4.63 ± 0.10 | 4.5 ± 0.2 | [95] |
| **Human DnaJB4 (II)** | 3.78 ± 0.10 | - | [95] |
| **Yeast Ydj1 (I)** | 4.57 ± 0.22 | - | [111] |
| **Yeast Sis1 (II)** | 3.46 ± 0.13 | 5.1 ± 0.2 | [111] |
| **Yeast SYS** | 4.65 ± 0.20 | - | [111] |
| **Yeast YSY** | 3.40 ± 0.19 | - | [111] |
| **Sis1_Δ124-174** | 3.4 ± 0.1 | 5.4 ± 0.1 | [128] |
| **Sis1_Δ121-257** | 3.0 ± 0.1 | 6.3 ± 0.1 | [128] |
| **Ydj1_Δ106-255** | 3.6 ± 0.1 | 5.9 ± 0.2 | [128] |

1, standard sedimentation coefficient at 0 mg/ml protein concentration in Sverdbergs (S); 2, diffusion coefficient

The hypothesis that the central domains specify both the quaternary structure and function, is strongly supported by experiments with chimeric mutants of yeast Hsp40s. Fan et al. [129] engineered two chimeric proteins in which the central part of Sis1 (residues 122–257) was exchanged with the central part of Ydj1 (residues 101–255) and vice versa, thereby generating the chimeras SYS and YSY, respectively. Surprisingly, these chimeras switch the specificity for binding substrates and in stimulating firefly luciferase refolding activity by DnaK/Hsp70 [129], which suggests that the central portion is important for chaperone activity specification. SAXS studies in solution showed that the switch of the central part of Sis1 and Ydj1 also induced an exchange in the structures of the chimeras SYS and YSY [111] (Figs. **6** and **7**). Therefore, these results indicate that Type I and Type II Hsp40s have different quaternary structures that are regulated by the Cys-rich region and Gly/ Met-rich region and are unique to each Hsp40 type. These conserved central modules have a major influence on determination of Hsp40 quaternary structure. Because Sis1 and Ydj1 have a similar structure at the C-termini and are highly identical at the J-domain, the difference in quaternary structure is likely to be coded by conserved regions of lower similarity in the central modules.

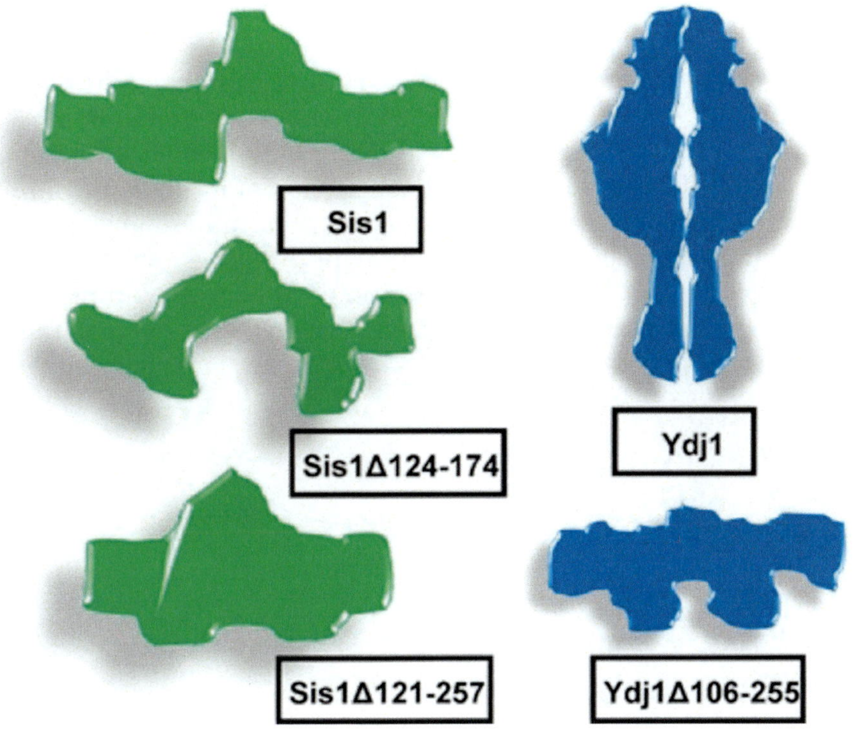


**Fig. (8).** Conformational effects of deleting the central regions of Types I and II Hsp40. Mutants are: Sis1_Δ124-174, where the G/M region has been deleted, Sis1_Δ121-257, where both the G/M and the CTDI has been deleted and Ydj1_Δ106-255, where both the Zinc finger-like region (ZFLR) and the CTDI had been deleted. Deletion of the central part of Sis1 did not result in a large change in its overall shape. However, deletion of the central part of Ydj1 largely altered its overall shape to a conformation similar to that observed for Sis1.

Further insights into how the central regions modulate the conformation of Hsp40s came from the study of deleted mutants in this region [128]: Sis1_Δ124-174, where the G/M region has been deleted, Sis1_Δ121-257, where both the G/M and the CTDI has been deleted and Ydj1_Δ106-255, where both the Zinc finger-like region (ZFLR) and the CTDI had been deleted. As expected from mutations in this region, the deletions had a decreased affinity for heated luciferase but were equally capable of stimulating ATPase activity of Hsp70. Although the deletion of the central part of Ydj1 largely altered its overall shape to a conformation similar to that observed for Sis1, deletion of the central part of Sis1 did not result in a large change in its overall shape [128] (Fig. **8**). Therefore, deletion of ZFLR-CTDI changed the relative position of the J-domains in Ydj1 in such a way that they ended up resembling those of Sis1. These results show that the central domains, previously appointed as important features for substrate binding, are also relevant for keeping the J-domains in their specific relative positions.

## FINAL REMARKS

The aim of this chapter is to present results on the conformation of proteins using small angle X-ray scattering and analytical ultracentrifugation. When high-resolution structures are not possible or provide only limited information, these techniques can provide helpful information on the conformation of protein. The fact that proteins with high molecular mass and multiple domains are the most difficult to obtain high-resolution structural information make them potential targets for study by small angle X-ray scattering and analytical ultracentrifugation techniques. The presented results with the Hsp70 system demonstrate how the results obtained by these techniques can advance the understanding of a given complex.

## CONSENT FOR PUBLICATION

Not applicable.

## CONFLICT OF INTEREST

The author declares no conflict of interest, financial or otherwise.

## ACKNOWLEDGMENTS

We are in great debt with FAPESP (2012/50161-8; 2014/07206-6; 2017/07335-9), CNPq (303129/2015-8; 305018/2015-9) and CAPES for financial support.

## REFERENCES

[1] Ramos CH, Ferreira ST. Protein folding, misfolding and aggregation: evolving concepts and conformational diseases. Protein Pept Lett 2005; 12(3): 213-22. [http://dx.doi.org/10.2174/0929866053587156] [PMID: 15777268]

[2] Dill KA, MacCallum JL. The protein-folding problem, 50 years on. Science 2012; 338(6110): 1042-6. [http://dx.doi.org/10.1126/science.1219021] [PMID: 23180855]

[3] Baldwin RL, Rose GD. Molten globules, entropy-driven conformational change and protein folding. Curr Opin Struct Biol 2013; 23(1): 4-10. [http://dx.doi.org/10.1016/j.sbi.2012.11.004] [PMID: 23237704]

[4] Englander SW, Mayne L. The nature of protein folding pathways. Proc Natl Acad Sci USA 2014; 111(45): 15873-80. [http://dx.doi.org/10.1073/pnas.1411798111] [PMID: 25326421]

[5] Goloubinoff P. Recent and future grand challenges in protein folding, misfolding, and degradation. Front Mol Biosci 2014; 1: 1. [http://dx.doi.org/10.3389/fmolb.2014.00001] [PMID: 25988142]

[6] Dodson G. Protein Folding: Deciphering the Second Half of the Genetic Code. Trends Biochem Sci 1991; 16: 76-7. [http://dx.doi.org/10.1016/0968-0004(91)90030-Y]

[7] Baldwin RL. The nature of protein folding pathways: the classical versus the new view. J Biomol

NMR 1995; 5(2): 103-9.
[http://dx.doi.org/10.1007/BF00208801] [PMID: 7703696]

[8] Shakhnovich E, Farztdinov G, Gutin AM, Karplus M. Protein folding bottlenecks: A lattice Monte Carlo simulation. Phys Rev Lett 1991; 67(12): 1665-8.
[http://dx.doi.org/10.1103/PhysRevLett.67.1665] [PMID: 10044213]

[9] Dill KA, Fiebig KM, Chan HS. Cooperativity in protein-folding kinetics. Proc Natl Acad Sci USA 1993; 90(5): 1942-6.
[http://dx.doi.org/10.1073/pnas.90.5.1942] [PMID: 7680482]

[10] Uversky VN, Fink AL. Conformational constraints for amyloid fibrillation: the importance of being unfolded. Biochim Biophys Acta 2004; 1698(2): 131-53.
[http://dx.doi.org/10.1016/j.bbapap.2003.12.008] [PMID: 15134647]

[11] Doyle SM, Genest O, Wickner S. Protein rescue from aggregates by powerful molecular chaperone machines. Nat Rev Mol Cell Biol 2013; 14(10): 617-29.
[http://dx.doi.org/10.1038/nrm3660] [PMID: 24061228]

[12] Knowles TP, Vendruscolo M, Dobson CM. The Amyloid State and Its Association with Protein Misfolding Diseases (Vol 15, Pg 384, 2014). Nat Rev Mol Cell Biol 2014; 15(7): 496.
[http://dx.doi.org/10.1038/nrm3826]

[13] Eisenberg D, Jucker M. The amyloid state of proteins in human diseases. Cell 2012; 148(6): 1188-203.
[http://dx.doi.org/10.1016/j.cell.2012.02.022] [PMID: 22424229]

[14] Tiroli-Cepeda AO, Ramos CH. An overview of the role of molecular chaperones in protein homeostasis. Protein Pept Lett 2011; 18(2): 101-9.
[http://dx.doi.org/10.2174/092986611794475093] [PMID: 21121892]

[15] Martin J, Hartl FU. Chaperone-assisted protein folding. Curr Opin Struct Biol 1997; 7(1): 41-52.
[http://dx.doi.org/10.1016/S0959-440X(97)80006-1] [PMID: 9032064]

[16] Douglas PM, Summers DW, Cyr DM. Molecular chaperones antagonize proteotoxicity by differentially modulating protein aggregation pathways. Prion 2009; 3(2): 51-8.
[http://dx.doi.org/10.4161/pri.3.2.8587] [PMID: 19421006]

[17] Hartl FU, Bracher A, Hayer-Hartl M. Molecular chaperones in protein folding and proteostasis. Nature 2011; 475(7356): 324-32.
[http://dx.doi.org/10.1038/nature10317] [PMID: 21776078]

[18] Kim YE, Hipp MS, Bracher A, Hayer-Hartl M, Hartl FU. Molecular chaperone functions in protein folding and proteostasis. Annu Rev Biochem 2013; 82: 323-55.
[http://dx.doi.org/10.1146/annurev-biochem-060208-092442] [PMID: 23746257]

[19] Morimoto RI. The heat shock response: systems biology of proteotoxic stress in aging and disease. Cold Spring Harb Symp Quant Biol 2011; 76: 91-9.
[http://dx.doi.org/10.1101/sqb.2012.76.010637] [PMID: 22371371]

[20] Fink AL. Chaperone-mediated protein folding. Physiol Rev 1999; 79(2): 425-49.
[http://dx.doi.org/10.1152/physrev.1999.79.2.425] [PMID: 10221986]

[21] Saibil H. Chaperone machines for protein folding, unfolding and disaggregation. Nat Rev Mol Cell Biol 2013; 14(10): 630-42.
[http://dx.doi.org/10.1038/nrm3658] [PMID: 24026055]

[22] Priya S, Sharma SK, Goloubinoff P. Molecular chaperones as enzymes that catalytically unfold misfolded polypeptides. FEBS Lett 2013; 587(13): 1981-7.
[http://dx.doi.org/10.1016/j.febslet.2013.05.014] [PMID: 23684649]

[23] Hendrick JP, Hartl FU. Molecular chaperone functions of heat-shock proteins. Annu Rev Biochem 1993; 62: 349-84.
[http://dx.doi.org/10.1146/annurev.bi.62.070193.002025] [PMID: 8102520]

[24] Bukau B, Horwich AL. The Hsp70 and Hsp60 chaperone machines. Cell 1998; 92(3): 351-66. [http://dx.doi.org/10.1016/S0092-8674(00)80928-9] [PMID: 9476895]

[25] Lindquist S, Craig EA. The heat-shock proteins. Annu Rev Genet 1988; 22: 631-77. [http://dx.doi.org/10.1146/annurev.ge.22.120188.003215] [PMID: 2853609]

[26] Hartl FU, Hayer-Hartl M. Molecular chaperones in the cytosol: from nascent chain to folded protein. Science 2002; 295(5561): 1852-8. [http://dx.doi.org/10.1126/science.1068408] [PMID: 11884745]

[27] Lee S, Tsai FT. Molecular chaperones in protein quality control. J Biochem Mol Biol 2005; 38(3): 259-65. [PMID: 15943899]

[28] Ramos CH. Prevention of Protein Misfolding and Aggregation by the Machinery of Molecular Chaperones. 2008.

[29] Borges JC, Ramos CH. Protein folding assisted by chaperones. Protein Pept Lett 2005; 12(3): 257-61. [http://dx.doi.org/10.2174/0929866053587165] [PMID: 15777275]

[30] Mayer MP. Gymnastics of molecular chaperones. Mol Cell 2010; 39(3): 321-31. [http://dx.doi.org/10.1016/j.molcel.2010.07.012] [PMID: 20705236]

[31] Parsell DA, Kowal AS, Singer MA, Lindquist S. Protein disaggregation mediated by heat-shock protein Hsp104. Nature 1994; 372(6505): 475-8. [http://dx.doi.org/10.1038/372475a0] [PMID: 7984243]

[32] Zietkiewicz S, Krzewska J, Liberek K. Successive and synergistic action of the Hsp70 and Hsp100 chaperones in protein disaggregation. J Biol Chem 2004; 279(43): 44376-83. [http://dx.doi.org/10.1074/jbc.M402405200] [PMID: 15302880]

[33] Shorter J. The mammalian disaggregase machinery: Hsp110 synergizes with Hsp70 and Hsp40 to catalyze protein disaggregation and reactivation in a cell-free system. PLoS One 2011; 6(10): e26319. [http://dx.doi.org/10.1371/journal.pone.0026319] [PMID: 22022600]

[34] Rampelt H, Kirstein-Miles J, Nillegoda NB, *et al.* Metazoan Hsp70 machines use Hsp110 to power protein disaggregation. EMBO J 2012; 31(21): 4221-35. [http://dx.doi.org/10.1038/emboj.2012.264] [PMID: 22990239]

[35] Mokry DZ, Abrahão J, Ramos CH. Disaggregases, molecular chaperones that resolubilize protein aggregates. An Acad Bras Cienc 2015; 87(2) (Suppl.): 1273-92. [http://dx.doi.org/10.1590/0001-3765201520140671] [PMID: 26312418]

[36] Mattoo RU, Sharma SK, Priya S, Finka A, Goloubinoff P. Hsp110 is a bona fide chaperone using ATP to unfold stable misfolded polypeptides and reciprocally collaborate with Hsp70 to solubilize protein aggregates. J Biol Chem 2013; 288(29): 21399-411. [http://dx.doi.org/10.1074/jbc.M113.479253] [PMID: 23737532]

[37] da Silva VC, Ramos CH. The network interaction of the human cytosolic 90 kDa heat shock protein Hsp90: A target for cancer therapeutics. J Proteomics 2012; 75(10): 2790-802. [http://dx.doi.org/10.1016/j.jprot.2011.12.028] [PMID: 22236519]

[38] Gava LM, Ramos CH. Human 90 Kda Heat Shock Protein Hsp90 as a Target for Cancer Therapeutics. Curr Chem Biol 2009; 3(1): 10-21. [http://dx.doi.org/10.2174/2212796810903010010]

[39] Wegele H, Müller L, Buchner J. Hsp70 and Hsp90--a relay team for protein folding. Rev Physiol Biochem Pharmacol 2004; 151: 1-44. [http://dx.doi.org/10.1007/s10254-003-0021-1] [PMID: 14740253]

[40] Garrido C, Paul C, Seigneuric R, Kampinga HH. The small heat shock proteins family: the long forgotten chaperones. Int J Biochem Cell Biol 2012; 44(10): 1588-92. [http://dx.doi.org/10.1016/j.biocel.2012.02.022] [PMID: 22449631]

[41] Haslbeck M, Vierling E. A first line of stress defense: small heat shock proteins and their function in protein homeostasis. J Mol Biol 2015; 427(7): 1537-48. [http://dx.doi.org/10.1016/j.jmb.2015.02.002] [PMID: 25681016]

[42] Graewert MA, Svergun DI. Impact and progress in small and wide angle X-ray scattering (SAXS and WAXS). Curr Opin Struct Biol 2013; 23(5): 748-54. [http://dx.doi.org/10.1016/j.sbi.2013.06.007] [PMID: 23835228]

[43] Petoukhov MV, Svergun DI. Applications of small-angle X-ray scattering to biomacromolecular solutions. Int J Biochem Cell Biol 2013; 45(2): 429-37. [http://dx.doi.org/10.1016/j.biocel.2012.10.017] [PMID: 23142499]

[44] Rambo RP, Tainer JA. Accurate assessment of mass, models and resolution by small-angle scattering. Nature 2013; 496(7446): 477-81. [http://dx.doi.org/10.1038/nature12070] [PMID: 23619693]

[45] Svergun DI. Determination of the Regularization Parameter in Indirect-Transform Methods Using Perceptual Criteria. J Appl Cryst 1992; 25: 495-503. [http://dx.doi.org/10.1107/S0021889892001663]

[46] Svergun DI. Restoring low resolution structure of biological macromolecules from solution scattering using simulated annealing. Biophys J 1999; 76(6): 2879-86. [http://dx.doi.org/10.1016/S0006-3495(99)77443-6] [PMID: 10354416]

[47] Petoukhov MV, Svergun DI. Global rigid body modeling of macromolecular complexes against small-angle scattering data. Biophys J 2005; 89(2): 1237-50. [http://dx.doi.org/10.1529/biophysj.105.064154] [PMID: 15923225]

[48] Volkov VV, Svergun DI. Uniqueness of Ab Initio Shape Determination in Small-Angle Scattering. J Appl Cryst 2003; 36: 860-4. [http://dx.doi.org/10.1107/S0021889803000268]

[49] Svergun DI, Petoukhov MV, Koch MH. Determination of domain structure of proteins from X-ray solution scattering. Biophys J 2001; 80(6): 2946-53. [http://dx.doi.org/10.1016/S0006-3495(01)76260-1] [PMID: 11371467]

[50] Kozin MB, Svergun DI. Automated Matching of High- and Low-Resolution Structural Models. J Appl Cryst 2001; 34: 33-41. [http://dx.doi.org/10.1107/S0021889800014126]

[51] García De La Torre J, Huertas ML, Carrasco B. Calculation of hydrodynamic properties of globular proteins from their atomic-level structure. Biophys J 2000; 78(2): 719-30. [http://dx.doi.org/10.1016/S0006-3495(00)76630-6] [PMID: 10653785]

[52] Ralston G. Introduction to Analytical Ultracentrifugation. Fullerton: Beckman Instruments Inc. 1993.

[53] Laue TM, Stafford WF III. Modern applications of analytical ultracentrifugation. Annu Rev Biophys Biomol Struct 1999; 28: 75-100. [http://dx.doi.org/10.1146/annurev.biophys.28.1.75] [PMID: 10410796]

[54] Schuck P. Size-distribution analysis of macromolecules by sedimentation velocity ultracentrifugation and lamm equation modeling. Biophys J 2000; 78(3): 1606-19. [http://dx.doi.org/10.1016/S0006-3495(00)76713-0] [PMID: 10692345]

[55] Harding SE, Winzor DJ. Sedimentation Equilibrium in the Analytical Ultracentrifuge.Protein-Ligand Interactions: Hydrodynamics and Calorimetry: Partical Approach. Oxford: Oxford University Press 2001; pp. 105-35.

[56] Winzor DJ, Harding SE. Basic Principles of Sedimentation Velocity.Protein-Ligand Interactions: Hydrodynamics and Calorimetry: Partical Approach. Oxford: Oxford University Press 2001; pp. 75-104.

[57] Laue T. Biophysical studies by ultracentrifugation. Curr Opin Struct Biol 2001; 11(5): 579-83.

[http://dx.doi.org/10.1016/S0959-440X(00)00250-5] [PMID: 11785759]

[58] Lebowitz J, Lewis MS, Schuck P. Modern analytical ultracentrifugation in protein science: a tutorial review. Protein Sci 2002; 11(9): 2067-79. [http://dx.doi.org/10.1110/ps.0207702] [PMID: 12192063]

[59] Schuck P, Perugini MA, Gonzales NR, Howlett GJ, Schubert D. Size-distribution analysis of proteins by analytical ultracentrifugation: strategies and application to model systems. Biophys J 2002; 82(2): 1096-111. [http://dx.doi.org/10.1016/S0006-3495(02)75469-6] [PMID: 11806949]

[60] Howlett GJ, Minton AP, Rivas G. Analytical ultracentrifugation for the study of protein association and assembly. Curr Opin Chem Biol 2006; 10(5): 430-6. [http://dx.doi.org/10.1016/j.cbpa.2006.08.017] [PMID: 16935549]

[61] Gabrielson JP, Arthur KK, Stoner MR, *et al.* Precision of protein aggregation measurements by sedimentation velocity analytical ultracentrifugation in biopharmaceutical applications. Anal Biochem 2010; 396(2): 231-41. [http://dx.doi.org/10.1016/j.ab.2009.09.036] [PMID: 19782040]

[62] Borges JC, Ramos CH. Analysis of molecular targets of Mycobacterium tuberculosis by analytical ultracentrifugation. Curr Med Chem 2011; 18(9): 1276-85. [http://dx.doi.org/10.2174/092986711795029537] [PMID: 21366535]

[63] Ghirlando R. The analysis of macromolecular interactions by sedimentation equilibrium. Methods 2011; 54(1): 145-56. [http://dx.doi.org/10.1016/j.ymeth.2010.12.005] [PMID: 21167941]

[64] Zhao H, Brautigam CA, Ghirlando R, Schuck P. Overview of Current Methods in Sedimentation Velocity and Sedimentation Equilibrium Analytical Ultracentrifugation. [http://dx.doi.org/10.1002/0471140864.ps2012s71]

[65] Johnson ML, Correia JJ, Yphantis DA, Halvorson HR. Analysis of data from the analytical ultracentrifuge by nonlinear least-squares techniques. Biophys J 1981; 36(3): 575-88. [http://dx.doi.org/10.1016/S0006-3495(81)84753-4] [PMID: 7326325]

[66] da Silva KP, Borges JC. The molecular chaperone Hsp70 family members function by a bidirectional heterotrophic allosteric mechanism. Protein Pept Lett 2011; 18(2): 132-42. [http://dx.doi.org/10.2174/092986611794475057] [PMID: 21121894]

[67] Young JC. Mechanisms of the Hsp70 chaperone system. Biochem Cell Biol 2010; 88(2): 291-300. [http://dx.doi.org/10.1139/O09-175] [PMID: 20453930]

[68] Craig EA. Essential roles of 70kDa heat inducible proteins. BioEssays 1989; 11(2-3): 48-52. [http://dx.doi.org/10.1002/bies.950110203] [PMID: 2686623]

[69] Mayer MP. Hsp70 chaperone dynamics and molecular mechanism. Trends Biochem Sci 2013; 38(10): 507-14. [http://dx.doi.org/10.1016/j.tibs.2013.08.001] [PMID: 24012426]

[70] Daugaard M, Rohde M, Jäättelä M. The heat shock protein 70 family: Highly homologous proteins with overlapping and distinct functions. FEBS Lett 2007; 581(19): 3702-10. [http://dx.doi.org/10.1016/j.febslet.2007.05.039] [PMID: 17544402]

[71] De Maio A. Extracellular Hsp70: export and function. Curr Protein Pept Sci 2014; 15(3): 225-31. [http://dx.doi.org/10.2174/1389203715666140331113057] [PMID: 24694368]

[72] Duncan EJ, Cheetham ME, Chapple JP, van der Spuy J. The Role of Hsp70 and Its Co-Chaperones in Protein Misfolding, Aggregation and Disease. In: Blatch GL, Edkins AL, editors. The Networking of Chaperones by Co-Chaperones. Subcellular Biochemistry. 78: Springer International Publishing; 2015. p. 243-73.

[73] Veinger L, Diamant S, Buchner J, Goloubinoff P. The small heat-shock protein IbpB from Escherichia

coli stabilizes stress-denatured proteins for subsequent refolding by a multichaperone network. J Biol Chem 1998; 273(18): 11032-7.
[http://dx.doi.org/10.1074/jbc.273.18.11032] [PMID: 9556585]

[74] Lee GJ, Vierling E. A small heat shock protein cooperates with heat shock protein 70 systems to reactivate a heat-denatured protein. Plant Physiol 2000; 122(1): 189-98.
[http://dx.doi.org/10.1104/pp.122.1.189] [PMID: 10631262]

[75] Bracher A, Verghese J. Grpe, Hsp110/Grp170, Hspbp1/Sil1 and Bag Domain Proteins: Nucleotide Exchange Factors for Hsp70 Molecular Chaperones. In: Blatch GL, Edkins AL, editors. The Networking of Chaperones by Co-Chaperones. Subcellular Biochemistry: Springer International Publishing; 2015. p. 1-33

[76] Langer T, Lu C, Echols H, Flanagan J, Hayer MK, Hartl FU. Successive action of DnaK, DnaJ and GroEL along the pathway of chaperone-mediated protein folding. Nature 1992; 356(6371): 683-9.
[http://dx.doi.org/10.1038/356683a0] [PMID: 1349157]

[77] Russell R, Wali Karzai A, Mehl AF, McMacken R. DnaJ dramatically stimulates ATP hydrolysis by DnaK: insight into targeting of Hsp70 proteins to polypeptide substrates. Biochemistry 1999; 38(13): 4165-76.
[http://dx.doi.org/10.1021/bi9824036] [PMID: 10194333]

[78] Szabo A, Langer T, Schröder H, Flanagan J, Bukau B, Hartl FU. The ATP hydrolysis-dependent reaction cycle of the Escherichia coli Hsp70 system DnaK, DnaJ, and GrpE. Proc Natl Acad Sci USA 1994; 91(22): 10345-9.
[http://dx.doi.org/10.1073/pnas.91.22.10345] [PMID: 7937953]

[79] Borges JC, Ramos CH. Characterization of nucleotide-induced changes on the quaternary structure of human 70 kDa heat shock protein Hsp70.1 by analytical ultracentrifugation. BMB Rep 2009; 42(3): 166-71.
[http://dx.doi.org/10.5483/BMBRep.2009.42.3.166] [PMID: 19336004]

[80] Schönfeld HJ, Schmidt D, Schröder H, Bukau B. The DnaK chaperone system of Escherichia coli: quaternary structures and interactions of the DnaK and GrpE components. J Biol Chem 1995; 270(5): 2183-9.
[http://dx.doi.org/10.1074/jbc.270.5.2183] [PMID: 7836448]

[81] Cantor CR, Schimmel PR. Size and Shape of Macromolecules.Biophysical Chemistry, Part Ii: Techniques for the Study of Biological Structure and Function. New York: W.H. Freeman Company 1980; pp. 539-90.

[82] Liberek K, Marszalek J, Ang D, Georgopoulos C, Zylicz M. Escherichia coli DnaJ and GrpE heat shock proteins jointly stimulate ATPase activity of DnaK. Proc Natl Acad Sci USA 1991; 88(7): 2874-8.
[http://dx.doi.org/10.1073/pnas.88.7.2874] [PMID: 1826368]

[83] Osipiuk J, Georgopoulos C, Zylicz M. Initiation of lambda DNA replication. The Escherichia coli small heat shock proteins, DnaJ and GrpE, increase DnaK's affinity for the lambda P protein. J Biol Chem 1993; 268(7): 4821-7.
[PMID: 8444859]

[84] Blattner FR, Plunkett G III, Bloch CA, *et al.* The complete genome sequence of Escherichia coli K-12. Science 1997; 277(5331): 1453-62.
[http://dx.doi.org/10.1126/science.277.5331.1453] [PMID: 9278503]

[85] Naylor DJ, Stines AP, Hoogenraad NJ, Høj PB. Evidence for the existence of distinct mammalian cytosolic, microsomal, and two mitochondrial GrpE-like proteins, the Co-chaperones of specific Hsp70 members. J Biol Chem 1998; 273(33): 21169-77.
[http://dx.doi.org/10.1074/jbc.273.33.21169] [PMID: 9694873]

[86] Borges JC, Cagliari TC, Ramos CH. Expression and variability of molecular chaperones in the sugarcane expressome. J Plant Physiol 2007; 164(4): 505-13.

[http://dx.doi.org/10.1016/j.jplph.2006.03.013] [PMID: 16687190]

[87] Borges JC, Peroto MC, Ramos CH. Molecular Chaperone Genes in the Sugarcane Expressed Sequence Database (Sucest). Genet Mol Biol 2001; 24(1-4): 85-92. [http://dx.doi.org/10.1590/S1415-47572001000100013]

[88] Harrison CJ, Hayer-Hartl M, Di Liberto M, Hartl F, Kuriyan J. Crystal structure of the nucleotide exchange factor GrpE bound to the ATPase domain of the molecular chaperone DnaK. Science 1997; 276(5311): 431-5. [http://dx.doi.org/10.1126/science.276.5311.431] [PMID: 9103205]

[89] Grimshaw JP, Jelesarov I, Schönfeld HJ, Christen P. Reversible thermal transition in GrpE, the nucleotide exchange factor of the DnaK heat-shock system. J Biol Chem 2001; 276(9): 6098-104. [http://dx.doi.org/10.1074/jbc.M009290200] [PMID: 11084044]

[90] Gelinas AD, Langsetmo K, Toth J, Bethoney KA, Stafford WF, Harrison CJ. A structure-based interpretation of E.coli GrpE thermodynamic properties. J Mol Biol 2002; 323(1): 131-42. [http://dx.doi.org/10.1016/S0022-2836(02)00915-4] [PMID: 12368105]

[91] Chesnokova LS, Slepenkov SV, Protasevich II, Sehorn MG, Brouillette CG, Witt SN. Deletion of DnaK's lid strengthens binding to the nucleotide exchange factor, GrpE: a kinetic and thermodynamic analysis. Biochemistry 2003; 42(30): 9028-40. [http://dx.doi.org/10.1021/bi0346493] [PMID: 12885236]

[92] Harrison C. GrpE, a nucleotide exchange factor for DnaK. Cell Stress Chaperones 2003; 8(3): 218-24. [http://dx.doi.org/10.1379/1466-1268(2003)008<0218:GANEFF>2.0.CO;2] [PMID: 14984054]

[93] Borges JC, Fischer H, Craievich AF, Hansen LD, Ramos CH. Free human mitochondrial GrpE is a symmetric dimer in solution. J Biol Chem 2003; 278(37): 35337-44. [http://dx.doi.org/10.1074/jbc.M305083200] [PMID: 12840016]

[94] Oliveira CL, Borges JC, Torriani IL, Ramos CH. Low resolution structure and stability studies of human GrpE#2, a mitochondrial nucleotide exchange factor. Arch Biochem Biophys 2006; 449(1-2): 77-86. [http://dx.doi.org/10.1016/j.abb.2006.02.015] [PMID: 16579957]

[95] Borges JC, Fischer H, Craievich AF, Ramos CH. Low resolution structural study of two human HSP40 chaperones in solution. DJA1 from subfamily A and DJB4 from subfamily B have different quaternary structures. J Biol Chem 2005; 280(14): 13671-81. [http://dx.doi.org/10.1074/jbc.M408349200] [PMID: 15661747]

[96] Summers DW, Douglas PM, Cyr DM. Prion propagation by Hsp40 molecular chaperones. Prion 2009; 3(2): 59-64. [http://dx.doi.org/10.4161/pri.3.2.9062] [PMID: 19535913]

[97] Kampinga HH, Craig EA. The HSP70 chaperone machinery: J proteins as drivers of functional specificity. Nat Rev Mol Cell Biol 2010; 11(8): 579-92. [http://dx.doi.org/10.1038/nrm2941] [PMID: 20651708]

[98] Cyr DM, Ramos CH. Specification of Hsp70 Function by Type I and Type Ii Hsp40. 2015. [http://dx.doi.org/10.1007/978-3-319-11731-7_4]

[99] Caplan AJ, Cyr DM, Douglas MG. Eukaryotic homologues of Escherichia coli dnaJ: a diverse protein family that functions with hsp70 stress proteins. Mol Biol Cell 1993; 4(6): 555-63. [http://dx.doi.org/10.1091/mbc.4.6.555] [PMID: 8374166]

[100] Cyr DM, Langer T, Douglas MG. DnaJ-like proteins: molecular chaperones and specific regulators of Hsp70. Trends Biochem Sci 1994; 19(4): 176-81. [http://dx.doi.org/10.1016/0968-0004(94)90281-X] [PMID: 8016869]

[101] Qian YQ, Patel D, Hartl FU, McColl DJ. Nuclear magnetic resonance solution structure of the human Hsp40 (HDJ-1) J-domain. J Mol Biol 1996; 260(2): 224-35. [http://dx.doi.org/10.1006/jmbi.1996.0394] [PMID: 8764402]

[102] Cheetham ME, Caplan AJ. Structure, function and evolution of DnaJ: conservation and adaptation of chaperone function. Cell Stress Chaperones 1998; 3(1): 28-36. [http://dx.doi.org/10.1379/1466-1268(1998)003<0028:SFAEOD>2.3.CO;2] [PMID: 9585179]

[103] Fan CY, Lee S, Cyr DM. Mechanisms for regulation of Hsp70 function by Hsp40. Cell Stress Chaperones 2003; 8(4): 309-16. [http://dx.doi.org/10.1379/1466-1268(2003)008<0309:MFROHF>2.0.CO;2] [PMID: 15115283]

[104] Walsh P, Bursać D, Law YC, Cyr D, Lithgow T. The J-protein family: modulating protein assembly, disassembly and translocation. EMBO Rep 2004; 5(6): 567-71. [http://dx.doi.org/10.1038/sj.embor.7400172] [PMID: 15170475]

[105] Li J, Qian X, Sha B. Heat shock protein 40: structural studies and their functional implications. Protein Pept Lett 2009; 16(6): 606-12. [http://dx.doi.org/10.2174/092986609788490159] [PMID: 19519518]

[106] Yochem J, Uchida H, Sunshine M, Saito H, Georgopoulos CP, Feiss M. Genetic analysis of two genes, dnaJ and dnaK, necessary for Escherichia coli and bacteriophage lambda DNA replication. Mol Gen Genet 1978; 164(1): 9-14. [http://dx.doi.org/10.1007/BF00267593] [PMID: 360041]

[107] Minami Y, Höhfeld J, Ohtsuka K, Hartl FU. Regulation of the heat-shock protein 70 reaction cycle by the mammalian DnaJ homolog, Hsp40. J Biol Chem 1996; 271(32): 19617-24. [http://dx.doi.org/10.1074/jbc.271.32.19617] [PMID: 8702658]

[108] Rüdiger S, Schneider-Mergener J, Bukau B. Its substrate specificity characterizes the DnaJ co-chaperone as a scanning factor for the DnaK chaperone. EMBO J 2001; 20(5): 1042-50. [http://dx.doi.org/10.1093/emboj/20.5.1042] [PMID: 11230128]

[109] Lee S, Fan CY, Younger JM, Ren H, Cyr DM. Identification of essential residues in the type II Hsp40 Sis1 that function in polypeptide binding. J Biol Chem 2002; 277(24): 21675-82. [http://dx.doi.org/10.1074/jbc.M111075200] [PMID: 11919183]

[110] Li J, Qian X, Sha B. The crystal structure of the yeast Hsp40 Ydj1 complexed with its peptide substrate. Structure 2003; 11(12): 1475-83. [http://dx.doi.org/10.1016/j.str.2003.10.012] [PMID: 14656432]

[111] Ramos CH, Oliveira CL, Fan CY, Torriani IL, Cyr DM. Conserved central domains control the quaternary structure of type I and type II Hsp40 molecular chaperones. J Mol Biol 2008; 383(1): 155-66. [http://dx.doi.org/10.1016/j.jmb.2008.08.019] [PMID: 18723025]

[112] Borges JC, Seraphim TV, Mokry DZ, Almeida FC, Cyr DM, Ramos CH. Identification of regions involved in substrate binding and dimer stabilization within the central domains of yeast Hsp40 Sis1. PLoS One 2012; 7(12): e50927. [http://dx.doi.org/10.1371/journal.pone.0050927] [PMID: 23227221]

[113] Szyperski T, Pellecchia M, Wall D, Georgopoulos C, Wüthrich K. NMR structure determination of the Escherichia coli DnaJ molecular chaperone: secondary structure and backbone fold of the N-terminal region (residues 2-108) containing the highly conserved J domain. Proc Natl Acad Sci USA 1994; 91(24): 11343-7. [http://dx.doi.org/10.1073/pnas.91.24.11343] [PMID: 7972061]

[114] Banecki B, Liberek K, Wall D, *et al.* Structure-function analysis of the zinc finger region of the DnaJ molecular chaperone. J Biol Chem 1996; 271(25): 14840-8. [http://dx.doi.org/10.1074/jbc.271.25.14840] [PMID: 8662861]

[115] Szabo A, Korszun R, Hartl FU, Flanagan J. A zinc finger-like domain of the molecular chaperone DnaJ is involved in binding to denatured protein substrates. EMBO J 1996; 15(2): 408-17. [PMID: 8617216]

[116] Suh WC, Lu CZ, Gross CA. Structural features required for the interaction of the Hsp70 molecular

chaperone DnaK with its cochaperone DnaJ. J Biol Chem 1999; 274(43): 30534-9. [http://dx.doi.org/10.1074/jbc.274.43.30534] [PMID: 10521435]

[117] Sha B, Lee S, Cyr DM. The crystal structure of the peptide-binding fragment from the yeast Hsp40 protein Sis1. Structure 2000; 8(8): 799-807. [http://dx.doi.org/10.1016/S0969-2126(00)00170-2] [PMID: 10997899]

[118] Caplan AJ, Tsai J, Casey PJ, Douglas MG. Farnesylation of YDJ1p is required for function at elevated growth temperatures in Saccharomyces cerevisiae. J Biol Chem 1992; 267(26): 18890-5. [PMID: 1527016]

[119] Caplan AJ, Douglas MG. Characterization of YDJ1: a yeast homologue of the bacterial dnaJ protein. J Cell Biol 1991; 114(4): 609-21. [http://dx.doi.org/10.1083/jcb.114.4.609] [PMID: 1869583]

[120] Lu Z, Cyr DM. Protein folding activity of Hsp70 is modified differentially by the hsp40 co-chaperones Sis1 and Ydj1. J Biol Chem 1998; 273(43): 27824-30. [http://dx.doi.org/10.1074/jbc.273.43.27824] [PMID: 9774392]

[121] Cyr DM. Cooperation of the molecular chaperone Ydj1 with specific Hsp70 homologs to suppress protein aggregation. FEBS Lett 1995; 359(2-3): 129-32. [http://dx.doi.org/10.1016/0014-5793(95)00024-4] [PMID: 7867784]

[122] Luke MM, Sutton A, Arndt KT. Characterization of SIS1, a Saccharomyces cerevisiae homologue of bacterial dnaJ proteins. J Cell Biol 1991; 114(4): 623-38. [http://dx.doi.org/10.1083/jcb.114.4.623] [PMID: 1714460]

[123] Terada K, Kanazawa M, Bukau B, Mori M. The human DnaJ homologue dj2 facilitates mitochondrial protein import and luciferase refolding. J Cell Biol 1997; 139(5): 1089-95. [http://dx.doi.org/10.1083/jcb.139.5.1089] [PMID: 9382858]

[124] Lopez N, Aron R, Craig EA. Specificity of class II Hsp40 Sis1 in maintenance of yeast prion [RNQ+]. Mol Biol Cell 2003; 14(3): 1172-81. [http://dx.doi.org/10.1091/mbc.E02-09-0593] [PMID: 12631732]

[125] Moriyama H, Edskes HK, Wickner RB. [URE3] prion propagation in Saccharomyces cerevisiae: requirement for chaperone Hsp104 and curing by overexpressed chaperone Ydj1p. Mol Cell Biol 2000; 20(23): 8916-22. [http://dx.doi.org/10.1128/MCB.20.23.8916-8922.2000] [PMID: 11073991]

[126] Linke K, Wolfram T, Bussemer J, Jakob U. The roles of the two zinc binding sites in DnaJ. J Biol Chem 2003; 278(45): 44457-66. [http://dx.doi.org/10.1074/jbc.M307491200] [PMID: 12941935]

[127] Lian HY, Zhang H, Zhang ZR, *et al.* Hsp40 interacts directly with the native state of the yeast prion protein Ure2 and inhibits formation of amyloid-like fibrils. J Biol Chem 2007; 282(16): 11931-40. [http://dx.doi.org/10.1074/jbc.M606856200] [PMID: 17324933]

[128] Silva JC, Borges JC, Cyr DM, Ramos CH, Torriani IL. Central domain deletions affect the SAXS solution structure and function of yeast Hsp40 proteins Sis1 and Ydj1. BMC Struct Biol 2011; 11(1): 40. [http://dx.doi.org/10.1186/1472-6807-11-40] [PMID: 22011374]

[129] Fan CY, Lee S, Ren HY, Cyr DM. Exchangeable chaperone modules contribute to specification of type I and type II Hsp40 cellular function. Mol Biol Cell 2004; 15(2): 761-73. [http://dx.doi.org/10.1091/mbc.E03-03-0146] [PMID: 14657253]

CHAPTER 4

# Structural Characteristics of the TPR Protein-Hsp90 Interaction: A New Target in Biotechnology

**Ana Cauerhff**[1,*] and **Mario D. Galigniana**[1,2]

[1] *Laboratory of Nuclear Receptors, Institute of Biology & Experimental Medicine, Buenos Aires, Argentina*

[2] *Instituto de Biología y Medicina Experimental (IBYME)-CONICET, Buenos Aires, Argentina*

**Abstract:** Nature employs multiple repeat protein scaffolds in order to promote protein-protein interactions. In this sense, TPR proteins participate in different natural pathways, especially in diverse processes of eukaryotic cells. An important aspect for cellular homeostasis is the maintenance of the folding of recently synthesized peptides as well as all mature proteins such as SHRs. Since, an aberrant protein folding drives loss of function, this effect induce the expression or modulate the function of molecular chaperones. Hsp90 and Hsp70 with the cooperation of cochaperones are involved in the stabilization of several proteins implicated in signaling, and in the tumor phenotype of various cancers. Therefore, cochaperones are essential component of the cytosolic Hsp90 folding pathway, since their function comprises targeting clients to Hsp90, modulating their conformational changes or Hsp90 ATPase activity. The scientific knowledge in the properties and structure of chaperones and the searching of compounds that can modulate their function on different cellular mechanism has became remarkably important in the treatment of diverse diseases specially those in which a protein mechanism is involved. A description of diverse structural aspects of Hsp90-TPR cochaperones interaction in the context of SHR, as well as a structural comparison of different isoforms of Hsp90 is presented in this chapter. Besides, the primary and new biotechnological approaches inhibiting Hsp90 interactions are also discussed, since Hsp90 and its interactions have become the main targets for inhibiting the growth of specific tumor types.

**Keywords:** TPR domain, TPR protein structure, Hsp90-TPR protein interaction, Hsp90 cochaperones, Protein folding, Steroid hormone receptor, Hsp90 inhibitors.

* **Corresponding author Ana Cauerhff:** Laboratorio de Biología Molecular y Celular, Departamento de Química Biológica, Facultad de Ciencias Exactas y Naturales, Universidad de Buenos Aires, Buenos Aires (1428), Argentina; Tel: 54+11/ 5285-8691; Fax: 4576-3342; Email: anacauer@qb.fcen.uba.ar

**Mario D. Galigniana (Ed.)**

# INTRODUCTION

The formation of protein-protein complexes is a cardinal requirement for many key factors of the living cell to acquire their specific biological functions. The study of protein-protein interactions comprises chemical, physical and structural aspects. In general, protein-protein interactions are highly specific resulting in medium to high affinity, however, in nature, there are proteins sufficiently flexible to interact also with other non-protein ligands. The different proteins found in living organisms have evolved to present multiple repeats to promote protein-protein interactions, which are components of various metabolic pathways. One of such repeat domains is the so called TPR (tetratricopeptide-repeats) motif, which participates in many natural mechanisms. The TPR motif (as other repeat motif) produces structural domains that function as interaction scaffolds comprising multiprotein complexes involved in several cellular processes such as transcription, cell cycle, protein translocation, protein degradation, and host defense against invading pathogens. Similar to other repeat protein scaffolds, TPR-containing proteins were found to intervene in diverse processes in eukaryotic cells, including peroxisomal targeting and import [1], synaptic vesicle fusion, and mitochondrial and chloroplast import. Additionally, TRPs proteins are essential for many bacterial pathways, such as bio-mineralization of iron oxides in magnetotactic bacteria [2], outer membrane assembly, and pathogenesis [3]. Besides, mutations in TPR proteins have been associated with several human diseases, such as Leber's congenital amaurosis and chronic granulomatous disease [4]. Accordingly, TPR domains can establish different types of interactions and thus perform multiple functions. On the other hand, the quality control of the proteins that carry on the cellular metabolism is essential so that the life continues inside the cells. In the same way, the folding of recently synthesized peptides to mature proteins as steroid receptors is crucial for cellular homeostasis. The Hsp90 and Hsp70 chaperones with the cooperation of other proteins as cochaperones constitute the Hsp70/Hsp90-based chaperone machinery which regulates that cellular homeostasis process [5]. Basically, cochaperones as component of the cytosolic folding pathway, target client proteins to Hsp90 and modulate its conformational changes and/or Hsp90 ATPase activity. Moreover, recent evidence shows that cochaperones promote or stabilize suitable conformations of Hsp90. Besides, they regulate the nucleotide status, and thus protein function of Hsp70 and Hsp90, and deliver non-native proteins to their respective polypeptide-binding domains for folding. The cochaperones which regulate Hsp70 include Hsp40, Hsp70-interacting protein (Hip), Hsp-organizing protein (Hop) and small glutamine-rich TPR protein (SGT). Instead, cochaperones which regulate Hsp90 comprise among others, Hop, p23, PP5, CyP40, Aha1, FKBP51, FKBP52, CHIP (C-terminal of Hsp70-interacting protein). Cochaperones when interacting with chaperones facilitate the processes of folding and assembly of proteins. Consequently, cochaperones are essential for cellular homeostasis, differentiation, proliferation, and protection from proteotoxic stress (protein toxic stress). Otherwise, the Hsp90 heterocomplex is comprised of two dimerized Hsp90 protomers surrounded by a set of cochaperones. Those cochaperones perform different tasks which include recruiting client proteins as well as modulating the ATPase cycle to enable the Hsp90 in the maintenance of protein folding. The aberrant folding of the protein provokes loss of function and reduces life expectancy as could be observed in neurodegenerative diseases. To avoid these detrimental effects it is necessary the development of compounds that may modulate the function or induce the expression of cochaperones. This research field has become very important for the advancement of scientific knowledge on different cellular mechanisms as well as in the searching for new drugs to increase the production of proteins properly folded. The chaperone Hsp90 is pivotal for cell physiology, because it is involved in the stabilization of several proteins related to cell signaling and in consequence, in the homeostasis of diverse cancer cells. Thus, Hsp90 heterocomplex is a central point of regulation of several cellular pathways. Recently, the strategy of generating compounds that interfere with the Hsp90-cochaperonas interaction has emerged as a possible application in the therapeutic field. Many of these cochaperones have TPR domains able to interact with a specific MEEVD motif located at the C-terminus of Hsp90. For this reason, the study and understanding of the Hsp90-TPR domain protein interaction is crucial to develop new drugs to attack several diseases. In this chapter, different aspects of Hsp90-TPR cochaperone interaction are discussed in the context to the steroid hormone receptor (SHR) complex stabilization to highlight the importance of these interactions in future basic and applied research. The first part of the chapter consists of a description of TPR protein characteristics: sequence, structure, ligand binding, folding, folding design and stability. The second part refers to the sequence, function and basic structure of Hsp90 protein, different isoforms and conformational changes of Hsp90 performed due to its chaperone function. The third part describes the several aspects of different Hsp90-TPR interactions in the chaperone cycle in the context of SHR. Finally, the fourth part of this chapter contains a review about biotechnology applications of the Hsp90 molecule as a target for the treatment of various diseases and the latest developments on drugs that target the co-chaperones Hsp90-interaction.

## TPR PROTEIN CHARACTERISTICS

### Definition and Prediction of the Sequence and Basic Structure of TPR Motifs

During the evolution of life the emergence of characteristic protein domains have fostered protein-protein interactions since the different protein domains are generally associated with certain functions and form part of a regulatory network. The TPR motif is a module that facilitates these protein-protein interactions. The TPR motif was discovered in 1990 within the genes of CDC23 and Nuc2, it is composed by an amino acid sequence present in either nine or ten inexact tandem repeats. It was first discovered in yeast cell cycle regulatory proteins [6, 7] but it has been identified in all living organism and a diverse cellular compartments. The TPR motif has been found in a variety of proteins which participate in several cellular process as mitochondrial and peroxisomal protein transport, NADPH oxidase activity, protein kinase inhibition, transcriptional regulation, immunity, mRNA processing, viral replication and protein folding and translocation [8 - 13]. The TPR motif is constituted by a 34 amino acid degenerate sequence arranged in tandem repeats [6, 7, 9]. This degenerate consensus sequence is organized by forming a packing pattern of small and large hydrophobic amino acids which results in a rather splayed arrangement of α-helices (Fig. **1**). It is notable that TPR motifs, despite having the same structural organization show different protein binding specificity and mediate also different protein-protein associations [14]. These protein-protein associations promote assembly of the TPR-protein complex into higher order structures [15, 16].

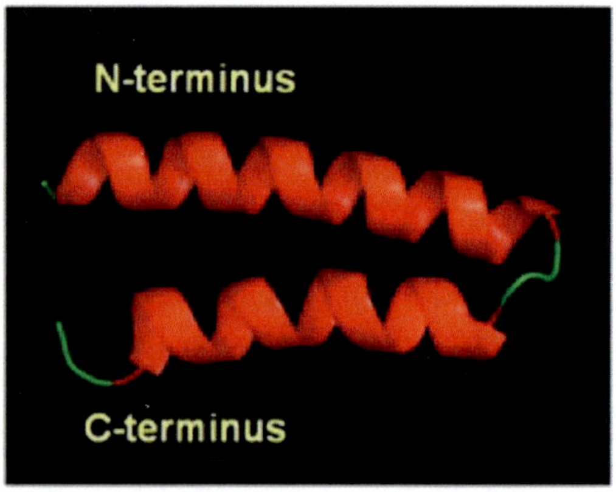


**Fig. (1).** TPR motif representation. TPR proteins possess a duplicated, degenerate 34 amino acid sequence termed the tetratricopeptide repeat (TPR) motif.

The X-ray crystallographic structure of most TPR domain-containing proteins showed that TPR motifs as two anti-parallel-helices packed generally in tandem arrays. In proteins which present this kind of organization, an extended and right-handed superhelical arrangement can be shaped and are completed by a C-terminal hydrophilic "capping-helix" that is thought to improve solubility [17, 18]. In this sense, TPR proteins are non globular repeat proteins that possess a duplicated TPR motif. The number of observed TPR units in diverse proteins ranges from 3 to 16 (Fig. **2**). After analyzing 3418 TPR-containing proteins,

Regan and colleagues [17, 18] distinguished the 34 amino acids with the highest propensity to create the TPR consensus sequence. Nevertheless, there are no fully invariant residues in such domains. This consensus conserved pattern includes positions 4, 7, 8, 11, 20, 24, 27, and 32, numbered relative to the motif N-terminal residue (for more details see reference [3]). Some residue types are highly conserved only at positions 8 (Ala or Gly), 20 (Ala), and 27 (Ala), whereas consensus positions 4, 7, 11, and 24 do not exhibit a preference for a particular amino acid but for a certain class of residue such as small, large or aromatic [4].

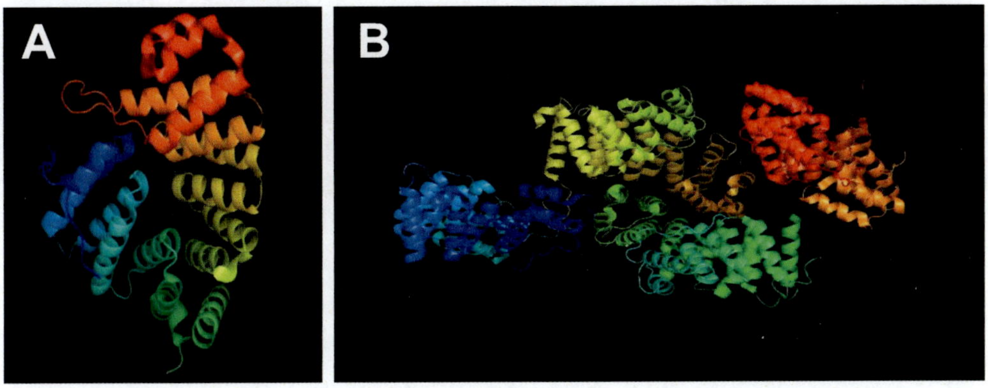


**Fig. (2). A)**-Peroxisomal targeting signal 1 receptor (PEXP5). Like other TPR proteins, PEXP5 has TPR motifs arranged in tandem arrays. **B)**-Structure of the homo-oligomerization tendency of a TPR protein.

Usually, between helical segments there are turns which in TPR motifs present residue conservations suggesting a crucial role in the TPR interaction and stability. Some authors suggest that those residues may play a key role in specifying the superhelical twist of the repeat and then the resultant structure may perform a determined function [4]. Consensus position 32, located in the turn between two TPR motifs, normally contains helix-breaking residues, such as proline [3, 4]. The study of the protein phosphatase 5 (PP5) [19] which is composed by three-TPR domains showed that the positions 8, 20 and 27 have strong preference for small amino acids and constituted the closest point between α-helices. This fact not only illustrates the role of residue type preference but the absolute consistency of connecting between small and large amino acids in a complementary manner. Only conservative mutations in these positions are tolerated, non-conservative mutations produce dysfunctional TPR domains that lead to disease [19]. However, other positions inside TPR motif or domain are better tolerated without loss of function.

Due to the increased awareness about the importance of TPRs, some sequence analysis programs, such as the Simple Modular Architecture Research Tool

(SMART) [20, 21] or the PROSITE dictionary of protein sites and motif patterns [22] can be excellent tools for identifying TPR consensus sequences. An additional specifically designed tool is TPRpred, which utilizes the profile representation of known repeats to detect TPR motifs and other patterns of protein repeats [23]. Consequently, the TPRpred Web server computes the statistical significance of the occurrence of protein repeats from a query sequence. Consequently, in the current genomic era, the TPR-containing protein distribution in diverse living organism can be predicted. In this way, TPR proteins were found in all forms of life, namely eukaryotes, bacteria, and archaea.

## Three Dimensional Structure

The canonical unit of the TPR motif adopts a basic helix-turn-helix fold without any long connecting loops (Fig. **1**). Adjacent TPR units form a series of repeating anti-parallel α-helices due to their parallel packing, yielding an overall superhelix structure that is influenced by the residue type positioned between adjacent TPR motifs. In this manner, the unique superhelix fold shapes a pair of concave and convex curved surfaces (Fig. **3**), where the groove is the proposed site for protein-protein interactions. In addition, the surfaces of TPR domains display amino acid variety and some degree of elasticity, allowing the binding of diverse ligands, usually via the concave surface. Thus, the regular and extensive surfaces found in repeat proteins facilitate them to mediate molecular recognition events. Therefore, repeat proteins as TPR-domain proteins fit out a central scaffold composed by a conserved sequence motif and a variable sequence in loops presented on the surface. Then, those loops may offer assorted interaction surfaces both in terms of geometry and in functional groups contents. For all the above, it must be considered that in nature there are a lot of repeats interaction motifs as TPR with different sequence and structure to perform different function [24, 25].

### *Curvature and Shape of the TPR Domain*

If we study the overall curvature angle and shape displayed by TPR-containing proteins, we can observe a considerable diversity due to many factors. Those factors include the arrangement as sequential motifs, the separation of repeating motifs into subdomains, the occupancy of residues types, helix breaking residues, among others. In spite of their apparent significance, there are no reported extensive studies about the overall configuration and curvature angle displayed by TPR-containing proteins. Until recently, TPRs were considered as rigid proteins that do not change their confirmation upon ligand binding, with their presented curvature angle being considered to be rigid as well. However, a study performed by Zarivach and coworkers has shown that some TPR-containing proteins, experience conformational changes upon binding of a putative ligand imitator [3].

Conformational changes and radial movements by the two N-terminal TPR motifs was noticed in the different crystal structures, suggesting that changes in the presented curvature angle might allow the proper binding of the ligand (*i.e.*, induced fit). Accordingly, the possible role of the curvature angle and its flexibility should be considered as an important component in ligand recognition and binding.

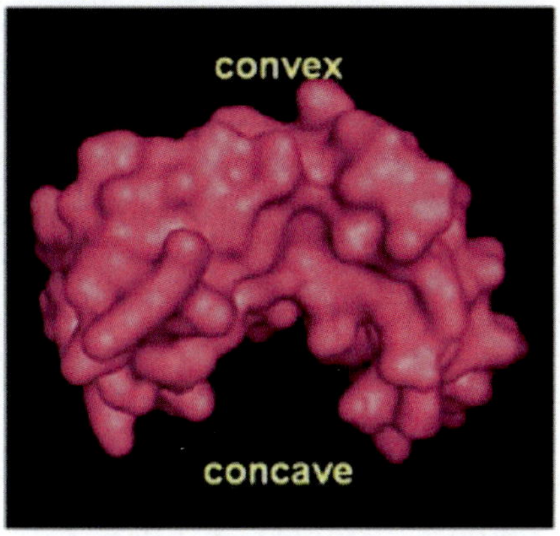


**Fig. (3).** Overall superhelix structure between adjacent TPR motifs. Surface representations of the superhelix fold forming a pair of concave and convex curved surfaces in Hop (1ELR).

## *Examples of TPR Protein Structures*

The first insights related to the structures of TPR proteins were acquired with the analysis of the crystal structure of the Ser/Thr-protein phosphatase PP5, which became an iconic client protein of Hsp90 (see below). PP5 is composed by three TPR repeats that constitute a right-handed superhelical structure with a continuous helical groove [19]. In such structure, the parallel arrangement of contiguous TPR motifs generates a regular series of anti-parallel α-helices, which is a distinctive feature of 14-3-3 proteins. Additionally, a right-handed α-helical conformation is generated and creates a channel that is capable to accommodate polypeptides from other substrate proteins [26, 27]. The TPR domain of PP5 exhibits an additional "capping helix" at the C-terminal end. This characteristic found in PP5 is not unusual since an equivalent helix is also observed in almost all TPR structures published to date. Some authors have conjectured that this "capping helix" is necessary for the solubility and/or stability of the TPR domains [4]. It is quite remarkable that TPR proteins, like other tandem repeat proteins, are also formed by other domains, which in turn assist to recruit specific substrates, as an example we can cite F-box proteins (see Itzhaki & Lowe [28] for an extensive and completed review of this family of proteins).

## Ligand Binding

The presence of a TPR domain inside a protein structure implies the existence of specific ligand interactions. Having said this, it should also be emphasized that some TPR proteins include TPR domains within certain structural arrangements to permit functional interactions with specific protein targets. The first reported X-ray crystal structures of three-TPR domains in complex with their peptide ligands revealed that the bound peptide is displayed in an extended conformation on the concave binding surface of the TPR superhelix [15, 29]. Later, the analysis of several reported structures of TPRs in complex with a peptide or protein ligand highlighted the adaptability of this helical domain. Structural modifications in the loop regions or in the α-helices will be allowed as long as the canonical TPR fold can be maintained. All these variations in structure, likewise adaptations result in TPR-containing proteins with different modes of interaction with distinct binding partners [4, 9].

Actually, not all TPR domains bind to ligands in the same manner. In this sense, several studies demonstrated that TPR-containing proteins bind diverse ligands without sequence or secondary structure similarity because the TPR binding pockets are composed for different kind of amino acids. Thus, different surface amino acids in each binding surface yield specific ligand interactions. TPR motifs utilize their distinct fold as an interaction scaffold resulting in wide-ranging types of binding. Generally, the binding of TPR domain proteins is highly specific, and these proteins are able to identify their ligands even within the crowded cellular environment. As it has been reported for the majority of the protein-protein complexes, binding specificity is not the result of a single bonding force. Rather, specificity is mostly achieved by a combination of factors: residue type, charge, electrostatics, hydrophobic pockets, among others. This feature also is observed in available TPR-ligand protein structures. Likewise, the secondary structure of TPR-bound ligand is an important factor to consider when TPR protein ligand binding is analyzed. Such secondary structure can differs between an extended coil to a α-helix conformation. An elongated conformation enlarges the ligand surface presented to the TPR domain and optimizes H- bonding formation, as well as the specific recognition of short amino acid fragments. Even more, several structural studies reviewed in Zeytuni *et al.* [3] indicate that exist a correlation between TPR binding surfaces and the amino acid sequences of their ligands. Nowadays, due to the deep knowledge about the structure of the TPR domain and with the help of bioinformatics tools we can identify the TPR-containing proteins by analyzing their amino acid sequences [20, 21, 23]. Due to their limitations, bioinformatics tools till now cannot predict TPR-interacting partners and regions of interaction. For this task to be performed successfully, the identification of TPR-interacting proteins by a genomic approach must be done firstly. D'Andrea and Regan [4] generated a list of 22 predicted TPR-containing proteins encoded by the *Saccharomyces cerevisiae* genome. Then, a protein-protein interaction databases scanning was performed to identify potential binding partners and 80 potential interacting proteins were predicted, some of which are known to contribute to multiprotein complex formation. Another approach consists of using the information derived from unbound TPR-containing proteins structure. Upon inspection of a TPR domain containing protein structure (mainly of the concave binding pocket) as to residue composition, dimensions, and electrostatic potential [30], a prediction of possible binding partners can be done. In this study, both concave and convex surface of the binding pocket should be considered in such analyses. Besides, the prediction of the TPR domain interaction region embraces first, the identification of the TPR structure within the protein and secondly the interacting partner binding sequence.

Over the past few years, several attempts have been made at docking ligand peptides onto TPR-containing proteins. The first trials mainly comprised manual ligand peptide docking onto available TPR domain-containing protein structures. However, the final models did not take into account side-chain conformational changes and flexibility attributable to peptide binding, and employed restricted energy minimization tools [15, 31]. Nonetheless, with the advance of modern bioinformatics tools, better and accurate models for TPR-containing protein-peptide interactions can be generated and in the near future, could contribute in the knowledge about molecular mechanisms of binding and recognition.

## Folding and Stability of TPR Proteins

The application of TPR proteins and domains for biotechnological purposes lead us to the need to study in depth the stability and folding dynamics of those molecular species. In general, TPR proteins present a globular shape; however, due to the lack of contacts between distant amino acids along the sequence in TPR proteins, these proteins show different characteristics from globular proteins [32]. Among these aspects, we will focus on the features of their folding processes. The traditionally accepted model of protein folding defines it as a two-state event where an extended chain reaches its native structure without the presence of stable intermediate species [33, 34]. This is the classical cooperative folding model for globular proteins. This model is also valid mostly for some repeat proteins [35]. Nevertheless, many authors have reported studies of repeat proteins for which this model is not valid [36 - 38]. It has also been assumed that the modularity of their native structures results in a modular stability of different repeats, enabling some of the sections of the protein to perform independently of the others [32]. All the work accumulated in the study of the TPR domain protein architecture showed that these TPR proteins can be designed with improved thermodynamic stabilities by virtue of their folding pathways. At the same time, according to advanced the folding pathway studies new consensus repeat proteins are been designed as platforms for engineering novel binding specificities. In this sense, the singular architecture of repeat proteins makes them a particular interesting target for biophysical analysis. As has been pointed out above, the modular nature and the lack of long-range interactions in TPR proteins are thought to contribute to the cooperative folding model. A correlation between folding rate and proportion of short-range contacts in the native structure has been observed, since the entropic cost is smaller for closing short loops [39]. Those short loops are located between TPR motifs. In protein folding studies, an important parameter is the m-value that can be defined as a measure of the size of the structural unit that is unfolding in a cooperative step. Diverse studies have shown that both the free energy of unfolding and the m-value increase with increasing number of repeats, signifying that there is a cooperativity of folding but interestingly, cooperativity is not always is related with the increase of the number of repeats [25, 40].

Regan and colleagues [36] using Differential Scanning Calorimetry (DSC) concluded that the folding of CTPRa2 and CTPRa3 proteins could be well fitted into a two-state process, because they contain 2 and 3 repeats, respectively. Nevertheless, two-state folding processes were not observed for the larger proteins of the series [36]. Then, Main and coworkers [41] also used DSC to study the family of CTPR proteins and compared these factors with the two smallest members of the CTPR series: CTPR2 and CTPR3. Both groups of proteins are pretty similar, but for substitution of two residues on each 34-amino-acid repeat module. CTPRa2 and CTPRa3 present a two-state unfolding system. However, as the number of repeats (*n*) is increased, the stability of the whole protein and partially unfolded intermediates increase (CTPRa6 to CTPRa8 for the CTP*Ran* group and CTPR2 onward for the CTPR group) [41]. In addition, the Levy laboratory [42, 43] found similar results for several members of the CTPR family and for other repeat protein [44]. The study concentrated on the folding kinetics and the relationship with the topology of the native state, as represented by the contact order [39] or other properties derived from this property. Further, the relationship between the unfolding cooperativity and the inter-repeat coupling was investigated using one dimensional Ising models [45]. One dimensional Ising model has not phase transition which, each repeat has an intrinsic equilibrium constant for folding. This model considers an additional equilibrium constant which describes the interactions or coupling between contiguous repeats. On the other hand, Kajander *et al.* [37] utilized the one dimensional Ising model to study the cooperativity of folding in designed TPR repeat proteins. The authors have studied several synthetic repeat proteins with varying numbers of consensus repeat domains. By fitting the data all at once for all the studied proteins, Regan and coworkers [46] afforded to extract the equilibrium constant for folding and coupling. The probability of partially unfolded conformations at the unfolding transition midpoint is high in TPR proteins that posses multiple copies of consensus sequence even more, the most of the population was observed to be in a partially unfolded state. Then, the authors hypothesized that natural proteins could present some degree of optimization of coupling for cooperative folding whereas this fact was not found in proteins containing multiple repeats of an identical sequence. In further studies, Wolynes and collaborators [47] used a model generated by molecular dynamics simulations (MDS) that, further would be enclosed in their Ising like model for the folding of CTPR family. But, they utilized only a single PDB structure of the CTPR3 protein to build native contact maps for different numbers of repeats. González-Charro and Rey [48] instead, capitalized the computational efficiency of a coarse-grained model to go beyond that analysis. They considered the structure of the thermodynamic intermediates detected, and tried to rationalize their populations and structures according to the experimental contact maps of both natural and consensus sequences proteins. In that work, performing a systematic study they demonstrated that CTPRa2 and CTPRa3 have no stable intermediate states yielding, again, a clear two-state thermodynamic transition. The presence of possible partially stabilized intermediate minima for both proteins has been observed. However, their free energy is very high in comparison with the native and unfolded states and, consequently, at the equilibrium temperature, those states are poorly populated. Similar results were found by Cortajarena and Regan [36]. Instead, the profile of CTPR4 is totally different. For this protein the free energy of the intermediate minimum, is close to the native state, and it is comparable with those of the folded and unfolded states. In another work, González-Charro and Rey [48] proved the existence of intermediates of folding. The native centric model provides intermediates which are similar to the native structures, or to a fraction of them in the repeat proteins studied by them. It would be interesting to check if these results are still obtained when interactions related to the real sequences are considered. Therefore, the structure of the native state may explain different properties of their folding-unfolding equilibrium. Consequently, new experiments are necessary to analyze not only the effect of the number of repeats on the global characteristics of the folding processes, but also the structure of the folding intermediates and the differences between proteins with the same number of repeats, depending on the actual type of sequence (*i.e.*, consensus or natural).

### *Oligomerization, Stability and Biological Functions*

A particular characteristic of TPR domains is their homo-oligomerization ability. Since, TPR-containing proteins are known to promote and be part of multiprotein complexes, within these complexes, TPR domains or proteins can be found as a single mediator or as homo-oligomers (Fig. **2**). The first approach to study the

TPR domain oligomerization was the characterization of a TPR-containing truncated form of *Schizosaccharomyces pombe* nuclear protein nuc2$^{+}$, which suggested that the structural assembled TPR domain functions may be considered as a protein interaction module and capable of forming large protein aggregates. At the same time, this feature facilitates the assembly of multi-protein complexes [6]. It is important to study the relationship between the TPR oligomerization and the function of the TPR proteins in life. In this sense, we can postulate that TPR proteins display a broad range of oligomerization states, including monomers [29, 49] and multimers [2, 50, 51]. Likewise, the self-assembly property of TPR proteins to give high order structures, can be intrinsic or in response to external stimuli [52 - 55]. This additional level of regulation allows for the fine-tuning of biological processes. Some examples can be mentioned as the TPR-containing protein rapsyn, which participates in the neuromuscular synapse [52, 53]; another example is the mitochondrial outer membrane protein Fis1 which is involved in mitochondrial fission [55]. These two examples emphasize the biological importance of TPR-mediated oligomerization. In addition, structural analysis on TPR domains deposited in the Protein Data Bank (PDB) has been performed using the program/server Protein Interfaces, Surfaces and Assemblies (PISA) [56] at the European Bioinformatics Institute on a PDB wide scale. PISA is an interactive tool for the exploration of macromolecular interfaces which revealed the possibility of self-association of TPR proteins in a number of cases. In order to elucidate the structural and solution data that explain TPR self-assembly for designing an oligomeric TPR domain, Kleanthous and colleagues have [57] identified key residues involved in oligomerization of YbgF, which is linked to the Toll system in Gram-negative bacteria. They also characterized this trimeric TPR-containing protein using analytical ultracentrifugation, size exclusion chromatography-multiangle light scattering (SEC-MALLS), circular dichroism (CD), X-ray crystallography, among other techniques. Their study laid the foundations for understanding the structural basis for TPR domain self-association and how such self-association can be regulated in TPR domain-containing proteins and also validated the involvement in TPR self-assembly by solution methods. On the other hand, Kleanthous and colleagues [57] have engineered residues that have been identified as relevant for the self-association and monomeric protein oligomerization in solution of the YbgF TPR into a three-repeat version of the consensus TPR protein (CTPR3,5). In such a way, conserved residues that deviate from the TPR consensus sequence are the fundamental to oligomerization. Moreover, the structure analysis of this engineered TPR domain can elucidate important principles for the formation of higher order TPR-oligomers. The authors of this study [57] previously subtracted the residues belonging to the TPR consensus sequence, which are required for folding stability by analyzing residues conserved in YbgF homologues. Then, manipulating those residues that are located in loops regions between TPR motifs, onto the CTPR3 (monomeric consensus protein TPR), the formation of oligomers can be induced. In this way, Kleanthous and coworkers [57] were be able identify residues involved in oligomerization of the C-terminal YbgF TPR domain. Analyzing the crystal structure of this engineered oligomer showed that the key of oligomerization are stacking interactions produced by introduced tyrosines and the displacement of the C-terminal hydrophilic capping helix. Even more, asymmetric trimerization of the YbgF TPR domain and CTPR3Y3 drives to the formation of higher order oligomers both in the crystal and in solution [57]. It is interesting to note that this self-association does not happen in full-length YbgF indicating that the N-terminal coiled-coil domain of the protein restrains further oligomerization. The described example illustrates the complexity of the TPR motif/protein oligomerization and the same time the possibility to be engineered for biotechnological purposes.

### *Novel TPR Protein Design*

It is important to remark that in nature, TPR-containing domains usually contain three or more tandem motifs. In the literature some examples of folding and stability studies can be cited, Kajander *et al.* [46] designed and determined the structure of a non-natural recombinant TPR-containing protein that includes 20 sequential TPR motifs. By addressing proteins with an ever-increasing number of repeats, these authors determined a positive correlation between protein thermo-stability and the number of TPR repeats. Also, Main *et al.* were able to design novel TPR proteins by consensus design. They constructed proteins containing 1.5-3.5 TPR motifs [18] and furthermore included an N-capping helix-stabilizing Gly-Asn-Ser (GNS) sequence at the N-terminus and an extra solvating helix at the C-terminus. Since, statistical studies have shown that Gly, Asn and Ser have the highest propensities to occur at the N″, N′, and N cap positions in α-helices. The addition of the solvating helix was the result of the structural study of the TPR-containing proteins where it was found that the last TPR motif was capped with another helix from the protein and was found to increase solubility. Besides, the 2.5- and 3.5-TPR repeat proteins folded into the correct structures and exhibited high thermal stabilities [25]. Nevertheless, Regan and colleagues were able to design a cooperatively folded polypeptide consisting of only one and a half TPR motifs that is a minimum required for a cooperative unit of structure. These findings ratify the estimation from contact maps of TPR proteins and suggest that the TPR module can be independently folded and that makes more contacts within the module than between modules [18]. Furthermore, it has been investigated the effects of increasing numbers of TPR repeats on the cooperativity in the protein-folding kinetic and the folding energy landscape. For this purpose, a large series of highly symmetric, designed TPRs comprising 2.5-10.5 repeats were used [58].

In this study, a multistate kinetic folding pathway for TPR proteins containing three or more repeats was determined and shown to involve thermodynamic interplay between the stability of individual repeats, as well as the interaction between repeats in a similar manner as it has been observed with other repeat proteins, such as ankyrins. However, since the repeats employed by Javadi and Main [58] were identical, there was no energetic favorability for the formation of a stable folding core by a single or by only a few repeats. Such a core would allow for stable protein folding around it without misfolded intermediates, as they were shown to exist in the earlier study [58]. As a conclusion, we can propose that a TPR motif-generated protein-protein interaction platform can support the binding of different ligands. The elegant superhelical fold of TPR-containing proteins exhibits several binding surfaces that can promote the formation of multiprotein complexes. TPR-containing proteins are presented as a great promise for protein engineering, therapeutics, and biotechnology, as the basic TPR scaffold can be redesigned to modulate binding specificity and/or affinity toward desired peptide ligands. Moreover, TPR-containing proteins can be inhibited by designed ligands with higher affinities, serve as platforms for protein presentation in nanotechnological applications, and more. Since the goal of this chapter is the description of TPR proteins-Hsp90 interactions, structural aspects of the Hsp90 molecule and a profound study of TPR client proteins-Hsp90 interactions will be described in the following items.

## SEQUENCE, FUNCTION, AND BASIC STRUCTURE OF HSP90 PROTEINS: HSP90 ALPHA AND BETA

### Introduction

Hsp90 is a molecular chaperone and one of the most abundant and conserved protein in nature. It is highly conserved from bacteria to eukaryotes with 50% sequence similarity between *Escherichia coli* and humans. More accurately, Hsp90 is an ATP-dependent molecular chaperone which is ubiquitously expressed in eukaryotes and comprises one of the most abundant proteins in the cytosol of cells. The function of Hsp90, in contrast to other chaperones, has remained an enigma for several decades. Hsp90 is a homodimeric ATPase [59] and its ATPase activity is indispensable to perform its function in vivo [60, 61]. Accordingly, in eukaryotes, Hsp90 is essential for housekeeping functions and is induced under various stress conditions, whereas in many prokaryotes, Hsp90 appears to be non-essential for most growth conditions. In eukaryotes, Hsp90 has dual chaperone functions participating both in the conformational maturation of the nuclear hormone receptors and protein kinases, and in cellular stress response [62, 63]. These two processes are likely to have in common the ability of Hsp90, in cooperation with Hsp70 and other factors to prevent protein aggregation and mediate the ATP- dependent refolding of heat-denaturated proteins *in vivo* and *in vitro* [64]. It is remarkable that, while the other ATP-dependent chaperones were shown to refold denatured model proteins in vitro (Hsp60 and Hsp70), or unfold proteins (Hsp100), no such activities could be found for Hsp90 [64, 65]. Furthermore, different from other known molecular chaperone like Hsp70 and GroEL/ES, Hsp90 does not appear to make an important contribution to *de novo* folding of proteins, another pivotal task of molecular chaperones [66]. Nevertheless, Hsp90 foments the final maturation of selected client proteins. Therefore, Hsp90 does not only function in protein folding, but it also contributes to various cellular processes such as signal transduction, protein transport and protein degradation.

It was noticed that Hsp90 establishes complexes with certain nuclear receptors and protein kinases, so it is thought that this protein acts as activity regulator of a limited number of protein substrates named clients. Until now, more than 400 Hsp90 clients have been reported [67]. These client proteins can be grouped into two major groups, for example a group would consist of: SHRs, transcription factors (such as p53) and kinases, most of which participate in signal transduction pathways of cell growth and differentiation. The other group of client proteins would be a more diverse and composed by DNA- and RNA-binding proteins (including polymerases), ribosomal proteins, small GTPases, cytoskeletal proteins, and ion channels. Overall, Hsp90 clients do not share the same sequence or structure. Moreover, the mechanisms by which Hsp90 binds to client proteins and regulates their activity, turnover, trafficking, cofactor insertion, membrane insertion, ligand binding, or covalent modification are yet to be elucidated. Another interesting aspect to keep in mind is that in eukaryotic cells, the ATPase cycle of Hsp90 is intimately coupled to the Hsp70 chaperone machine. Both systems rely on a large number of specific cochaperones, which enter and leave the machine in a defined order, thus regulating client maturation. In addition, the cochaperone requirement varies and depends on the actual client protein. While in bacteria, a homologous protein similar to Hsp90 called HtpG was found for example in *Escherichia coli*, no Hsp90 gene has been found in Archaea. Even more, bacterial Hsp90 seems to be non essential and its precise function remains to be investigated. However, novel studies postulate that this Hsp90 isoform collaborates with the DnaK (Hsp70) system in substrate remodeling and may function as a protective defense against oxidative stress. In yeast, there are two Hsp90 isoforms in the cytosol, Hsc82 and Hsp82, in which Hsp82 is up-regulated up to 20 times under heat stress conditions.

In summary, Hsp90 participates in the regulation of the stress response. It functions in complex with other cochaperones in stabilizing and refolding denatured proteins after stress, preventing misfolding and aggregation of unfolded or partially folded proteins. Besides, Hsp90 assists in correctly folding of newly synthesized proteins and in protein transport across the endoplasmic reticulum (ER) and organellar membranes.

## Hsp90 Isoforms

Approximately forty years ago, several laboratories addressed the expression of Hsp90 on the cell surface, either as a tumor antigen or a helper protein in the context to antigen presenting cell. Furthermore, it was found that there is an accumulation of Hsp90 inside cells to be secreted during injury. The early studies about the steady-state storage of great amounts of Hsp90 in cells led us to think that this fact would be responsible for a fast preserving response to environmental damage, such as heat, hypoxia, reactive oxygen species (ROS), injury-released growth factors, UV, gamma-irradiation, among others [68]. Therefore, originally, Hsp90 was proposed to increase upon heat stress. Nonetheless, it is remarkable that already in unstressed cells, the protein amounts reach about 1-3% of the total soluble cytosolic protein pool [69]. In vertebrates, two distinct isoforms are found: the mostly inducible Hsp90α isoform that represents the major form and the constitutively expressed Hsp90β [70]. In 2002, another isoform of Hsp90 was discovered, the Hsp90N, that participates in cellular transformation [71]. Other Hsp90 analogues were also found out: Grp94 in the endoplasmic reticulum, Hsp75/Trap1 in the mitochondrial matrix [72] and ch-Hsp90 in the chloroplast. It is remarkable that Trap1 presents a unique LxCxE motif [73], which is not found in all other Hsp90 family members [74]. In summary, there are four human Hsp90 isoforms namely two cytosolic partners (Hsp90α and Hsp90β), one mitochondrial form (Trap1) and one specific form located at the endoplasmic reticulum (Grp94). The whole mapping of Hsp90α and Hsp90β to the human genome and the genomic location were revised in Sreedhar *et al.* [70]. The nucleotide sequences of Hsp90α and Hsp90β have low similarity, especially in their 5′ and 3′ noncoding regions, the introns, and the regulatory 5′-flanking sequences [75]. On the contrary, the Hsp90N nucleotide sequence shares a high similarity with that of Hsp90α, and represents a very recent gene rearrangement [71]. The mammalian Hsp90 isoforms α and β emerged by gene duplication approximately 500 million years ago [76]. The "signature sequences" found in Hsp90 isoforms are highly conserved regions: three are situated at the N-terminal domain and comprise amino acids 38-59, 106-114, 130-145 of human Hsp90α and the remaining two at the middle domain containing amino acids 360-370 and 387-401 [76]. Several phylogenetic distribution studies have been reported but the majority of these studies about Hsp90 focused on the cytoplasmic isoforms [76, 77]. Thus, the conservation of protein sequences between species can be studied by means phylogenetic trees, using this approach. Those phylogenetic studies show that Hsp90α and β clusters present high conservation between mammals and suggest that Hsp90 α isoform could be the result of beta gene duplication [78].

In vertebrate Hsp90α and Hsp90β are encoded by genes Hsp90AA and Hsp90AB, with seven and one signature sequences, respectively. Both genes were clustered into a single clade [17, 23, 24, 27]. Even more, recent study performed by Chen and coworkers [79] suggests that the clade Hsp90A is monophyletic with high bootstrap support and 11 signature motifs. Nevertheless, the difference between the Hsp90AA and Hsp90AB classes was markedly smaller than the difference among HSP90A sequences across animals, fungi, plant and protists. Therefore, this result would signify that an additional gene duplications event took place early in the origin of vertebrates and would explain the presence of Hsp90AA and Hsp90AB classes [79].

After several years of research, it was concluded that there are profound differences between extracellular (Hsp90α) and intracellular (Hsp90β) Hsp90. In such manner, the first difference is the cellular location. The second difference comprises their function, the intracellular isoform of Hsp90 acts as molecular chaperone whereas the extracellular isoform functions as a pro-motility factor with other possible unknown properties. It is necessary to make a clear identification of these two isoforms by nomenclature, especially when we refer to the extracellular Hsp90, this protein can be named as "surface bound", released, or secreted. Few studies in the past few years have disclosed the mechanism and functional reason why Hsp90α is exported by both normal and tumor cells. Lately, several studies have demonstrated that secreted Hsp90α is a novel pro-motility factor in cell in response to tissue injury. The clinical importance of Hsp90α is due to tumor cells need this isoform to promote tumor invasion, so blocking this protein would result in a significant inhibition of tumor metastasis [80]. Currently it is known that under stress such as hypoxia, UV light, ionizing radiation, free radicals and tissue injury, normal cells secrete Hsp90α. On the contrary, cells from several tumor types secrete Hsp90α and Hsp90β [68].

## Sequence and Basic Structure of Hsp90 Proteins: Hsp90 α and β

As was mentioned above, sequences of all Hsp90 isoforms have been subjected to phylogenetic analysis. Chen and colleagues [79] performed a comparative genomic study and evolutionary analysis of the Hsp90 family of genes which includes 32 complete genomes across all kingdoms of life. They found 87 putative functional genes, and based on functional motif/domain prediction Chen and coworkers divided the gene subfamilies into five subfamilies: HTPG, TRAP, Hsp90C, Hsp90B, Hsp90A, at the same time Hsp90A was divided into Hsp90AA (Hsp90α) and Hsp90AB (Hsp90β) genes. Besides, these authors proposed a new nomenclature for the Hsp90 gene family which was accepted by the Human

Genome Organization (HUGO) Gene Nomenclature Committee [81]. In this nomenclature the root names Hsp90B means ER Hsp90 and Hsp90C designs chloroplast homologues of Hsp90 [82]. The whole analysis performed by Chen and coworkers indicated that the Hsp90 family of genes (with the exception of HTPG, and some HSP90A and HSP90B genes) contains a large number of introns. In addition, their genome investigation identified Hsp90-like proteins that exist in Archaea and Bacteria. Other interesting result was that all Hsp90A+B+C and TRAP sequences share 2 signature motifs that originated within HTPG Group A. In addition to the sequence of the different Hsp90 isoforms, the Hsp90 structure was extensively investigated and numerous crystal structures were published in the PDB database. At this moment, more than 400 entries are found for Hsp90 structures. In December 2008 Hsp90 was declared the molecule of the month and a related article was written by David Goodsell [83]. The Hsp90 is a member of a special class of structurally related, evolutionarily conserved split ATPases as the so-called gyrase, histidine kinases, MutL (GHKL) domain ATPases, which contain a Bergerat ATP-binding fold [59]. The Hsp90β isoform was the first to be discovered and the first to be extensively studied.

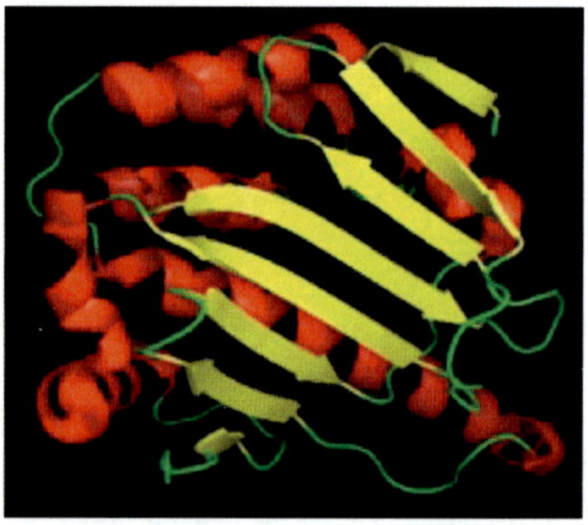

**Fig. (4).** N-terminal domain (NTD) of Hsp90. The spatial organization of the NTD consists of two layers composed by α and β structures. SCOP defines the fold of this domain as a α/β structure, the α-secondary structure is represented in red and the β-sheet is in yellow.

Basically, Eukaryotic Hsp90 consists of four structural domains [84 - 87]:

-a highly conserved N-terminal domain (NTD) of ~25 kDa
-a "charged linker" region (CL), that connects the N-terminus with the middle domain
-a middle domain (MD) of ~40 kDa
-a C-terminal domain (CTD) of ~12 kDa

All Hsp90 homologues have the same domain architecture with a NTD, a MD and a CTD site. Except for the connecting loop (CL) located between the NTD and MD in eukaryotic Hsp90, this domain organization is conserved from bacteria to *Homo sapiens*. Since the Hsp90 is a homodimeric protein, CTD comprises the dimerization domain (DD) and the NTD, the nucleotide-binding domain (NBD). NTD of Hsp90 is a ~25kDa fragment, whose structure was determined after proteolysis analyses. NTD is composed by a α/β structure arranged in two layers (Fig. **4**).

Using the SCOP databases for the structural classification, NTD can be defined as protein with α/β structure. Functionally, this domain appears as part of the ATPase domain of Hsp90 chaperone/DNA topoisomerase II/histidine kinase and comprises eight stranded mixed β-sheet and nine mixed helices (five α-helices and four $3_{10}$-helices). The spatial organization of the NTD consists of two layers composed by α and β structures, as it is shown in Fig. **4**.

Several sections in the Hsp90 sequence are homologous to MutL mismatch repair proteins and type II topoisomerases, which alter DNA with the aid of ATP. This fact made the researchers think that Hsp90 could bind ADP/ATP. In this sense, the binding of adenine nucleotides is by means a pocket in the NTD and this binding site is essential for Hsp90 to fulfill its ATPase function [88].

An important feature of the ATP binding region consist in that several conserved amino acid residues form a "lid" that closes over the nucleotide binding pocket in the ATP bound state but is open during the ADP bound state. The NTD is also the binding site of the specific inhibitors as geldanamycin (GA) and radicicol (RD) [89] among others (see the section Inhibitors of the N-terminal domain of Hsp90). In turn, the middle region or MD includes the catalytic loop and the binding site to most client proteins. MD consists of many short α-helices arranged in a compact coil that connects the NTD to the CTD.

The crystallographic structure of the MD of human Hsp90, resolved at 2.3Å resolution (PDB Id: 3PRY), shows that this domain comprise from Lys 284 to Asn 546 (263 amino acids). The fold classification in the SCOP database place it into the class of α and β structure, its fold is assorted as ribosomal protein S5 domain 2 like superfamily (core: β(3)-α-β-α; 2 layers: α/β; left-handed crossover). MD presents 13 helices and five β-strands, which correspond to five $3_{10}$-α-helices and eight right-handed α-helices. The human Hsp90 MD is represented in the Fig. **5**.

Client proteins could bind to Trp 300 by means hydrophobic interactions in addition to the region comprised by 327-340 residues. Also in the MD, Arg 380 residue and the catalytic loop are important for the reaction with ATP/ADP and ATPase function [90] (Fig. **5**). It should be noted that both ND and MD are involved in ATP hydrolysis, but NTD contains the ATP binding domain.

The CTD also named dimerization domain (DD) is composed by the α/β structure, this particular domain give the name of the fold considered in SCOP database. This fold consists of two layers α/β: α-β(3)-α(3) with crossing loops.

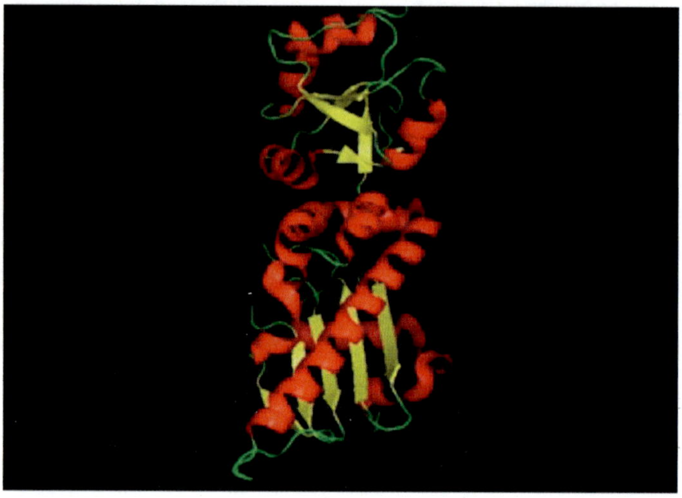

**Fig. (5).** Middle domain of Hsp90. MD consists of a tight coil of many α-helices that are small in length

Unfortunately, the structure of the human Hsp90 CTD is still unknown, but the structure of HtpG of *E. coli* has been resolved at 2.60Å by Agard and colleagues [91] in 2004 (PDB Id: 1SF8). Analyzing this structure, we can observe six α-helices (one belonging to $3_{10}$ type) and 3 β-sheets. In the structure can also be found a crossing loop that is composed by Ala 557-Tyr 566, between the α-helix 3 and the β-sheet 2, the N-terminal of this loop comprise a turn. So far, no secondary structure was assigned by the rest of the loop (Fig. **6**). CTD region can open and close rapidly, for this reason is critical in Hsp90 dimerization process. The dimerization is carried out by the union of two bundles of 2 α-helices present in each Hsp90 monomer [91] and it is shown in Fig. **7**. This region is the most divergent in Hsp90 due to few deletions and a lower sequence similarity (Fig. **8**). Moreover, the association between Hsp90 and cochaperones with a TPR domain is performed by a conserved MEEVD motif at the C-terminal end of CTD [29]. Interestingly, the opening of the CTD in eukaryotic Hsp90 is anti-correlated to the closing of the NTD [92]. Nevertheless, prokaryotic and eukaryotic Hsp90 shows structural differences indicative of mechanistic differences. In eukaryotes, both Hsp90 proteins possess a charged flexible linker between the NTD and MD. Hainzl *et al.* [93] showed that the linker is necessary for cochaperone regulation and client activation but not for ATPase activity. The linker may provide additional freedom for the relative orientations of the NTD and MD, which could be required for the mechanism of regulation through cochaperones, and also to create a flexible interface for client interactions. Furthermore, HtpG (prokaryotic isoform) possesses only two substrates and no cochaperones suggesting that the EEVD motif and the linker present in eukaryotic isoforms of Hsp90 (among other characteristics) evolved to create a flexible system to reclute more client proteins.

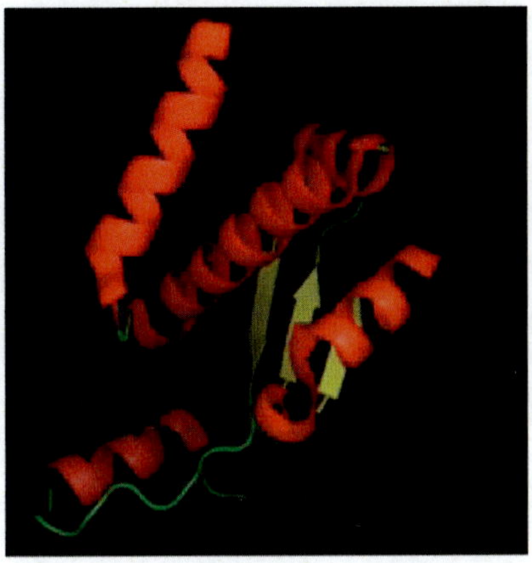

**Fig. (6).** C-terminal domain (CTD) structure of HtpG of *E. coli*, resolved at 2.60Å [86]. According to SCOP database this fold consists of two layers α/β: α-β(3)-α(3) with crossing loops.

Although the complete structure of the eukaryotic Hsp90 has not yet resolved, a realistic model from the combination of the structures of regions from different organisms could be obtained. However, a problematic region could be the link between the MD and the NTD due to its low sequence conservation between organisms [94]. It is worthy to note that due to the great importance of Hsp90 in maintaining the integrity of the proteins carrying out the basic processes of the cell under normal conditions and in disease, structural studies of Hsp90 in various organisms are necessary. At the end of 90s Pearl and Pavletich laboratories [87, 95] have determined the first crystal structures of the NTD. Then, in 2006 the Pearl group [96] determined the structure of Hsp90 in complex with Sba-1 and ATP as ligand. One year later, Agard's group [97] determined the structure of the apo-protein and that bound to ADP. The structural plasticity of Hsp90 makes almost impossible to resolve the whole protein structure by X-ray crystallography. That flexibility led to conformational changes that are explained bellow.

## *Difference in Structure of Hsp90 α–and β–isoforms*

Although it is really difficult to separate both isoforms of Hsp90 is worth the attempt, since they have important functional differences. One of those differences is the dimerization capability, Hsp90α can easily dimerize while Hsp90β is less efficient. Both homogeneous and heterogeneous dimerization process are observed in Hsp90, when homogeneous the result is a homodimer (αα or ββ), whereas when heterogeneous a heterodimer (αβ) is composed. However, higher oligomers of both isoforms can be found in nature. The dimerization capability depends of the quality of the C-terminal 190 amino acids of Hsp90 [70].

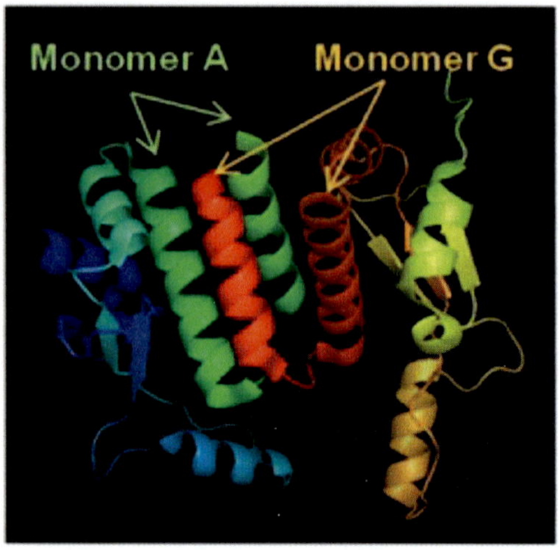


**Fig. (7).** CTD dimerization in Htpg in *Escherichia coli* (PDB Id: 1SF8 [86],). Two α-helices from one monomer (green) interact with two α-helices from the other one (red and orange).

On the other hands, there are differences in the sequence of the amino acid residues of particular areas in α and β isoforms, which favor specific functions, such as differential binding to client proteins. In this sense, for example, Hsp90N lacks the 25 kDa N-terminal domain although presents high sequence homology with Hsp90α and β [71]. Another example is the case of the prokaryotic homologue, HtpG which lacks the highly charged hinge region of the N-terminal section although is similar to the mitochondrial isoform Trap1 [72, 73]. Functional differences between Hsp90 isoforms and clinical implications are reviewed in Sreedhar *et al.* [70].

## Conformational Changes in Hsp90

Due to its function of chaperone, dimeric Hsp90 displays a large degree of conformational freedom, as it can be observed from its X-ray crystallography structure, electron microscopy (EM) images, and small angle X-ray scattering (SAXS) data for Hsp90 different homologues [96 - 99]. The NTD, MD and CTD of Hsp90 show high levels of structural conservation, however, the domain boundaries suffer large structural changes because of DD rearrangements. All three domains interact with different cochaperones and substrates, although the MD appears to play a central role in these interactions. Nevertheless, the subdomains show significant differences in their relative orientations, and the protomers perform wide shearing and twisting motions around the DD interaction, ranging from wide open to very compact closed states with dimerized NTDs.

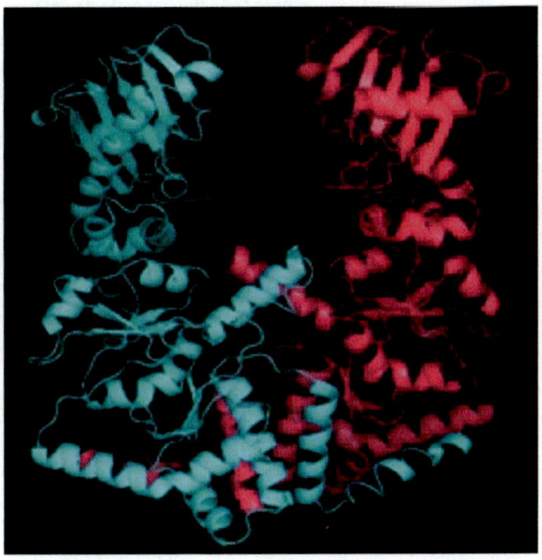

**Fig. (8).** Dimerization of Hsp90. Dimerization is carried out by the interaction of the CTD from two monomers of Hsp90 forming a four-helix bundle (two from each monomer). CTD and MD are shown in each Hsp90 monomer (cyan and magenta, respectively).

Furthermore, Hsp90 appears to fluctuate continuously between two open and two closed conformations in the apo, ATP- and ADP- bound states, as shown for the yeast homolog, Hsp82 [100]. The open conformation (apo-state) adopts a "V"-configuration whereas ATP binding triggers a series of conformational changes such as the NTD-MD orientation and a repositioning of the NTD lid region. In the apo-state, Hsp90 adopts a "V"-configuration, termed "open conformation". On the other hand, ATP binding triggers a series of conformational changes including repositioning of the NTD lid region and a dramatic change in the NTD-MD orientation. Finally, Hsp90 reaches a more compact state, termed "closed conformation" in which the NTDs are dimerized [96, 97].

In 2007, the crystal structure of the Grp94 (endoplasmic reticulum homologue) was resolved [98], researchers found that is very similar to closed state. From the comparison with other structures it can be inferred that the Hsp90 apo-state displays several conformations due mainly to rigid-body rotation at the MD-CTD interface, thus monomer arms may form a variable-size cleft. Then, these dramatic conformational changes produce 50Å translations and 50x rotations upon the open/close transition in Grp94. Besides, SAXS measurements of Hsp90 in the presence of osmolytes indicate that the Hsp90 conformation could be energetically related to folding steps [101]. It is notable, that all the studied Hsp90 homologues as HtpG, Hsc82, Hsp90α, Grp94 and TRAP [102], present high conformational heterogeneity in the apo-state. This means that flexibility in all Hsp90 isoforms is important for their function as chaperone and their interaction with client proteins. Therefore, the client proteins may induce that the different isoforms of Hsp90 to present certain conformations which in turn affect the client folding/unfolding [103]. Thus, the conformational equilibrium of Hsp90 plays a main role in client proteins interactions and depends on the demand of each organism. Due to Hsp90 has a relatively complex structure with several domains and also presents several conformations in the inter-domain junctions, Hsp90 can offer different binding surfaces and geometries in order to interact with structurally diverse clients and cochaperones [103]. In addition, nucleotide binding changes the relative abundance of different conformations by altering the forward and backward transition rates. ATP binding restructures an N-terminal helical lid region, in turn produces profound changes in the NTD/MD orientation and this event is related to a β-strand swapping across monomers to stabilize N-terminal dimerization [103]. The distribution of conformations at the N-terminal also varies between species [104], explaining the different ATPase rates since only the closed, N-terminally dimerized conformation, can hydrolyze ATP. The first conformational feature is represented by the fact that some region in the ATP binding site forms a lid that closes over the nucleotide binding pocket in the ATP-bound state. On the contrary, the lid is open during the ADP-bound state [96]. On the other hand, the helix-loop-helix motif is found neighboring to the nucleotide-binding pocket and makes extensive contacts between monomers [87, 105]. Isolated NTD structures in the apo, AMP-PNP, and ADP bound states display that the lid suffers important structural changes due to nucleotide binding. Biochemical studies highlight the role of the lid in ATP hydrolysis and in the regulation of NTD dimerization. Although the removal of the lid fully abolishes the ATPase activity, when only one of the NTD lacks it, the closure and ATPase hydrolysis rate are accelerated [106, 107]. It was observed that nucleotide binding, lid restructuring and β-strand swapping are all coupled, since the deletion of the first 24 amino acids of the N-terminus, which participate in β-strand swapping and stabilizes NTD dimerization, provides plasticity to the lid [107]. The MD of Hsp90 also is involved in ATP hydrolysis since contains crucial catalytic residues for forming the complete ATPase site. Moreover, the MD also contributes to interaction sites for client proteins and some cochaperones. Thus, the second principal structural change upon ATP binding is located in the NTD/MD interface. This change orients the highly conserved Arg 380 (yeast numbering) which is present in the MD catalytic loop to bind the ATP c-phosphate. Consequently, the cooperativity of the ATP hydrolysis is favored by the numerous contacts in the NTD and MD regions of both protomers [108 - 110]. Another factor to analyze is the NTD/MD orientation which is variable in distinct crystallographic structures. A regulation of the alignment of the Arg 380/ATP c-phosphate by the cochaperones and the client proteins is possible because of the freedom of rotation between the NTD and MD domains. The Arg 380 on the catalytic loop constitutes a hydrophobic network with a loop formed by Thr 22, Val 23 and Tyr 24 (Hsp82 numbering) present on the opposing monomer in the closed state. This interaction network is indicative of a cross-monomer coupling of closure and hydrolysis. The cross-monomer modulation of hydrolysis in yeast was confirmed by means site directed mutagenesis in residues that participate in that network [108]. Truncation experiments where Hsp90 deletion constructs lack CTD still present basal ATPase activity when the two monomers are joined by a linker. However, when Hsp90 constructs imply only cross-linked NTDs without MD exhibit a 100-fold decrease in the ATPase activity in comparison with WT [103]. In a similar manner, for the *Escherichia coli* HtpG in the ATP-induced state the majority conformation seems to be the most compact. Similarly, for the *Escherichia coli* homologue, HtpG, the most compact conformation prevails in the ATP-induced state [99, 111, 112], but some authors found different results [104]. Graf and coworkers showed that ATP binding induces slow stepwise conformational changes in the secondary structure that start at the nucleotide-binding pocket, proceed with dimerization of NTDs, and finally lead to the docking of ND and MD. In their study, they found that in HtpG for ATP binding and hydrolysis conformational changes are needed, but the same does not occur in Hsp82. Moreover, MDS experiments using full-length Hsp90 proved that there is an increase of molecular interactions between residues in the NTD and CTD upon ATP binding [113]. Interestingly, it has been shown that the binding of both ATP and ADP produces interactions over a distance of 80Å, however each nucleotide produces interactions between different residues located in NTD and CTD. These dissimilar interaction patterns could explain the different conformations of Hsp90 in the presence of ATP or ADP [103].

### *Sequence of Conformational Changes Induced by ATP Binding to Hsp90*

After fast ATP binding, Hsp90 reaches the first intermediate state (I1), this is a slow process in which the ATP lid is closed and the NTDs are still open. The NTD dimerization drives to the formation of the second intermediate state (I2), in this state MD change its position and interact with the NTD. Later, Hsp90 reaches a fully closed state in which ATP hydrolysis occurs. Then, NTDs dissociate, release ADP as well as inorganic phosphate (Pi), and Hsp90 returns to the open conformation again [106]. It is notable that the conformational rearrangement of Hsp90 is influenced by nucleotide binding and by the interaction with cochaperones and client proteins [101, 114]. As was pointed out above, during ATP binding to Hsp90 there may be a dynamic equilibrium between the different conformations of Hsp90, this conformational plasticity may allow Hsp90 to adapt to different client proteins. Next, we will develop this feature with more detail.

## HSP90-TPR PROTEIN INTERACTIONS

### Introduction

The particular objective of this section is to describe the cochaperones containing TPR as well as the structural basis of the interaction between Hsp90 and these cochaperones in the context of the SHR complex. The importance of the study of the Hsp90-TPR proteins interaction is due to the fact that several cochaperones that interact with Hsp90 are defined as TPR proteins. Moreover, Hsp90 complexes are essential in the mechanism of folding of recently synthesized peptides towards the generation of correctly folded and functional proteins such as SHR and besides, the complete mechanism is crucial to maintain cellular homeostasis. The overall process is actively regulated by Hsp70 and Hsp90, with the cooperation of cochaperones belonging to the Hsp70/Hsp90-based chaperone machinery [5]. To properly fulfill its function Hsp70/Hsp90 machinery directs misfolded proteins towards the ubiquitin proteasome, in this cellular compartment misfolded proteins are degraded. Thus, Hsp70/Hsp90 machinery in concert with ubiquitin proteasome system plays an important role in protein quality control [115 - 117].

The regulatory function of cochaperones comprises the regulation of nucleotide status and non-native proteins delivery to Hsp70/Hsp90 binding domains for correctly folding. In this context, cochaperones regulate the function of both Hsp90 and Hsp70. The cochaperones regulation is an evolutionarily conserved characteristic of the eukaryotic Hsp90 system [118, 119]. Also, they regulate the inhibition and activation of the ATPase of Hsp90, as well as recruitment of specific client proteins to the cycle. Interestingly, cochaperones work in coordination to promote the maturation of Hsp90 client proteins [94, 120]. The cochaperones that modulate Hsp90 include: Hop, p23, PP5, CyP40, FKBP51 and FKBP52 among others and those which regulate Hsp70 comprise: Hsp40, Hsp70-interacting protein (Hip), Hsp-organizing protein (Hop) and small glutamine-rich TPR protein (SGT). However, some cochaperones regulate both Hsp90 and Hsp70; this is the case of C-terminal of Hsp70-interacting protein (CHIP). It is noticeable that the composition of cochaperone complexes seems to depend to some grade on the presence of a specific client protein. Consequently, cochaperones modulate progression through the cycle by inhibiting (Sti1/Hop, Sba1/p23, Cdc37) or stimulating (Aha1) the ATPase activity of Hsp90, or by competing for binding to the EEVD motifs at the C-termini of Hsp70 and Hsp90 (*i.e.* TPR-containing proteins such as Sti1/Hop, FKBP51, FKBP52, CyP40, Cpr6, PP5/Ppt1, Tah1, CHIP [119]). Some cochaperones function as a platform for the early Hsp70-Hsp90 complex (Sti1/Hop) or as client-specific adapters and targeting factors, e.g. Cdc37 for protein kinases, androgen receptor (AR), hepatitis

B virus reverse transcriptase, GCUNC45 for myosin and SHRs, Tah1 for snoRNA maintenance machinery, etc [86].

## Chaperone Cycle in SHRs

SHRs belong to the large nuclear receptor superfamily; they are hormone-regulated transcription factors [121, 122]. SHRs respond to hormonal signals, these hormones bind to specific binding sites on the surface of SHRs to provoke conformational changes that are necessary for the particular SHR function. SHRs comprise receptors for: glucocorticoids (GR), mineralocorticoids (MR), estrogens (ER), progesterone (PR) and androgens (AR). Their function implies an obligate interaction with Hsp90 and other molecular chaperones before hormone-dependent activation (reviewed in reference 5). Historically, around 1985 the laboratories of Toft, Baulieu, and Pratt [123] discovered almost simultaneously that Hsp90 establishes essential complexes with SHRs. Soon thereafter, the Faber's laboratory [123] discovered an extra Hsp90-associated protein in several SHRs complexes and it was defined as a TPR-domain cochaperone. For teaching purposes from here on we will refer to the SHRs as a set, *i.e.* the SHR.

In an ATP-dependent assembly process, high affinity hormone binding is achieved through the direct interaction of SHR ligand binding domain with Hsp90 and specific Hsp90-associated chaperones. During maturation process various different complexes can be originated in the context of the Hsp90-SHR assembly [124, 125]. Reconstitution experiments with a reticulocyte lysate system or with purified proteins showed that SHRs must pass through three complexes with different cochaperones to achieve their functional conformation. Then, the activation cascade of SHR begins with the binding of SHR to Hsp40 and to Hsp70 to form the "early complex" [124, 126]. In other words, after synthesis, SHR enter to the Hsp90 chaperoning pathway by first association with Hsp40 followed by incorporation of Hsp70 and Hip. Later, SHR is then transferred to Hsp90 by means the help of the cochaperone Sti1/Hop, in this process an "intermediate complex" is generated [127 - 129]. This transference is accompanied by conformational changes in Hsp90 to a closed conformation. Then, Hsp70 and Hop dissociate from the assembly to allow the recruitment of cochaperones like Sba1/p23 and prolyl isomerases (PPIases) such as FKBP51/52 or CyP40. In this moment, a third complex constituted by a PPIase and the cochaperone p23 have been created [5, 130, 131]. At the last step of the cycle, Sba1/p23 and PPIases are released and the activation of SHR is completed [130]. Notably, these types of heterocomplexes are found from yeast to *Homo sapiens* even in the absence of a client protein [124].

Biophysical studies by fluorescence resonance energy transfer (FRET), analytical ultracentrifugation (AUC), nuclear magnetic resonance (NMR) spectroscopy, and EM, have provided additional insights about the regulation of the exchange of cochaperones [131] besides, an alternative model was postulated. Therefore, Hop/Sti1 binds first to the Hsp90 dimer which stabilizes its open conformation and inhibiting its ATPase activity. It has been postulated the existence of a potential (transient) secondary TPR acceptor site which might be preferentially occupied by a PPIase protein, conducting to an asymmetric Hsp90 intermediate complex. A conformational change that suffers Hsp90 after binding of ATP and p23/Sba1 (closed conformation) weaken the binding of Hop/Sti1 and consequently provoke its release from the complex. Another PPIase or TPR cochaperone could bind to assemble the final complex along with Hsp90 and p23/Sba1. Following ATP hydrolysis, p23/Sba1 and the folded client are released from Hsp90 [131]. In the following sections, we will describe and discuss structural characteristics of TPR-cochaperones as well as their different modes of binding to Hsp90 in the context of the maturation of SHRs.

## Hop/Sti1

The first cochaperone to describe is Hop. Hop stands for "Hsp70/Hsp90 organizing protein" since it is required to bring together both chaperones, which are unable to associate one another in the absence of Hop. It was first discovered in yeast and homologues (and named Sti1, or stress inducible protein 1), and then was also identified in humans, mouse, rat, insects, plants, parasites, and virus. It is a modular protein that can function as linker protein between Hsp70 and Hsp90 [128, 132]. GR and PR cannot bind hormone without the action of Hop [89] because Hop performs two important functions in SHR complex association one of them is the recruitment of Hsp90 to preformed Hsp70-receptor complexes, and the other comprises the inhibition of the ATPase activity of Hsp90 [133, 134]. Besides, it was demonstrated that Hop interferes in the binding of geldanamycin in the binding pocket at NTD of Hsp90 [95]. Hop belongs to the TPR cochaperones group; all of them present a conserved TPR-clamp domain. This domain is made up by three TPR motifs and recognizes the C-terminal MEEVD motif in Hsp90 [29].

### *Overall Structure*

Hop is composed (almost exclusively) by various TPR domains [135, 136] and it has been questioned its ability to function as a chaperone by itself [137, 138]. Hop binds Hsp70 and Hsp90 simultaneously via its TPR domains. According to structural predictions in Hop sequence, this protein would contain nine TPR motifs which would constitute two TPR domains. However, three dimensional structural studies showed that Hop contains three TPR domains (TPR1, TPR2A and TPR2B) and two domains of unknown structure (DP1 and DP2). TPR1 is the N-terminal TPR domain, succeeded by an aspartic acid/proline (DP)-rich region, and two more contiguous TPR domains (TPR2A and TPR2B) in contact with a second DP-rich region (Fig. **9**). Both TPR domains (TPR1 and TPR2A) form a bend of seven α-helices which are arranged in a head to tail manner similar to the TPR domain of PP5 (see below) [19].

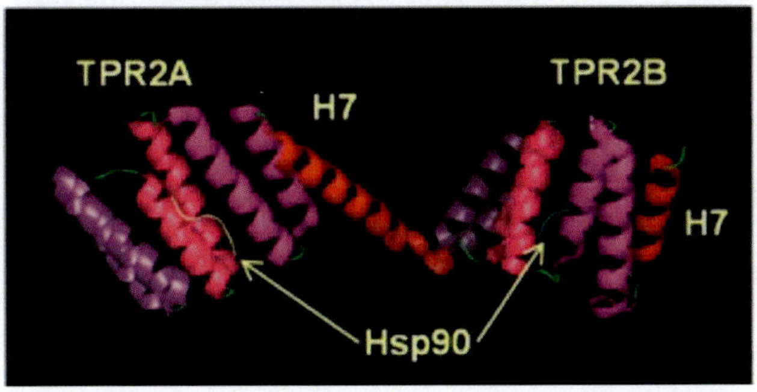


**Fig. (9).** TPR2A and TPR2B domains of HOP. Structure of TPR2A and TPR2B domains in yeast Sti1 (PDB Id=3UQ3 [139],), α-helices forming TPR motif are colored in violet, pink and magenta from N-terminal to C-terminal, respectively in both domains. α-Helix seven is shown red, Hsp90 peptide interacting with TPR2A is yellow and the corresponding peptide that interacts with TPR2B is cyan. Note that the binding sites of Hsp90 peptide in both TPR domains (TPR2A and TPR2B) are oriented in opposite directions.

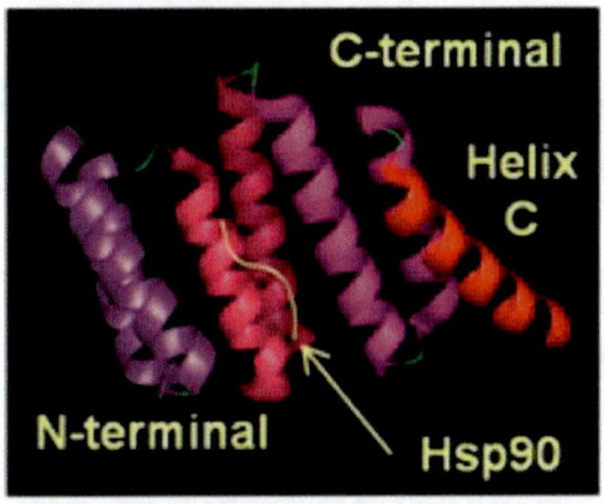


**Fig. (10).** Hop TPR2A-Hsp90 complex. HOP TPR2A domain representation showing TPR motifs in violet, pink and magenta, helix C is in red and Hsp90 peptide (MEEVD) in yellow.

However, TPR1 and TPR domain of PP5 are slightly different from TPR2A, since the latter contains an insertion between repeat two and three which leads to stretching of α-helices B2 and A3 by one turn each. In Hop and in PP5, C-terminal α-helix named α-helix seven or α-helix C, located at the end of the three TPR consensus blocks, is essential for TPR domain function [29] (Figs. **9**, **10** and **11**). On the other hand, Buchner and colleagues [139] have determined the structure of the TPR2A-TPR2B segment bound to the pentapeptide MEEVD (C-terminal end of yeast Hsp90). As was observed in other TPR domains, both

TPR2A and TPR2B consist of three TPR motifs (Fig. **9** and Fig. **10**). In the overall structure, a right-handed α-helical structure with an amphipathic channel is created by the interaction of seven α-helices. A well defined linker orients the two TPR domains in an S-shaped conformation with their binding grooves pointing in opposite directions. The structural analysis and other additional experiments published by Buchner and colleagues [139] ratified that the architecture of both the TPR domains and the linker is physiologically relevant. Because of its capacity to bind Hsp70 and Hsp90, some investigators previously proposed that Hop is a dimer [140, 141], however later conclusive data indicated that Hop is indeed a monomer [142].

### *Hop-Hsp90 Interaction: Structural and Biophysical Aspects*

Deletion mutants analysis of Hop confirmed that TPR1 domain contains three TPR motifs and interacts with the C-terminus of Hsp70, while the C-terminal TPR2 domain, composed by six TPR motifs, principally interacts with Hsp90 [127]. Like for Hsp70, the binding site for Hop has been outlined in the CTD of Hsp90 [143]. However, as it is expected, it is necessary the conservation of C-terminal EEVD motif in Hsp90 for the interaction with Hop [29, 132] and (as was pointed out before) the CTD of Hsp90 also binds other TPR containing cochaperones as CyP40, FKBP51, FKBP52, and PP5. Competition experiments showed that Hsp90 CTD contains only one TPR acceptor site [143, 144]. Scheufler and coworkers [29] have described the structure of the N-terminal TPR domain of Hop (TPR1) bound to a C-terminal peptide of Hsc70 composed by 12 amino acids. Additionally, they described the structure of TPR2A in complex with a five residue peptide located at the C-terminal of Hsp90 (PDB Id: 1ELW and 1ELR respectively) (Fig. **10** and Fig. **11**).

The crystal structures of the TPR-peptides (Hsc70 and Hsp90) complexes show that the peptides are in an extended conformation, spanning a groove in the TPR domains. Electrostatic interactions have been found between EEVD motif from Hsp90 and TPR domain since the C-terminal aspartate acts as a two-carboxylate anchor. Hydrophobic interactions, between TPR domain and residues upstream of EEVD are also found and are crucial for specificity. It is noticeable that the peptide residue Val employs conserved hydrophobic and van der Waals contacts with TPR domain.

With the purpose to measure the $K_A$ and to improve our knowledge about the thermodynamics of the Hsp90/Hsp70-Hop interaction, isothermal titration calorimetry (ITC) was employed. To date, this is the best quantitative method to determining thermodynamics parameters of protein-protein interactions, since it measures directly the heat developed upon association of a protein with its binding partner. Scheufler and coworkers made ITC experiments with expressed and purified fragments corresponding to the N-terminal (aa 1-118) and the MD TPR2A (aa 223-352). ITC measurements [29], demonstrated that a 12 kDa fragment of Hsp90 (C90) that includes the dimerization domain and the genuine C-terminus of Hsp90 (aa 629-732) bound to TPR2A with a $K_D$ of 6 μM ($K_A$=1.6 x $10^5$ $M^{-1}$) and a stoichiometry factor (N) close to one. On the other hand, a five residue peptide from Hsp90 was be able to bind TPR2A with a $K_D$ of 11 μM ($K_A$=9.09 x $10^4$ $M^{-1}$). Also ITC measurements confirmed that a 25-kDa fragment from human Hsp70 (C70), which includes the C-terminus of Hsp70 and the substrate binding domain, bound to TPR1 with an affinity of 15 μM ($K_A$=6.6 x $10^4$ $M^{-1}$) and a N close to one.

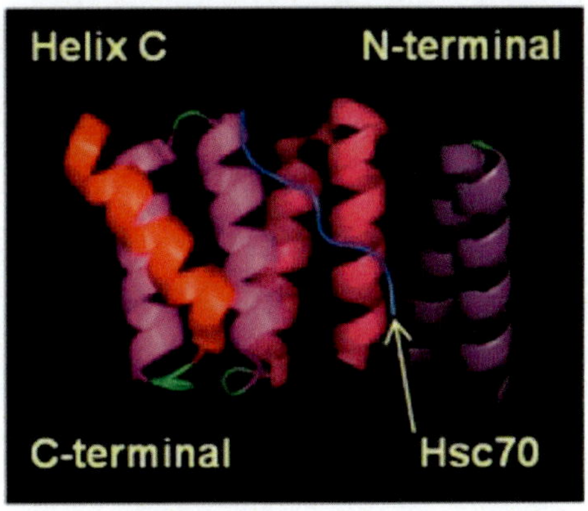


**Fig. (11).** Hop TPR-Hsc70 interaction. Representation of HOP TPR1 domain in complex with Hsc70 peptide (1ELW), three TPR motifs from N-terminus to C-terminus are shown in violet, pink and magenta, respectively, α-helix C and the Hsc70 octa-peptide (GPTIEEVD) are represented in red and marine blue respectively.

Additionally, the binding of TPR1 to C70 can be studied by measurements of the interaction between TPR1 and a 12-mer C-terminal peptide of Hsp70. Simultaneously; it was observed that only five C-terminal residues present in Hsp90 interact with TPR2A. The affinities measured are comparable to those determined for the interaction of SH3 domains with their peptide ligands [145]. In order to better understand the significance of the results of ITC, it must be taken into account that in the TPR2A complex (Hsp90) the bound peptide is considerably shorter than in the TPR1 complex (Hsp70) (Fig. **10** and **11**). It is surprising that the buried surface area for the structured octamer peptide in the TPR1 complex comprises 1330Å$^2$, but only 650Å$^2$ for the EEVD motif. The overall buried surface area in the TPR2A/MEEVD complex comprises 930 and for 750Å$^2$ for the residues EEVD, respectively. Anyways, the $K_A$ value of the Hsp90 MEEVD pentapeptide for TPR2A is comparable to the $K_A$ of the Hsp70 GPTIEEVD peptide for TPR1 but, longer peptides cannot bind with higher affinity. In their work, Schleufer and colleagues [29] showed that both the thermodynamic data and the high level of sequence conservation predicted that the consensus sequence MEEVD is necessary for TPR binding and comprise the unique TPR acceptor site in the CTD region [143, 144].

Recently, another structure of the Hop-Hsp90 complex was published [139] (PDB Id: 1ELR) by the Buchner's group who, by a combination of *in vivo* and *in vitro* experiments, defined the function of basic modules of Sti1 and determined their structures. Hop/Sti1 presents three TPR domains for two different binding partners, for that reason it is necessary to examine the peptide-binding specificity of the Sti1 TPR domains. The structural organization of yeast Sti1 was studied by dissecting the protein into its respective domains. ITC measurements for binding between Sti1 TPR domains and peptides of different length from CTD of yeast Hsp70 or Hsp90 respectively were performed. The authors found that for the TPR2A interaction higher affinities were measured for the peptides DTEMEEVD and TEMEEVD ($K_A$=3.3 x $10^6$ $M^{-1}$ and 1.0 x $10^6$ $M^{-1}$, respectively) whereas for TPR2B best binders were GPTVEEVD, DTEMEEVD and TEMEEVD ($K_A$=2.5 x $10^5$ $M^{-1}$ for the three peptides) and finally for TPR1 GPTVEEVD and VEEVD peptides were found to be associated with $K_A$=1.0 x $10^6$ $M^{-1}$ and 5.0 x $10^5$ $M^{-1}$, respectively. These results suggest that TPR1 domain from Sti1 preferentially binds Hsp70 peptides while TPR2A domain binds Hsp90 peptides, similar to mammalian Hop [29]. However, the TPR2B domain from Sti1 binds to Hsp70 peptides and Hsp90 peptides but, with relatively low affinity ($K_D$ ≈ 4 μM). Moreover, the maximum affinity constant value was measured using the heptapeptides, another difference with TPR1 and TPR2A. Schmid *et al.* performed a detailed analysis showing the structural basis for the weaker peptide affinity of the Sti1-TPR2B domain [139]. Overall, these peptide binding data imply that Sti1 has one high affinity Hsp90 peptide-binding site (TPR2A), one preferential Hsp70 peptide-binding site with slightly lower binding affinity (TPR1) and a third, less selective Hsp70/Hsp90 peptide-binding site with even lower affinity compared to the other TPR domains (TPR2B). In addition, binding of Sti1 to Hsp90 is accompanied by an inhibition of the Hsp90 ATPase activity [131, 141, 146]. In order to identify the minimum fragment of Sti1 responsible for this effect, the authors analyzed the Hsp90 ATPase activity in the presence of different Sti1 constructs. Their results demonstrated that TPR2A-TPR2B is the central unit of Sti1 necessary to inhibit Hsp90 activity, especially the peptide binding groove of TPR2A. The analysis of the affinity values for Hsp90 and for the Hsp90 C-terminal peptide showed that the affinity of Sti1 for Hsp90 ($K_D$=40 nM or $K_A$=2.5 x $10^7$ $M^{-1}$) [146] is much higher than the affinity for the Hsp90 C-terminal peptide ($K_D$=300 nM or $K_A$=3.3 x $10^6$ $M^{-1}$). Thus, the interaction between Sti1-Hsp90 is carried out not only by the Hsp90 C-terminus. Experiments with immunophilins made by Smith's laboratory [147] propose the same hypothesis, it means that outside the core TPR domains, Sti1 domains present alternative conformations which could be improve the binding with Hsp90 (see below). Furthermore, the ATPase inhibition by Sti1 is provoked by the interaction of an extra binding site. In order to identify this site, the interaction between Sti1 and individual Hsp90 domains was studied by NMR spectroscopy. The results indicated that TPR2A and TPR2B shape a joint binding site for Hsp90 MD. Therefore, the most interesting results that were found in the crystal structure reported by Buchner and colleagues [139] suggest that TPR2A and TPR2B are joined via a rigid linker orienting their peptide-binding sites in opposite directions allowing the simultaneous interaction with the Hsp90 CTD (TPR2A) and Hsp70 C-terminus (TPR2B). The two TPR domains also interact with the MD from Hsp90. Moreover, the authors proposed that the TPR1-DP1 module might function as an Hsp70-client delivery system to the TPR2A-TPR2B-DP2 segment that is necessary for client protein activation *in vivo*.

Another relevant aspect related to the interaction of Sti1/Hop-Hsp90 is the inability to support GR activity observed for TPR2B variants *in vivo*. To investigate if this inability was due to the loss of Hsp70 binding to this domain, Buchner and coworkers labelled Hsp70 (Ssa1) with Fluorescein (Hsp70*) and performed AUC experiments in presence and absence of Sti1 variants. Then, by those means they were be able detect the ternary complexes Hsp70*-Sti1-Hsp90 and Hsp70*-TPR2A-TPR2B-Hsp90. These conclusive results reinforced that TPR2A-TPR2B binds Hsp70 and Hsp90 simultaneously. Notably, simultaneous binding of both Hsp90 and Hsp70 can be accomplished via TPR2A and TPR2B or via TPR2A and TPR1. Otherwise, GR activation *in vivo* requires the integrity of TPR1 and the association of Hsp90 and Hsp70 via the TPR2A-TPR2B module [139]. Several works demonstrated that the interaction Hsp70-Hop-Hsp90 result in an active complex in which client proteins are passed from Hsp70 to Hsp90 to complete their folding and maturation, for more details, readers can see Fig. **1** in the reference [140]. Kundrat and Regan [140] focusing in Hsp70, Hsp90, Hop and CHIP, they investigated if a complex arrangement of chaperones and cochaperones were able to participate in both protein folding and degradation as Hsp90 associated functions. They found that folding and degradation machineries are mutually exclusive (see below in the CHIP item). Recently, the interaction between Hop and Hsp90 was described in *Plasmodium falciparum* [148]. In this organism, Hop was predicted to contain nine TPR motifs in its sequence; it means that Hop accommodates three TPR domains, each having three TPR motifs and two DP domains; DP1, between TPR1 and TPR2A; and DP2, between TPR2B and the C-terminal end of Hop. DP2 comprehends five α-helices, which are structured in an elongated V-configuration. DP1 is composed of similar five α-helices to DP2, but with a short extra α-helix adjacent to the N-terminus, and it is structure seems more globular than DP2. In addition to the structural studies already described, later studies about *Plasmodium falciparum* Hop (PfHop) proposed that PfHop with PfHsp70 and PfHsp90 constitute a complex in the trophozoite infective stage [148]. Using protein bioinformatics, it was observed [148] that different individual domains in Hop present distinct evolutionary rates within the protein. Structural differences were identified between human Hop (HsHop) and PfHop by motif analysis. Besides, homology modeling experiments using PfHop and HsHop in complex with their cytosolic Hsp90 were performed and indicate a high conservation of the concave surfaces formed by TPR motifs when are bound to the C-terminal regions of partner proteins. Analyzing the conservation of different Hop domains it was reported that the most conserved region was the TPR2B domain, both at sequence and at motif levels. The representation of DP2 as a single motif indicates the functional importance of this domain. On the contrary, DP1 domain and the region which connects DP1 with TPR2A (LR region) constitutes the less conserved region. Multiple sequence alignment (MSA) suggested that this region may perform a distinct structural function that has to be further explored in PfHop. Comparative interaction studies in both Pf and HsHop proposed that the residues which constitute the concave surface of TPR domains that interact with C-terminal residues from protein partners are better conserved than those forming the TPR2 convex surface that contact with PfHsp90 MD. These results constitute the basis to design inhibitors for PfHsp90 in the treatment against malaria.

## CyP40, FKBP51 and FKBP52

### *Function*

The CTD of Hsp90 also binds other cochaperones as CyP40, FKBP51 and FKBP52, and the serine-threonine phosphatase PP5 [149 - 151]. FKBP51, FKBP52 and CyP40 are known as the large immunophilins, their structure is composed by TPR domains and peptidyl-prolyl isomerase (PPIase) domains. The PPIase domain can bind immunosuppressive drugs such as FK506 as the case of FKBP51, FKBP52 and cyclosporin A (CsA) as the case of CyP40. PPIases are enzymes that catalyze the trans/cis isomerization of proline-peptide bonds that is often a rate-limiting step in protein folding.

To date, three subfamilies of this enzyme class (EC 5.2.1.8) have been identified: FKBPs (FK506 binding proteins), cyclophilins and parvulins. FKBP51, FKBP52, CyP40 and its yeast homologues Cpr6 and Cpr7 were found forming complexes with Hsp90, and due to their higher relative molecular mass (40-54 kDa) they are called large PPIases. Immunophilins as other TPR domain cochaperones bind the C-terminal EEVD peptide motif in Hsp90. The binding of PPIases to Hsp90 occurs via TPRs, which are arranged in tandem. But, although all of these proteins are TPR proteins and bind Hsp90, the TPR domain alone does not fully dictate the Hsp90-binding properties.

It was demonstrated that after binding with Hsp90 large immunophilins can hold or trigger conformational changes in client proteins which previously were bound to Hsp90. In this way, immunophilins can actively participate in Hsp90-regulated signal transduction pathways [152]. The knowledge about the molecular basis of the Hsp90-immunophilin interaction can lead to engineer inhibitors with potential use in the therapy against diverse diseases. Notably, CyP40 and the two FKBPs possess a similar structural organization. Each protein comprises an N-terminal binding site for the immune-suppressants CsA or FK506, respectively, and a C-terminal TPR domain. This structural pattern was confirmed by the crystal structures for CyP40 [152] FKBP51 [153] and FKBP52 [154]. Structurally, FKBP51 and FKBP52 contain two PPIase domains, FK1 and FK2 but only FK1 can bind FK506. Immune-suppressants when bound to the cyclophilin or FK1 domains inhibit the PPIase activity. As was mentioned above, *in vitro*, immunophilins act as PPIase, but their detailed biochemical mechanism in the Hsp90 chaperoning cycle has to be extensive studied. The selection of the PPIase present in the Hsp90 complex seems to depend on the substrate bound. For example, mature PR and GR the favored immunophilin to form complexes assembled in vitro is FKBP51 over CyP40 and FKBP52 [155, 156] but not in ER complexes [156].

As outlined above the binding of PPIases to Hsp90 occurs via TPRs [29] that in large PPIases are organized at the C-terminal part of the protein (12 kDa domain) [143] whereas PPIase domains are disposed at the N-terminal part [154]. FKBP51, FKBP52 and CyP40 apparently compete for a single binding site on Hsp90 [157]. FKBP51 and FKBP52 recognize the motif MEEVD at the Hsp90 C-terminal but with different affinities [158].They assemble differentially into various SHR complexes [156] and in this sense FKBP52 has been amply studied in relation to SHR signaling [159, 160].

### *CyP40 Overall Structure*

The structure of CyP40 has been analyzed in two crystalline forms: the monoclinic and the tetragonal form, it is interesting that both crystalline structures displayed some differences [152]. However, both monoclinic and tetragonal forms of CyP40 [161 - 163] show pretty similar cyclophilin domain conformations which was consistent with previously characterized single-domain cyclophilins (Fig. **12** and Fig. **13**). The active site of the cyclophilin family is composed 13 amino acids found to be essential in the binding of the drug CsA to hCypA (Fig. **14**).

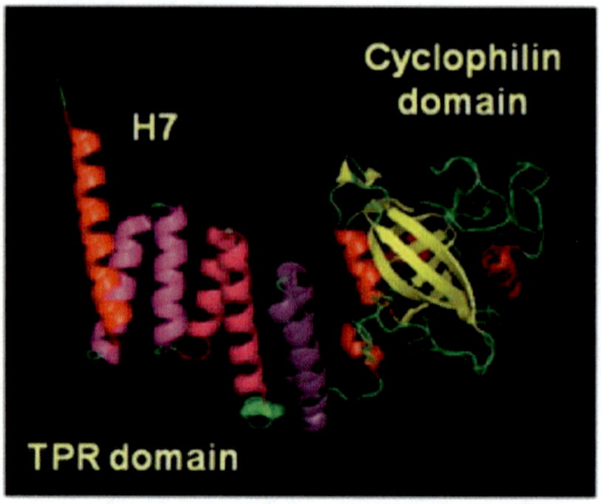


**Fig. (12).** Monoclinic form of the CyP40 is represented showing the cyclophilin domain (N-terminal PPIase) and TPR domain (C-terminal). The cyclophilin domain is colored showing the secondary structure as green for loops, red for α-helix and yellow for β-structures. In TPR domain the TPR motifs are represented in violet, pink and magenta, form N-terminal to C-terminal, the α-helix seven which contains the CBM is represented in red.

The active site residues are identical in both structures of bovine CyP40, while the PPIase activity has a $k_{cat}/K_m$ value of 1.9 x $10^6$ $M^{-1}s^{-1}$ (164) compared with the value of 1.4 x $10^7$ $M^{-1}s^{-1}$ for hCypA [165]. This CyP40 contains a "divergent loop" which comprises an insert of 8 residues ($^{60}$PTTGKPLH$^{67}$) (Fig. **14**). Similar to other divergent loops cyclophilins, CyP40 presents two cysteines in close proximity, enabling forming a disulphide bond, which in an oxidizing environment induces a conformational change that provides a signaling mechanism in response to oxidative stress. The TPR domain structure in the monoclinic crystal contains seven α-helices of variable length forming three TPR motifs. This TPR domain provide a concave binding surface to accommodate the Hsp90 C-terminal MEEVD sequence. In CyP40, a linker of 30 amino acids connects the TPR domain with the cyclophilin domain (Fig. **14**).

This acidic linker includes 11 aspartate and glutamate residues that configure a well-defined hydrogen-bonded structure with two β-turns. In addition, the positively charged surface of helix Q forms salt bridges with the linker. The seven α-helices labeled P, Q, R, S, T, U, and V are structured in an extended helix-tur--helix pattern. The first three helices (P, Q and R) are longer than expected for predicted TPR sequences. The comparison between the sequences of the C-terminal TPR-containing domains of large immunophilins with other Hsp90 binding proteins, including PP5, Hop, and CNS1, showed a consensus sequence pattern of small and large hydrophobic residues. Small residues are located at positions 8, 20, and 27 [152].

The calmodulin binding site (CBM) in the monoclinic crystal structure of CyP40 is represented by the C-terminal portion of the V (7) α-helix (Fig. **12** and Fig. **14**), which is extended beyond the shorter TPR helices. The α-helix V is mainly positive charged and it is protruding from the TPR domain. The calmodulin binding requires an amphiphilic α-helix that can provide a net positive charge. Specific binding between large immunophilins and calmodulin have been studied for bovine CyP40 [157], maize FKBP66 and rabbit FKBP52.

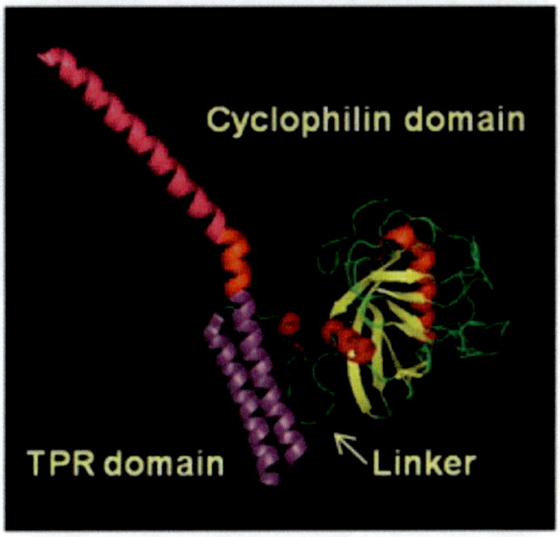


**Fig. (13).** Tetragonal CyP40 structure. The cyclophilin domain (N-terminal) colored by secondary structure is shown. Only the first TPR motif is represented (violet), α-helices of TPR domain are colored following the same pattern like monoclinic structure (1IHG). The residues in TPR motifs are colored as TPRPred.

In the tetragonal crystal form only the first TPR domain (α-helices P and Q) is similar to that found in monoclinic crystal; the second TPR motif (α-helices R and S) has sprung out to constitute an elongated helix (Fig. **13**). Furthermore, residues comprising α-helices T, U, and V are not visible in tetragonal form and two of the TPR α-helices have rearranged to constitute one extended α-helix, leading to a dramatically dissimilar conformation of the molecule. Accordingly, this structure can be considered as a trapped intermediate in the folding pathway of the α-helix domain. All these observations demonstrated the flexibility of such α-helical domains and their position toward the end of a α-helical domain may occur to give rise to protein-protein recognition events. The structural arrangement in the tetragonal lattice allows compensatory intermolecular helix-helix interactions which mimic the inter-molecular interactions found in the monoclinic crystal. The conservation of the quality of TPR-TPR interactions is indicative of a high-fidelity recognition mechanism. The structural analysis of monoclinic and tetragonal crystal forms suggests that possibly, partially folded forms of TPR intervene in domain swapping and protein recognition. At the same time, different conformations found in both X-ray structures (PDB Id: 1IHG y 1IIP) demonstrate the multifunctionality of CyP40 as a signaling protein that participate in several protein-protein associations [152].

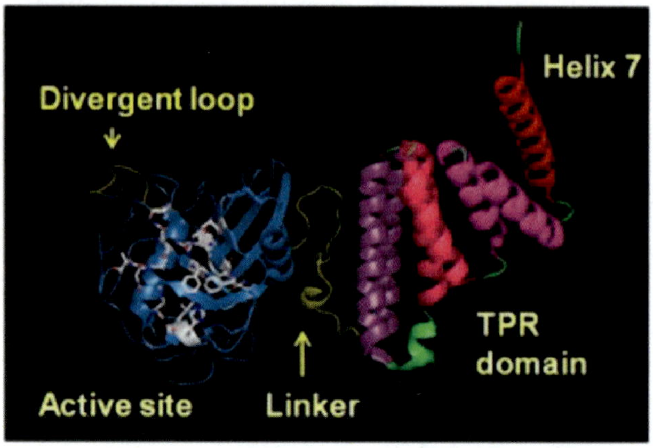


**Fig. (14).** CyP40 crystal structure. The cyclophilin domain of CyP40 (residues 1-183) is colored in blue. The side chains of the 13 residues essential in defining the active site of the cyclophilin family of proteins are shown in CPK. The divergent loop ($^{60}$PTTGKPLH$^{67}$) is colored in olive and the linker loop (residues 184-213) joining the cyclophilin and the TPR domains are represented in yellow. The TPR domain is formed by seven α-helices P, Q in violet, R, S in pink, T, U in magenta and V in red. The last α-helix is termed α-helix seven, which comprises the CBM.

## *Cyp40-Hsp90 Interaction: Structural Studies*

In a study conducted by Chinkers and coworkers using site-directed mutagenesis of PP5 it has been shown that Lys 97, Arg 101, Lys 32 and to a lesser extent Arg 74 are required for Hsp90 binding [166].

Since no crystal structure of CyP40-Hsp90 complex has been published, the sequence alignment of TPR sequence showed that residues Lys 97 and Arg 101 are conserved at positions 2 and 6 of the TPR3 motif in all the proteins that bind Hsp90. Moreover, structural analysis found that the C-terminal residues EKAAK fitting tightly into a groove on the surface shaped by the three TPR motifs. The groove is composed by the residues: Lys 227, Phe 234, Asn 278, Lys 308 and Arg 312 (Fig. **15**). It is remarkable that, the three charged residues are the most important residues in the interaction with Hsp90 [166]. Later, Walkinshaw and colleagues utilized these inter-molecular interactions as backbone template to model the interaction between CyP40 and the C-terminal MEEVD sequence from Hsp90 [152]. This model showed that the MEEVD peptide binds in opposite direction compared with that found in the Hop-peptide complex [29]. These findings suggest that the peptide binding groove of TPR domains may present a broad specificity. Thus, the X-ray structure of the monoclinic CyP40 TPR domain provided a model binding surface for Hsp90.

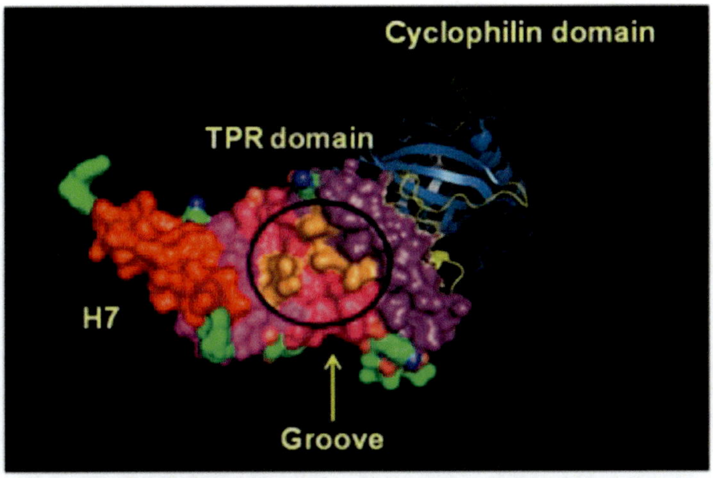


**Fig. (15).** Cyp40-Hsp90 interaction groove. Cyp40 TPR domain surface is represented; the residues of interaction with Hs90 defined by Russell *et al.* [160] are colored in orange. The groove is comprised by Lys 227, Phe 234, Asn 278, Lys 308 and Arg 312 (circled in black).

## *FKBP51: Overall Structure and Implications*

Reynolds *et al.* [167] observed that lymphocytes from squirrel monkey have 13 times higher FKBP51 protein levels and incorporate more FKBP51 into GR complexes than human lymphocytes. Using transfected cells and cell-free approaches, researchers have demonstrated that squirrel monkey FKBP51 reduces GR hormone-binding affinity and expression of a reporter gene [167]. The expression of human FKBP51 to comparable levels produced cortisol resistance, but to only one-fifth the extent induced by monkey FKBP51.

Sequence alignment showed that the human and squirrel monkey proteins are 94% identical (97% similar), and a superposition of both crystal structures (see Fig. **6** from Sinars *et al.* [153]) displayed that the overall architecture is the same. In order to study the structural basis of the biological function of FKBP51, Sinars and coworkers [153] resolved the three-dimensional structures from human and Bolivian squirrel monkey (*Saimiri boliviensis*) of FKBP51 to 2.7 and 2.8Å resolution, respectively (PDB Id: 1KT0 and 1KT1). As was expected, both structures are almost identical since they comprehend three domains, two FKBP-like domains: FK1 and FK2 and a three-unit TPR domain with a CBM (Fig. **16** and Fig. **17**).

Comparing both crystal structures 15 residue-substitutions were found but most of them are conservative. However, non conservative substitutions also have been found, six (of 32 aa) are located at the N-terminal and five (of 30 aa) are in the C-terminal. Although in the three-dimensional structural alignment no differences are observed, these substitutions may indirectly influence to FK2 or other regions that, in the context of Hsp90-SHR complexes, modulate receptor-hormone-binding affinity.

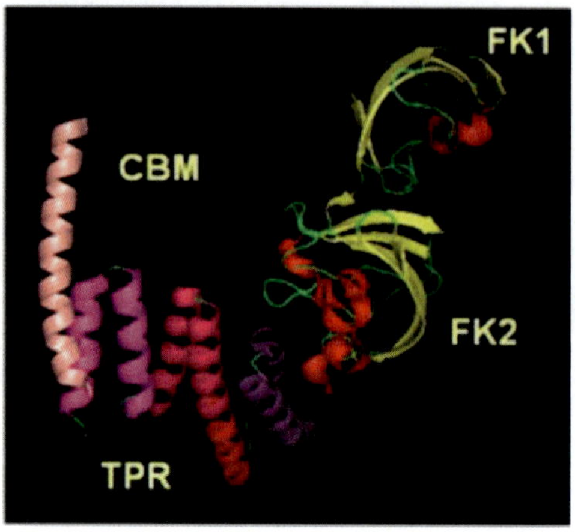


**Fig. (16).** Human FKBP51 domains. The two FKBP-like domains are represented and colored according to secondary structure (FK1 and FK2), TPR domain motifs are colored following the identification of TPR motif of TPRPRED software (from N to C in violet, pink and magenta), and CBM is represented in salmon at C-terminal of the structure.

### *FKBP-like Domains*

The first two domains, FK1 and FK2 (residues 33-138 and residues 147-251, respectively) display the typical FKBP domain structure, they are constituted by an anti-parallel five-stranded β-sheet wrapped around a central α-helix [168]. The spatial orientation of the binding groove of the TPR domain and the binding pockets of the FKBP domains indicates that FKBP51 participates in multiprotein complexes. The binding pocket of the first FKBP domain, FK1, as observed in Sinars *et al.* [153] is oriented ~ 180° from the putative binding pocket of the second FKBP domain, FK2. Analyzing the connection between FK1 and FK2, a short link of only eight residues can be observed. It is expected a rigid short link due to extensive interactions inside the molecule [153]. The first FKBP domain (FK1) involves five anti-parallel β-strands which curve around a central α-helix shaping the typical FK folds (Fig. **18**). In addition, it was found that several loops are disordered, specially the loop between β1 and β4 and the "40s" loop [153]. Comparing the structures of FKBP51 and FKBP12, the major difference was seen in the "80s" loop (FKBP51 residues 110-125), a region with the most variations among the studied FKBP domains [169]. FK1 (FKBP51) and FKBP12 share 48% sequence identity (60% similarity), the structural similarity is observed in the overall domain structure and in the binding pocket. The measurement of the PPIase activity of FKBP51 is comparable to that of FKBP12. FKBP51 $K_i$ values for FK506 and rapamycin are ~ 10 and > 5 nM respectively, compared with ~ 1 nM for FKBP12.

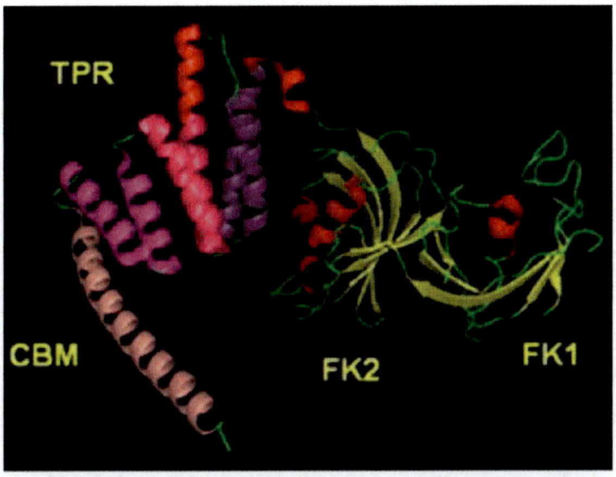


**Fig. (17).** Squirrel monkey FKBP51 domains. The two FKBP-like domains are represented and colored according to secondary structure (FK1 and FK2), TPR domain motifs are colored following the identification of TPR motif of TPRPRED software (from N to C in violet, pink and magenta), and CBM is represented in salmon at C-terminal region.

In order to know which of the two domains: FK1 or FK2 shows higher PPIase activity, site-directed mutagenesis experiments were performed, demonstrating that FK1 domain is the measurably active PPIase domain since the double point mutation in that domain decreases the enzymatic activity > 90% . On the other hand, FK2 domain is also structurally similar to FKBP12 despite having only 26% sequence identity and 44% similarity. Three of the 12 residues, forming the binding pocket, are identical (Phe 181, Asp 182, and Glu 194) and three are conserved (Val 55/Ile 198, Tyr 82/Phe 225, and Phe 99/Tyr 243). It was found that an insertion of 3 residues between β5 and the α-helix pushes into the binding pocket. Analyzing the result set, both the lack of sequence conservation and the binding pocket insertion are in concordance with previous work that proposed FK2 lacks measurable PPIase activity [156]. Even more, Sinars *et al.* [153] using FKBP51 truncation mutants lacking FK1 have confirmed those results (See below).

### *TPR Domain*

As was defined in previous item, usually, TPR domains are entirely α-helical structures composed by 2-16 motifs that possess a consensus sequence formed by 34 amino acids [19]. In FKBP51 structure [153] a single TPR motif consists of two serial α-helices of 12-15 residues each crossing at an angle of ~ 20° to each other. The third domain in FKBP51 is represented by the TPR domain [153] with three repeats, it comprises the residues 261 to ~ 384 and presents a similar structure to the Hsp90-binding TPR domains of other cochaperones as protein phosphatase 5 (PP5) [19], Hop [29], and CyP40 [152]. FKBP51 TPR domain comprehends besides, three repeats, one extra α-helix (Fig. **16**, **17** and **19**), named the seventh helix that extends beyond the final TPR motif, and constitutes the CBM (Fig. **16** and **17**). Similar structures located at the end of TPR domain shaping the CBM were found in PP5 and CyP40 (Fig. **16** and **17**).

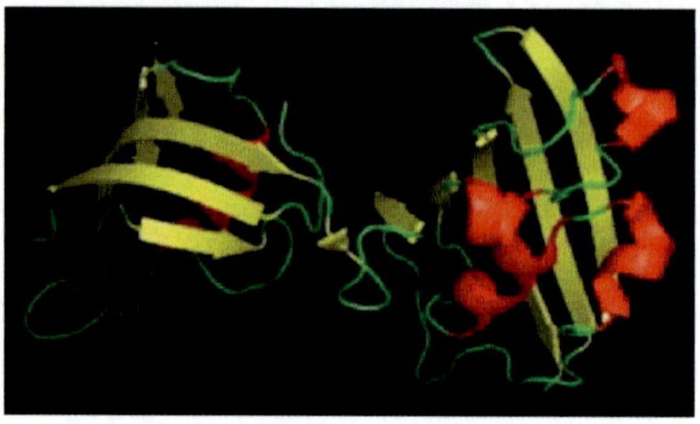

**Fig. (18).** FK1 and FK2 domains of human FKBP51. The orientation of the equivalent secondary structure in both domains is ~ 180° suggesting that the orientation of the binding pocket of FK1 is diametrically opposite to the putative binding pocket in FK2.

The α1- and α3-helices located at 21 and 25 residues of the FKBP51 molecule, exhibit a longer length than the typical α-helices composed by 12-15 residues. However, α2 and α4-α6 helices exhibit the average size for TPR α-helices. (Fig. **16**, **17** and **19**). The resolution (2,7-2,8 Å) of the whole FKBP51 structure reported in Sinars *et al.* [153] was limited with some non resolved residues nevertheless, represents the only structure in which the relative orientations of the FK1, FK2, and TPR domains can be appreciated; and a structure-function relationship can be inferred. Although FK2 does not appear to exhibit PPIase activity, it was surprising that the FK2-binding pocket is just near the binding groove of the TPR domain. In order to elucidate the functional role of the FK2 insertion loop, FKBP51 and FKBP52 molecules lacking this loop was tested for Hsp90 binding and receptor association [156]. The deletion reduced FKBP51 recovery in PR complexes to a similar level that FKBP52. However, the lacking of FK2 insertion loop did not show any critical change in the FKBP52 function [153].

Those results propose some hypotheses: one of them consists in considering that FKBP51 and FKBP52 have distinct interactions with PR and, the one takes into account that FK2 in FKBP51 present a dissimilar orientation and consequently, a different mechanism to create the assembly with Hsp90 and PR. The exceptional interaction between receptor and the FK2 domain of FKBP51 could explain the PR preference by FKBP51 over FKBP52. Probably, FK2 has been originated from an FK domain duplication event and has evolutionarily lost its PPIase activity although seems to have gained protein interaction ability.

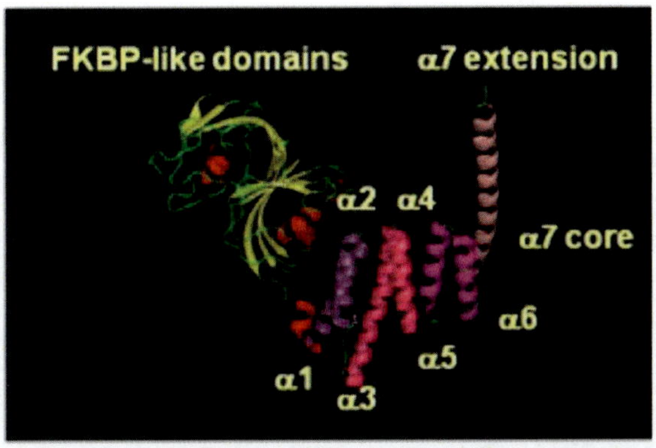


**Fig. (19).** FKBP51 TPR domain structure. The three repeat motifs in the TPR domain form six α-helices (H1-H6). The center of the seventh α-helix interacts with H6 and a region that extends beyond the TPR domain. The complete extent of H7 is unknown because the observed structure terminates at position 421 of 457.

## *FKBP51-Hsp90 Interaction: Structural Studies*

The reported FKBP51 structures, gives no indication about the TPR domain orientation upon the binding of Hsp90. MSA of Hsp90 combining sites of PP5, FKBP51, FKBP52, CyP40, Tom 34, Tom 70 and CNS1 [29] showed that the Hsp90 putative binding site in FKBP51 could involve Ala 268, Lys 272, Ala 318, Asn 322, Met 325, Lys 352 and Arg 356 (Fig. **20**). Additional structural studies are necessary to define exactly the Hsp90 binding site in TPR domain of FKBP51.

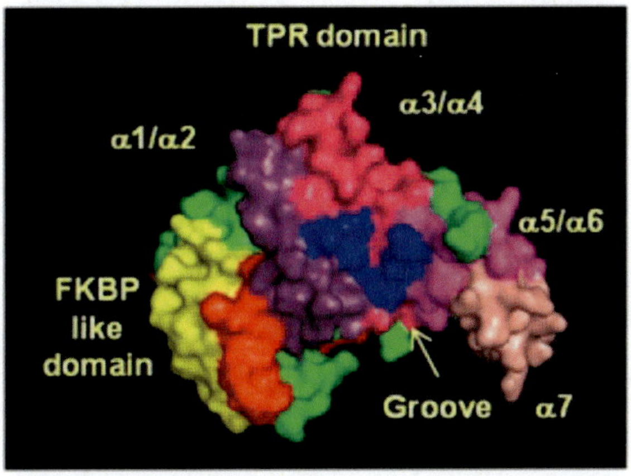


**Fig. (20).** Hsp90 putative binding site in FKBP51. The surface showing the putative binding site for Hsp90 is shown in blue. From MSA results, it comprises Ala 268 and Lys 272, from α1 (violet); Ala 318, Asn 322 and Met 325, from α3 (pink); Lys 352 and Arg 356 from α5 (magenta).

All those residues shape a groove in the concave face of TPR domain and they come from different motifs. However, an extensive study of the role of different part of FKBP51 structure in the binding of Hsp90 was performed by Smith's laboratory [156]. Although the authors took into account the role of the TPR domain in the binding with Hsp90, they also extended that role to amino acids outside the TPR domain. In that study, using truncated FKBP51 products expressed *in vitro* they demonstrated that only full-length FKBPs co-precipitated with Hsp90. This result suggests that residues upstream and downstream from the TPR domain are necessary for Hsp90 binding. Other experiments using FKBP51 truncated C-terminal mutants (51/N446 and 51/N431), demonstrated that the C-terminal end is important to maintain the 3-D structure of the TPR domain since N/414 failed to bind Hsp90 but 51/N431 retained Hsp90 binding. Unfortunately, there is not structure of the region beyond CBM since the structures corresponding to 1KT0 and 1KT1 comprise until residue 411. In this way, FKBP truncation mutants lacking sequences either up- or downstream from the TPR domain failed to bind Hsp90, it means that the whole domain is essential for binding to Hsp90. Previous mutation analysis and crystal structures of TPR domains have identified key residues for binding Hsp90, the conservation of such residues in Hop, CyP40, and PP5 suggests a functional similarity between these proteins [19, 29, 152, 166, 170]. However, so far no conclusive results that clearly define the whole picture about the FKBP51-Hsp90 interaction were found, what requires the performing of experiments with novel technologies.

### *FKBP52 Structure*

Similar to FKBP51, FKBP52 structure comprises three domains: two FKBP like domains, and a TPR domain with an extra α-helix that corresponds to CBM. The N-terminal PPIase domain, the FK1 domain accommodates the binding pocket for FK506 [158, 171]; the middle FKBP-like domain, the FK2 is the domain as yet with undiscovered function and finally the C-terminal domain, the TPR-clamp recognizes the C-terminal peptide of Hsp90 with high affinity, providing the major binding site for Hsp90.

Most FKBP52 structures deposited in the PDB database include FKBP-like domains: FK1 and FK2 (PDB Id: 1Q1C) (Fig. **21**) only two structures comprise the TPR domain (PDB Ids: 1P5Q and 1QZ2) (Fig. **22** and **23**).

### *FKBP-like Domains*

Both FKBP domains of FKBP52 consist of a five- to six-stranded ant parallel β-sheet, which is wrapped around a short α-helix with a right-handed twist [154]. FKBP domains are similar to those of FKBP51, FKBP12, and macrophage infectivity potentiator protein [168, 172]. The crystal structure can be accurately solved as far as Ala 425, suggesting a disordered structured from Lys 426 to Ala 459. Structural alignments of both regions N and C result in the entire structure of FKBP52 (Fig. **24**). In the structure of FKBP52 reported by Rao's group [154]

FK1 and FK2 domains are joined by a highly hydrophilic hinge (residues 139-148), and the FK506-binding pocket of FK1 is oriented ~ 180° from the hypothetical FK506-binding pocket of FK2 (Fig. **21**). Because of the hinge region is held by numerous contacts it does not appear to be a flexible region. The relative orientation of FK1 and FK2 in FKBP52 is stabilized by several factors as hydrogen bonds, hydrophobic interactions and the rigid hinge region. At the same time, all these interactions restrained the possible conformational changes that would affect such FK1-FK2 orientation. Nonetheless, a different fact was found by Hausch and colleagues [171].

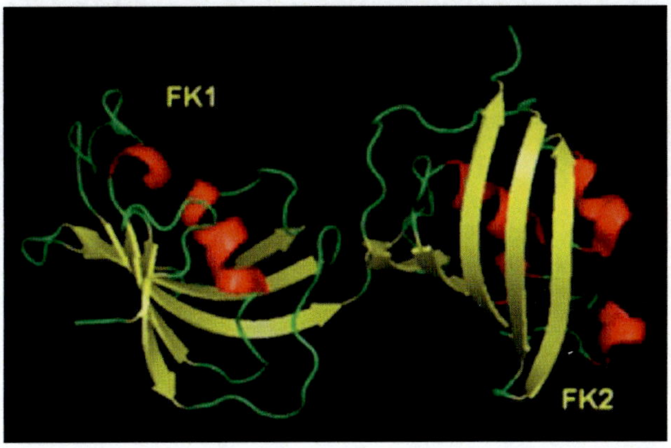


**Fig. (21).** PPIase domains FK1 and FK2 of FKBP52. The first domains of FKBP52 (residues 1-260, PDB=1Q1C) are colored by secondary structure. Note that α-helices structure in FK1 and FK2 are diametrically opposed. Thus, the FK506-binding pocket of FK1 is oriented approximately 180° from the similar pocket found in FK2.

In 2013, Hausch's group [171] reported the crystallographic structure of the FK1 and FK2 domains of the immunophilin FKBP52. In that crystal structure the authors disclosed that the short linker that joins FK1 with FK2 seems to be a flexible hinge. The authors further postulated that this improved flexibility and its modulation by phosphorylation could be the cause of the functional antagonism between FKBP51 and FKBP52. In addition, those authors presented two co-crystal structures of FKBP52 in complex with FK506 and a synthetic analogue. They found definite differences comparing this FK506-FKBP52 complex structure and the equivalent from FKBP51 and FKBP12. This fact opened up the possibility of using these differences to rationally design differential ligands for FKBP52 and other immunophilins. As happened in other immunophilins the FK1 domain shows PPIase activity in FKBP52, whereas the FK2 domain lacks of this enzyme activity. FK2 shares only 32% sequence with FK1, nonetheless that low similarity; both domains show an analogous basic structure although FK2 lacks the large bulge-splitting strand α4 of FK1. This domain comprehends 14 residues directly related to substrate binding, five of which are preserved in FK2. From the structural analysis it can be postulated that the loss of PPIase activity of FK2 could be the result of alterations in loops nearby the binding pocket and important amino acids substitution also occurred during the domain evolution. In this sense, the evolution of FK2 domain is an aspect that is analyzed in Wu *et al.* [154]: the authors hypothesized that FK2 domain in FKBP proteins may result from a duplication event during evolution (see above).

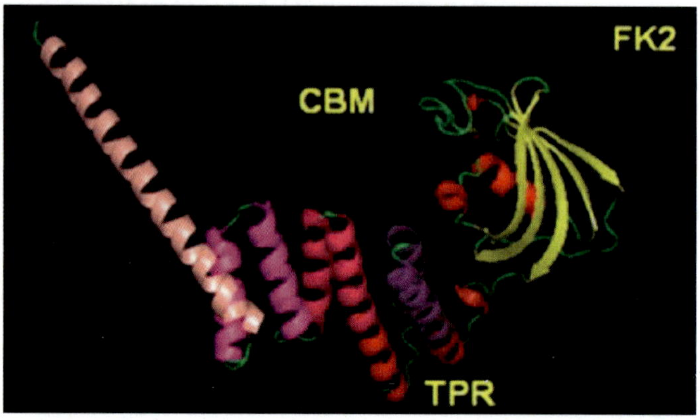


**Fig. (22).** C-terminal region of human FKBP52 domains at 2.80Å resolution. FK2, TPR and CBM are shown. FK2 consists of a five- to six-stranded ant parallel β-sheet, which is wrapped around a short α-helix with a right-handed twist (α-helices are in red, β-sheets are in yellow and loops are in green). TPR domain consists of three TPR motifs shown in violet, pink and magenta, according to the results of the TPRPred software (N-terminal TPR motif: 270 STIVKER GTVYFKEGKYKQALLQYKKIVSWLEYE 303; Middle TPR: 319 LASHLNLAMCHLKLQAFSA AIESCN-KALELDSNN 352 and C-terminal motif: 353 EKGLFRRGEAHLAVNDFELARADFQK VLQLYPNN 386). CBM is represented in salmon.

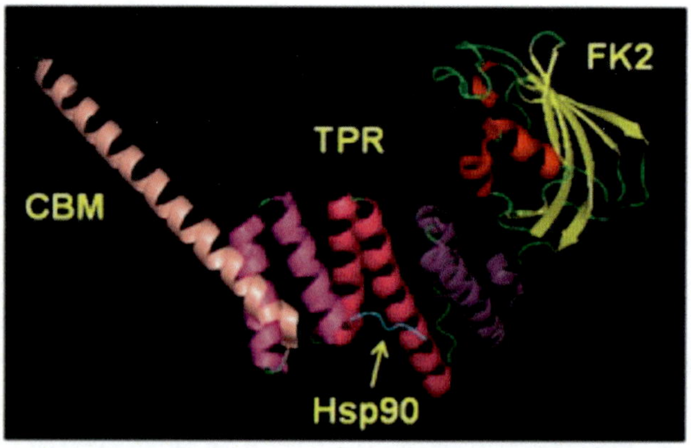


**Fig. (23).** Crystal structure of human FKBP52 C-terminal domain complex. Secondary structure of FK2 and TPR domain is represented (residues E 1 to A 125). TPR domains is comprised by three TPR motifs colored in violet, pink and magenta from N to C-terminal, the seven α-helix or CBM is colored in salmon.

An allosteric activation mechanism similar to that for CBM has been proposed for FK2 in FKBP52. Some authors discussed the improbability of the activation of a cryptic PPIase activity in FK2, even if FKBP52 could suffer conformational changes to permit that function. In early works, the structures of: a single FK1 domain, an FK1-FK2 fragment and an FK2-TPR construct of FKBP52 have been determined by NMR or by X-ray crystallography [154, 173]. In these structures, the orientation of the FKBP domains seems to be closely similar to that found in full-length FKBP51 crystal structures [153]. On the other hand, analyzing the conformation adopted by apo-N II FKBP52 structure [171] inside the crystal lattice, it can be observed that the PPIase pocket and the tip of the β4-β5 loop of the FK1 domain face the opposite side of the β-sheet of the FK2 domain. This loops, that was defined as functionally relevant for the improvement of SHR signaling [174] might be better aligned with Hsp90 binding site in the modeled TPR domain. Consequently, this conformational change observed in NMR and X-ray structures suggest that Hsp90 heterocomplex is highly dynamic [175]. The results obtained by Hausch and coworkers [171] demonstrate that FKBP52 may adopt multiple conformations during the chaperoning cycle, allowing reorientation of the FK1 domain toward Hsp90 and the bound client.

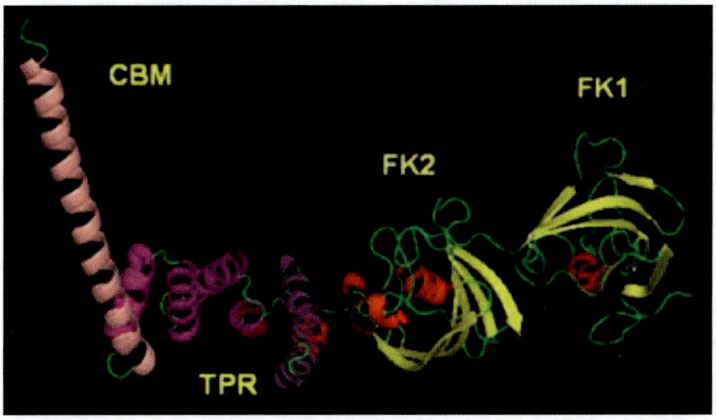


**Fig. (24).** FKBP52 overall structure. The three domains of FKBP52 are represented by means a superposition of structures of the N and C-terminal of FKBP52 solved by Rao's group (1Q1C and 1QZ2, respectively). Both structures share the FK2 domain, where the overlap is complete.

## *TPR Domain*

The TPR domain of FKBP52 similar to FKBP51 comprises three units TPR motifs (consensus 34-aa motif). Each motif consists of two consecutive α-helices containing 12-15 residues (except α1- and α3-helices which be composed by 21 and 23 residues, respectively) that overpass at an angle of ~ 20° to each other. The canonical structure of the FKBP52 TPR domain is also exhibited in FKBP51, CyP40, PP5, and Hop [19, 29, 152, 153]. As can be observed in FKBP51, there is an additional α-helix (α7) at the C- terminus, this α-helix contains the CBM and it is situated beyond the final TPR motif (Fig. **22**, **23**, **24**). At this point, it is obvious that the overall structure of FKBP52 described by Rao's group is very similar to that of FKBP51 except for their relative domain orientations [153] (Fig. **25**).

### *Comparison Between FKBP51 and FKBP52*

One difference between FKBP51 and FKBP52 appeared from the reported crystal structures of FKBP52 by Hausch's laboratory [171], as was discussed above they defined as flexible the short linker that joins FK1 and FK2 from FKBP52. This linker is basically identical in FKBP51, but the contacts which stabilize the different conformations tend to differ between FKBP51 and FKBP52. Nevertheless, alternative conformations seem to be more stable in FKBP52. In this point it is necessary to emphasize that the plasticity of the linker depends on the interactions with its surroundings. Thus, the distinct flexibility and its modulation by phosphorylation might be the reason of the functional antagonism between FKBP51 and FKBP52.

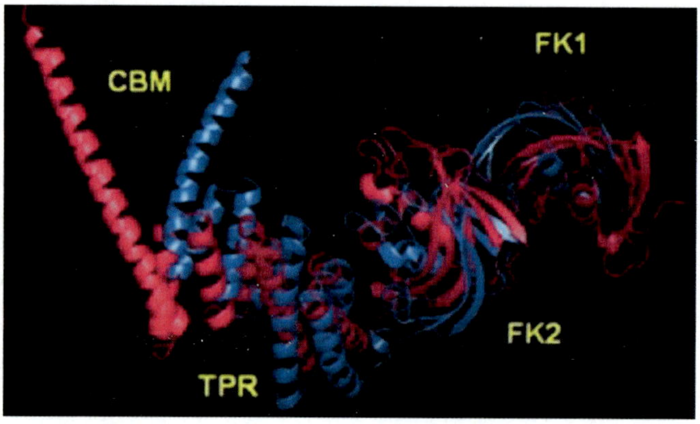


**Fig. (25).** Superposition of FKBP51 and FKBP52. The overall structure of FKBP52 (pink) and FKBP51 (blue) are very similar in the four domains, except for their relative domain orientations, as can be noticed in CBMs.

The linker plasticity connecting FK1 with FK2 in FKBP52 appears to be important for the mechanism of action of rapamycin as immunosuppressant anti-cancer drug. However, except the structural function that must be studied in depth, FK2 does not possess PPIase activity and neither binds FK506 nor rapamycin [158, 176]. In addition of the inter-domain hinge region, the most prominent difference to FKBP51 is the absence of the N-terminal α-helix in the FK1 domain of FKBP52, and the location of the so-called β3-bulge, a discontinuity in β-strand 3 in the same domain. Interestingly, the FK1 domain of FKBP52 has been used as a negative construct to retard the nuclear transport of selected Hsp90 clients (e.g. GR or p53) [177, 178], denoting that this domain includes the major interaction site with dynamitin, although this interaction was not studied in depth [179]. Considerable structural differences between FKBP51 and FKBP52 are also found at the tip of the β4-β5 loop. Depending on the structural context in the crystal lattice, both cis and trans-conformations of the Pro119-Pro120 peptide bond in FKBP52 were observed. Besides, the structural differences mentioned above, residue 119 was reported to be a significant determinant for the opposing effect of the FKBP51 and FKBP52 on SHR maturation [174]. Experiments using FKBP52 P119L mutant demonstrated that is much less effective in enhancing the signal transduction of GR or AR in the yeast model or in mammalian cells, respectively. This result leads to the hypothesis that the ability to adopt a trans-conformation in this position might contribute to the intensifying effects of FKBP52.

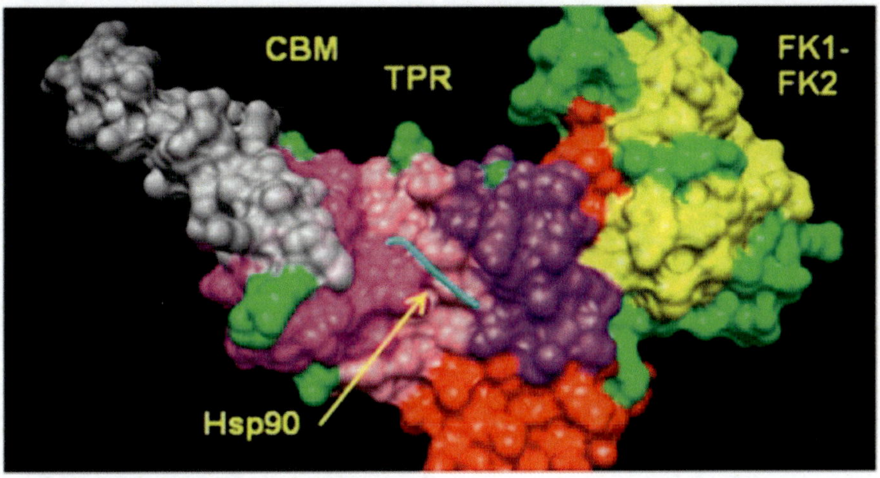


**Fig. (26).** FKBP52-Hsp90 interaction surface. In the representation of the interaction surface of FKBP52-Hsp90 (1QZ2) complex, this is colored according to the domains and the motifs showed in previous figures. Note that the cyan MEEVD Hsp90 peptide is accommodated into the binding groove surface from the TPR domain.

To perform a detailed analysis on the differences in conformational changes of FK1 domain in FKBP51 and FKBP52, Hausch's laboratory [171] performed MDS experiments. The authors used free and FK1 bound to FK506 for the simulations. No major conformational rearrangement or signs of unfolding were observed. As expected the residual mean square deviation (r.m.s.d.) were larger for the unbound FKBPs in comparison with ligand-bound FKBPs. Cluster analysis of the MD trajectories of the ligand-free FKBPs showed that the fluctuation of FKBP52 can be attributed to movements of the β4-β5 loop and the β3 bulge. Simulation experiments started from the FKBP52 structure with the trans-conformations for the Pro 119-Pro 120 peptide bond and showed higher variability for of the β4-β5 loop and the β3 bulge. On the contrary, the corresponding loops of FKBP51 seem to be comparatively more rigid. In the case of FKBP51, part of the conformational variation is due to reorientation of the additional N-terminal α-helix, this fact does not happen in FKBP52. Even more, the lack of this helix may however contribute to the larger conformational fluctuations of the β4-β5 loop in FKBP52 [171].

### *FKBP52-Hsp90 Complex*

As was pointed out above, Rao [154] and collaborators reported the crystallographic structures of two overlapped segments from FKBP52: the N fragment comprised of residues 1-260 and C segment comprehended by residues 145-459. Those fragments were crystallized in complex with the C-terminal segment MEEVD from Hsp90. It is significant that the structure of the peptide-bound to FKBP52 is similar found in free Hsp90. This peptide interacts with the cavity comprised of α1, α3, and α5-helices of the TPR domain (see Fig. **23**). In the asymmetric unit of the complex crystal, two of the three molecules bind the MEEVD peptide. The binding pockets are composed of exactly the same residues, namely Lys 282, Asn 324, Met 327, Lys 354, and Arg 358 (Fig. **26**).

The structural analysis of the two complexes in the asymmetric unit shows that the interactions that stabilize the binding of the peptide are conserved, in particular, the hydrogen bond between Lys 282 and Met 1, and the hydrogen bonds made by Lys 354, Arg 358, and Glu 2. Nevertheless, the orientations of these two complexes are different. Apart from the amino acids located at the N-terminal region, other residues also present in the structure of Hsp90 contribute to the binding and the interaction specificity with TPR domain [29]. Thus, the binding between the C fragment of FKBP52 (145-459) and MEEVD peptide from Hsp90 is not so strong and this fact could be the reason of different binding modes and protein conformations observed in performed structural analysis. As an example, the binding pocket of molecule C is blocked by the FK2 domain of another molecule C present in the same unit cell. However, preceding mutagenesis experiments and structural analysis have identified conserved residues in the TPR domain of diverse cochaperones as Hop, CyP40, and PP5 suggesting that those amino acids are crucial for Hsp90 binding [19, 29, 166, 152, 170]. Interestingly, those conserved residues also are present in FKBP52 structure proving the functional similarity of all these cochaperones.

### *FKBP52-FK506 Complex*

As was mentioned above, Hausch and coworkers [171] have solved the structures of the free FKBP52 cochaperone, the FKBP52-FK506 complex and other complex formed by I63, a small synthetic analogue. FK506 (the typical ligand of FK1) shares with I63 the pipecolyl-α-keto amide motif. Analyzing both complex structures, this moiety adopts similar positions, it binds into a deep surface pocket included in the FK1 domain that is composed by the residues: Tyr 57, Phe 77, Val

86, Ile 87 and Trp 90 (For more details, see Fig. **4c** and **d** of reference [171]). Almost identical interactions have been noticed with FKBP12, FKBP12.6 and FKBP51 in complex with various pipecolate-based ligands [180, 181]. Although the interactions between PPIases and FK506 analogues are similar, they are not completely identical, which implies that the minimal structural differences found in all these complexes can be harnessed for the rational design of selective enhancers or inhibitors of FKBP52 function.

## *Binding Studies of Immunophilins*

After several studies, it is known that the only presence of a TPR domain in one PPIase does not fully dictate the Hsp90-binding properties. In this sense, FKBP51 and FKBP52 bind Hsp90 with different affinities [158] and assemble differentially into various SHR complexes [156]. Mutagenesis experiments results showed some differences located downstream from the TPR domain [147]. Then, Smith's laboratory trials [147] indicated that there is an interaction within the region from α6 through α7 extended which is differently necessary by FKBP51 to bind Hsp90. In addition, experimental results with FKBP52 and FKBP51 mutants pointed out that the extended portion of α7 (amino acids 400-420) are significant for Hsp90 binding and FKBP function. Therefore, the Hsp90 binding properties of FKBP51 and FKBP52 reside in the combination of the core TPR domain and the influence of C-terminal amino acids sequences outside this domain. The existence of alternative modes of interaction with Hsp90 was postulated after performing point mutations experiments in the C-terminal region of the chaperone. Those mutations produced dissimilar effects on the binding of particular TPR cochaperones. The analysis of all these experiments proposes the hypothesis that cochaperones might interface with different parts of Hsp90 molecule in addition to a common interactions with the C-terminal MEEVD.

Based on the crystal structure for FKBP51 where α7 (salmon) extends beyond the core TPR domain (see Figs. **16, 17, 19**) Smith's laboratory identified a named Y charged motif within α7 which is essential for maximal Hsp90 binding. They raised two possibilities: the first one is that the charged Y motif contacts Hsp90 at a site separate form the core of TPR domains. The second possibility is that α7 could break in such way that an eighth helix containing the charged Y motif becomes part of the core TPR domain and enhances binding to Hsp90. However, the crystal packing did not stabilize the FKBP51 conformation intended to exist in solution or be induced by binding to Hsp90. Different conformations of TPR domains due to crystal packing was also described for CyP40 [152]. All the results from the analysis of several TPR domain structures suggest that probably certain TPR domains may eventually assume alternative conformations following the "jacknife" model. This model implies the TPR motif rearrangement through extension or folding back the α-helices.

### *CyP40, FKBP51 and FKBP52 Interaction with Hsp90: Physico-chemical Binding Studies*

In early works, large PPIases-Hsp90 interaction was investigated using cross-linking, but to obtain a value of affinity constant or to improve knowledge about the thermodynamics of the interaction, ITC was employed. All binding curves from ITC experiments presented a sigmoidal shape indicating that the binding follows the simple two-state association model. The binding constant values ($K_D$) derived from these measurements are 226 nM, 174 nM and 55 nM for CyP40, FKBP51 and FKBP52, respectively ($K_A$=4.42 x $10^6$ $M^{-1}$; 5.75 x $10^6$ $M^{-1}$ and 1.81 x $10^7$ $M^{-1}$). According to these values, FKBP52 displayed the higher affinity for Hsp90. In addition, binding reactions exhibited different enthalpic and entropic contributions. It was found that FKBP51 association displayed the most favorable enthalpy of binding, but it also showed the most unfavorable entropic contribution. On the contrary, CyP40 and FKBP52 exhibited lower binding enthalpies, but especially the entropic contribution of FKBP52 binding to Hsp90 is very favorable [158]. All these results demonstrated that the different PPIases have a unique form of Hsp90 binding. From the midpoint of the binding reaction the molar stoichiometry of interaction between the PPIases and Hsp90 could be calculated. Taking into account the union of a PPIase monomer per Hsp90, it should be expected a stoichiometry value of 0.5. The Buchner's laboratory determined (for the three PPIases) a molar ratio of approximately 1.0 indicating that one PPIase monomer binds per Hsp90 monomer; it means two PPIase per dimer of Hsp90 are bound to form the complex. Although, previous chemical cross linking experiments have shown that a PPIase monomer binds per Hsp90 dimer [182]. Further evidence of the binding of two functional TPR binding sites per Hsp90 dimer comes from the finding that Cns1, another TPR-containing cochaperone, was selectively immunoprecipitated together with Hsp90 and the yeast PPIase Cpr7 [150, 183]. Comparing the binding properties of CyP40, FKBP51 and FKBP52 again, in the three cases two PPIase monomers bind to an Hsp90 dimer, but with different affinities: FKBP52 the strongest and CyP40 the weakest. For further analysis, the PPIase activity, the influence on the refolding of RCM-T1 and their chaperone activity were assayed for CyP40, FKBP51 and FKBP52 also in Buchner's laboratory [158]. Nevertheless, the distinct binding types of the CyP40, FKBP51 and FKBP52 with Hsp90 were already noted by other authors. In previous works, binding assays using Hsp90 mutants have demonstrated that not all TPR-containing proteins interact with Hsp90 in an identical manner [132, 156]. The values of $K_D$ from ITC of the three large PPIases agree with earlier findings that CyP40 was easily washed off from Hsp90 immune pellets with low salt buffers [184] and acts in the same manner that the yeast

Cyp40 homologues, Cpr6 and Cpr7 [134, 185], and Sti1 [134]. However, in the mammalian system the differences in association constants of the three large PPIases to Hsp90, especially between FKBP52 and CyP40, are more notable than in the yeast system [185]. The absolute value of binding association constant of PPIases infers that *in vivo* they would be found predominantly in complex with Hsp90. Besides, it should be noted that Hsp90 is present in excess over any of the large PPIases with ratios of 1:10, 1:15 and 1:75 for CyP40, FKBP52 and FKBP51, respectively [149]. Consequently, the PPIase that is present in the highest relative concentration displays the weakest binding affinity. In addition it has been proposed that the interaction between the large PPIases and Hsp90 may be stabilized by the substrate bound.

In summary, the three PPIases, CyP40, FKBP51 and FKBP52 exhibit similar stoichiometry of interaction with Hsp90, but with different association constants and thermodynamically, with different enthalpic and entropic contributions. These results show that they play distinct roles in the Hsp90 chaperone complex.

### *Secondary Structure and Stability of CyP40, FKBP51 and FKBP52*

The basic structure of the PPIase domain in both large and small PPIases is conserved. In this manner, structural studies related with the stability of large and small PPIases can be compared. In this regard, a comparative study about the secondary structure of the large (FKBP51, FKBP52 and CyP40) and small human PPIases (FKBP12 and CyP18) was performed by the scientists Pikrl and Buchner [158] recording far UV-CD spectra. For CyP18, the far UV-CD spectrum showed two minima at 208 and 222 nm characteristic of α-helix secondary structures. But, due to the high content of β-strands the overall intensity of the signal was relatively low. For FKBP12 displayed only one distinct minimum at a wavelength of 212 nm with a mean residue ellipticity of ~5000 deg $cm^2$ $dmol^{-1}$. The spectra for all three large PPIases displayed a similar pattern; it means the typical CD spectrum of a α-helix-containing protein, though Cyp40 displayed a slightly higher ellipticity. The minima values were seen at 208 nm and 222 nm and the maximum value was below 200 nm. Also, the α-helical content was calculated from the UV-CD spectra. The overall shapes of the spectra and the ellipticities found for large PPIases are suggestive of a higher content of α-helical structure compared to the small PPIases, since FKBP12 and CyP18 exhibited 10% and 15.5% of α-helical content, respectively.

On the other hand, large PPIases displayed 30% for CyP40 and about 25% of α-helical content for both FKBP51 and FKBP52. Differences in the far UV-CD spectra are mostly based on the α-helical content of the C-terminal domains of CyP40, FKBP51 and FKBP52 comprising the TPR regions, the chaperone site and the CBM. Especially, the latter region contains a large amount of such secondary structure.

### *Stability of the CyP40, FKBP51 and FKBP52*

To understand the relationship between the α-helical content and the protein stability Pikrl and Buchner [158] monitored temperature-induced unfolding transitions by CD. In this point, FKBP52 demonstrated to be the most stable protein with a midpoint of denaturation of 49°C compared to 45°C for CyP40 and 46°C for FKBP51. However, in comparison with small PPIases, all three proteins showed decreased thermal stability. The midpoints of thermal denaturation between Cyp40 on the one hand and FKBP51 and GST-CyP40 (see Table 1 in Pikrl and Buchner [158]) on the other differ only by 1-2 degrees. Although, it should be considered that CyP40 aggregates under the experimental conditions whereas the other two proteins are stable. The authors discussed that this fact could be due to the thermal denaturation of CyP40 are highly cooperative compared to that FKBP51 and FKBP52.

All exposed on CyP40, FKBP51 and FKBP52 demonstrates that although they display similar structure, it does not mean that these proteins present the same affinity constants to Hsp90, the same quantity and quality of secondary structure, or exhibit the same stability. They constitute a clear example of the need for detailed structural studies in relation to the protein function to establish a whole picture about a molecular mechanism.

## PP5

### *Introduction*

PP5 or protein phosphatase 5 is a cochaperone with phosphatase activity that dephosphorylates serine and threonine residues on target proteins. It is a ubiquitous protein belonging to the PPP family of protein phosphatases [186] composed by a regulatory TPR domain situated at the N-terminal with a C-terminal phosphatase domain. A certain conformation of the TPR domain and the catalytic C-terminal domain causes its phosphatase activity to remain self-inhibited.

**Fig. (27).** PP5 structure representation. PP5 contains two domains: the N-terminal TPR domain and the C-terminal phosphatase domain, both domains are joined by a partially disordered linker region. TPR domain, as expected, is composed by α-helices, colored in violet, pink and magenta from N- to C-terminal and by a seventh helix (α7), represented in salmon.

Since PP5 works in response to stress, glucocorticoids and DNA damage, participates in several stresses activated cellular signaling pathways that regulate growth arrest, apoptosis and response to ionizing radiation-induced DNA damage [187 - 189]. The property of responding to these stimuli is also a feature of others members of PPP family as PP1, PP2A and PP4. Potential inhibitors of PP5 may include tumor promoters as okadaic acid, microcystin, cantharadin, calyculin A and tautomycin [190]. PP1, PP2A and PP4 exist as dimers or trimers with a catalytic subunit bound to regulatory subunit(s), instead PP5 only contains a regulatory N-terminal TPR domain joined to a C-terminal phosphatase catalytic domain (reviewed in [191]).

Since PP5 is a TPR-containing cochaperone, it interacts with the C-terminal fragment of the Hsp90 and competes with TPR-containing immunophilins in joining to GR complexes [192, 193]. It was described that proteins that interact with the TPR domain also stimulate phosphatase activity, as was notably observed for the C-terminal domain of Hsp90 [194] and this stimulation is also observable with arachidonic acid.

### *PP5 Overall Structure*

The crystal structure of the whole PPI molecule without the presence of ligands or binding partners showed the structural assembling of the autoinhibited form of human PP5 [188]. The structure published by Barford's laboratory (PDB Id=1WAO) [188] comprises two distinct domains: the N-terminal TPR domain and the C-terminal phosphatase domain, both domains are fused by a partially disordered linker region (Fig. **27**). The asymmetric unit contains four independent copies of PP5 with the same conformation of the catalytic domain however, small dissimilarities in the relative orientations of the TPR domains relative to the phosphatase domains were found. These differences probably are produced due to the linker flexibility. Another similar crystal structure of PP5 but from rat (*Rattus norvegicus*) and with higher resolution (2.0Å, PDB Id=4JA7) was published very recently by Richter and colleagues [195]. There, the authors have identified five specific PP5 activators that enhance the phosphatase activity up to 8-fold.

### *TPR Domain*

On the other hand, the TPR domain of PP5 is composed of three tandem TPR motifs organized in parallel [19]. Each TPR motif of 34 residues is the product of the joining of two anti-parallel α-helices since, the TPR domain is formed by a series of anti-parallel α-helices. Besides, each α-helix is termed A or B and the two anti-parallel α-helices are connected by shorts turns with a relative orientation of 24°. This architecture creates a right-handed superhelical structure generating a notable concave surface on one side and a convex surface on the other. Basic residues located at the concave surface interact with acidic residues present at the C-terminal of Hsp90 [166, 196] (Fig. **27**).

### *Catalytic Phosphatase Domain*

The Barford's laboratory [188] proposed that the inhibition of phosphatase activity can be mediated by the TPR domain and by Hsp90, and its stimulation is induced by arachidonic acid. Prior to this study, the crystal structure of the catalytic domain of PP5 (residues 169-499) was reported [197]. The tertiary structure of the phosphatase domain of PP5 is basically the same as of PP1 [198] and similar to that of calcineurin /PP2B [199], other members of the PPP phosphatase family [198], in the latter case PP5 shares approximately 40% sequence identity. The structural organization of PP5 and other members of PPP family consist in a central β-sandwich surrounded on one side by seven α-helices and on the other, by a subdomain composed by three α-helices and a three-stranded mixed β-sheet. The catalytic function is carried out by residues that function to coordinate two metal ions at the centre of the catalytic site. The spatial structure of catalytic residues is stabilized by loops connecting β-strands from the β-sandwich with the flanking α-helices. The two metal ions situated at the central part of the catalytic domain mediate the dephosphorylation reaction by coordinating the scissile phosphate group of the phosphorylated substrate contiguous to a metal-activated nucleophilic water molecule [188].

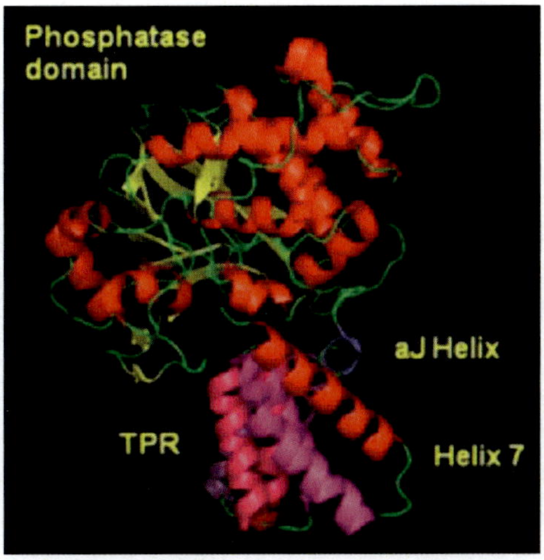


**Fig. (28).** PP5 domains. At the C-terminal to the phosphatase domain, a subdomain of 20 amino acids is observed. This subdomain (colored in blue) ends in a short two-turn α-helix (αJ) that stabilizes the autoinhibited conformation of PP5 and is found contiguous to TPR domain.

The catalytic centre is buried in a deep channel originated by the association of the two β-sheets of the β-sandwich, which probably favors the attachment with the phosphoprotein substrate. At the C-terminal to the phosphatase domain, a subdomain of 20 amino acids can be observed. This subdomain ends in a short two-turn α-helix (αJ) that stabilizes the autoinhibited conformation of PP5 and is found contiguous to TPR domain. This feature is distinctive of the PP5 structure (Fig. **28**).

## *Relationship Between Phosphatase Activity and Hsp90 Binding*

The complex regulatory mechanism of PP5 is unique among protein phosphatases, which ensures that substrates are dephosphorylated when recruited to the Hsp90 chaperone network [195]. The crystallographic structure of the human PP5 autoinhibited conformation [188] revealed that TPR domain interacts with the catalytic domain forming a vast interface, besides the αJ (short two-turn α-helix) blocks entrance of the phosphatase site. In other words, in the inactive conformation, the TPR domain associates with the phosphatase domain in such a way that restricts the substrate entrance to the catalytic site, being also this conformation stabilized by the interactions with the C-terminal subdomain of the protein [163]. In order to unblock PP5, it is necessary to disrupt the TPR-catalytic domain interface, this conformational change can occur due to the binding of polyunsaturated fatty acids or Hsp90 to the TPR domain [194, 200]. Thus, the catalytic activity of PP5 is controlled by a complex structural mechanism in which both the TPR domain (N-terminal) and the αJ-helix (C-terminal) participate [188, 192]. Moreover, PP5 is regulated by binding to the molecular chaperone Hsp90

[194] what produces that in the absence of Hsp90 PP5 exhibits low basal activity. The interaction of the C-terminal αJ-helix with the N-terminal TPR domain of PP5 blocks the enzyme activity in a latent state [188]. Upon binding of Hsp90, the C-terminal MEEVD from the chaperone stabilizes the TPR domain of the phosphatase [192, 193, 201], the interaction between TPR domain and αJ-helix is disrupted and the catalytic site is free to substrate entrance. Likewise, long-chain fatty acids activate PP5 by disrupting TPR domain contacts with the phosphatase domain and thus, stabilizing an alternate conformation of the TPR domain.

In the study performed by Richter and colleagues [195] were reported five compounds named PP5 small-molecule activators (P5SAs) that regulate the activity of PP5. An interesting aspect about this work is that the P5SAs moderate the enzymatic activity of PP5 following a molecular mechanism independent of the TPR domain. This mechanism is also dissimilar from other PP5 activators described so far as arachidonic acid and its derivatives, which interact to the TPR domain [194, 202]. Although novel P5SAs do not share a lot of structural likeness, each individual P5SA may increase significantly the turnover rate of the phosphatase. The discovery of P5SAs can have concrete application when PP5-activity is reduced or where increased dephosphorylation of its substrate proteins is desirable. The phosphorylation mechanism of PP5 is also relevant in the ATPase driven cycle to activate SHRs and in the organization of the complexes between Hsp90-SHR and PP5 [184, 193].

### *Structural Aspects of PP5-Hsp90 Interaction*

As was described above, early studies showed that PP5 has been found in complex with Hsp90 and Hsp70 [189, 192, 193]. Molecular interactions in this complex imply the TPR domain from PP5 and C-terminal EEVD residues from both Hsp70 and Hsp90 [166, 189, 196]. In order to elucidate the molecular basis of these interactions, biochemical and mutagenesis studies of PP5-TPR domain interactions with Hsp90 [166, 196] have been performed. This experimental approach was combined with the structural analysis of the Hop-TPR domain in complex with peptides from both Hsp90 and Hsp70 [29]. Some residues as Arg, Lys and Asn are conserved for Hop-Hsp70/90 interactions as well as in other Hsp70/90-binding TPR domains, including PP5 [29]. Interaction studies about PP5-Hsp90 carried out by Chinkers and coworkers [166] demonstrated that Lys 32, Lys 97 and Arg 101 are required for Hsp90-PP5 interactions. Lys 97 and Arg 101 are located on adjacent helical turns of the αA-helix of TPR 3, in the vicinity of Lys 32 on αA-helix of TPR 1. The association of these residues with Arg 74 of αA-helix of TPR 2 shapes a basic surface inside the TPR peptide-binding channel where an acidic C-terminal region from Hsp70/90 can interact. All the analysis suggests that the PP5-TPR domain interacts with Hsp70/90 in a similar way than that seen in the Hop-peptide complexes (Fig. **30**). Nonetheless, Chinkers and co-workers [202], performing site-directed mutagenesis study of PP5, found that the replacement of basic residues inside the peptide-binding groove (shaped by the TPR domain) did not produce changes in the basal activity of PP5. This result infers that the residues which participate in TPR-Hsp90 interaction are not the same than those necessary for autoinhibitory interactions with the phosphatase domain. In summary, the PP5-Hsp90 interaction is carried out by means acidic residues of the EEVD motif from C-termini of Hsp90 and Hsp70; those chaperones interact with PP5 and Hop following analogous mechanisms [196]. In addition to EEVD residues, an acidic region comprised between residues 9 to 13 (relative to the Hsp90 C-terminus) is also important for the Hsp90-PP5 binding [196], evidencing that significant interaction between Hsp90 and PP5 overlaps the TPR-αJ helix contacts. Further, Cliff and coworkers [201] presented NMR studies of the wild-type (WT) PP5-TPR domain and G83N PP5, the mutant resultant of single point mutation experiments in complex with the C-terminal MEEVD peptide of Hsp90. The election of Gly 83 is because it possesses high accessible solvent area, it is situated on the opposite face of the TPR domain that interacts with Hsp90 and besides, this point mutation improved the spectral quality. It is remarkable that Gly 83 is not conserved in PP5 from non mammalian species. However, the effect of this mutation was only observed in the improvement of the stability of the major conformation of the native sequence. No effects were seen in binding thermodynamics with Hsp90 pentapeptide. The structure published by Cliff and coworkers [201] using NMR techniques was the first structure from a TPR-pentapeptide complex that was determined in solution. Consequently, it was confirmed that the consensus superhelical structure constitutes the binding site of Hsp90 C-terminal pentapeptide. Besides, the structural basis of PP5-Hsp90 studied in Cliff *et al.* [201] showed that the peptide binds at the same site as that noticed in the homologous TPR2A domain from Hop (Fig. **29**).

The Hsp90 peptide recognition mechanism in PP5 and Hop imply the "clamp" however, in PP5 is different because it lacks differential contacts with Hsp70 and shows that both the binding site in the TPR domain and the peptide-ligand present different conformations in comparison with Hop-peptide interaction. In PP5 the peptide orientation is such that the C-terminal aspartate is making salt bridges with two highly conserved residues Lys 32 and Lys 97 [166, 196] and a hydrogen bond between its backbone amide and N36. This manner of binding was named "two-carboxylate clamp" and was first observed in the Hop structure. In PP5 the spatial conformation of the residues Lys 32, Asn 36, Asn 67, Lys 97, and Arg 101 conserved the binding motif in the TPR domain (Fig. **30**). In the three dimensional structure of PP5 can be noticed the presence of a secondary and largely apolar pocket constituted by Phe 39 and Leu 70, these residues contact with CH2 group from the asparagine side chains from both conserved residues: Asn 36 and Asn

67. This pocket fit with Val from the Hsp90 peptide in the Hop, CHIP, and PP5 structures. Interestingly, the F39A mutation in the CyP40 protein notably decreases the binding to Hsp70/90 peptides. The presence of a hydroxyl group of the conserved Tyr 51 at the base of the pocket could be the reason of the mobility of the Val in PP5 since this polar group would act weakening that hydrophobic interaction (Fig. **30**). Moreover, the structure of PP5 published by Barford's laboratory [188] proposes that high affinity TPR-Hsp90 interactions cannot be carried out because the autoinhibited conformation observable in free PP5. Although a region of the Hsp70/90 peptide-binding site was accessible in the autoinhibited structure, high affinity binding needs a fast and easy access to the TPR domain that, in this case is blocked by the catalytic domain.

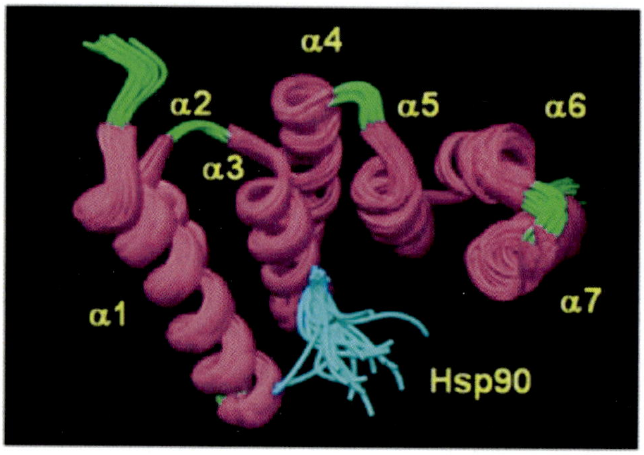


**Fig. (29).** NMR structure of the combining site of Hsp90 in PP5. NMR technique confirmed that the consensus superhelical structure of the TPR domain constitutes the binding site of Hsp90 C-terminal pentapeptide. This is the first structure from a TPR-pentapeptide complex that was determined in solution.

As a consequence, the phosphatase domain is predicted to antagonize Hsp90 binding, by blocking the combining site in TPR groove. In this way, the TPR domain would require dissociating from the phosphatase domain and then, both domains would be accessible to perform their specific functions. In order to assess these suggestions, the authors performed diverse kinetic and ITC experiments.

### *PP5-Hsp90 Binding Studies*

Barford's laboratory [188] tested the ability of the whole Hsp90 molecule versus its C-terminal pentapeptide to activate the catalytic capacity of PP5. They found that the whole Hsp90 molecule is a discrete activator of PP5, inducing ~7-fold stimulation of the enzyme with a half maximal activation (AC50) of ~2 μM. However, the TSRMEEVD peptide from human Hsp90α induces a comparable fold activation but with a reduced AC50 constant value of 57 μM. These results signify that the binding between PP5 and the peptide is weaker and are consistent with previous findings that a 12 kDa C-terminal domain of Hsp90 provoked a 10-fold stimulation of PP5 with an AC50 constant of 6 μM [194]. Further, using ITC, those authors measured the binding affinity of Hsp90 peptide for both the isolated TPR domain and for the whole PP5 molecule. As expected, Hsp90 peptide binds the isolated TPR domain firmly, with a $K_D$ of 40 nM ($K_A$=2.5 x $10^7$ $M^{-1}$), 500-fold higher than its affinity for the whole protein ($K_D$ of 20 μM or $K_A$=5.0 x $10^4$ $M^{-1}$). Other studies employing ITC and CD corroborated the high binding affinity of the Hsp90 peptide-TPR domain union: $K_{Dobs}$ ~50 nM ($K_A$=2.0 x $10^7$ $M^{-1}$) [203]. These experiments demonstrates that to unlock PP5 and to expose it enzymatic activity, it is necessary that the entire Hsp90 molecule interacts with the N-terminal TPR domain of the cochaperone. The hypothesis that supports that whole Hsp90 molecule is a more powerful activator of PP5 than the C-terminal Hsp90 peptide, also supports that regions located at the N-terminal to the MEEVD motif in Hsp90 and outside of TPR domain in PP5 could be involve in the complete interaction and in the enzymatic activation of PP5. Other results (explained above) show that a segment, composed by residues -9 to -13 relative to the Hsp90 C-terminus, is essential for strong binding affinity of PP5-Hsp90 union [197].

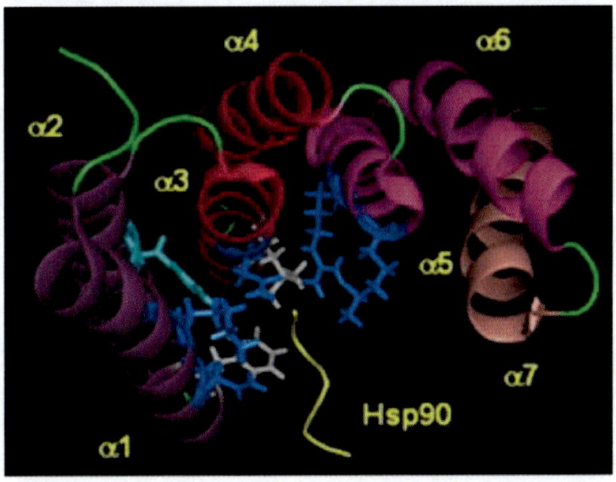


**Fig. (30).** Combining site of Hsp90 in PP5 (TPR domain). The groove of the Hsp90 combining site is constituted by residues from different α-helices: Lys 32, Asn 36 and Phe 39 (α1); Tyr 51 located at the base of the pocket colored in cyan (α2), Asn 67 and Leu 70 (α3) and Lys 97 and Arg 101 (α5). Basic and apolar residues are illustrated in blue and white, respectively.

## *Structural Considerations*

It is important to highlight that contacts between the catalytic and TPR domain involve the intra-repeat turns of the TPR protein, an unusual characteristic not observed previously. The autoinhibited state of PP5 implies that TPR domain blocks the phosphatase activity; however, the interactions between the C-terminal αJ helix and the groove of the TPR domain stabilize this state. In this sense, both

TPR domain and C-terminal αJ helix collaborate to abrogate phosphatase activity. At the same time these interactions partially block the TPR domain groove preventing the formation of high affinity contacts between this domain and the C-terminal of Hsp90. Thus, Hsp90 provokes a strong activation of PP5 since it competes for TPR groove. Then, the Hsp90-TPR interaction unblocks the PP5 phosphatase site. According to Bardford's group results [188] the mechanism of PP5 auto-inhibition by its N-terminal TPR domain, and the activation by Hsp90, shares analogies to how SH2 domains regulate the activities of Src family kinases [239], and the protein tyrosine phosphatases SHP1/2 [204].

However, the structure of the PP5-Hsp90 peptide complex published by Cliff *et al.* does not seem to be in concordance with this hypothesis [203]. Diverse conformations of the peptide found in the TPR pocket means that Hsp90 (or Hsp70) can associate avoiding the autoinhibitory interface (for better understanding see Fig. **7** in Cliff *et al.* [203]). Instead, structural alignment studies contemplate the possibility of an allosteric mechanism for relief of auto-inhibition. Binding of the peptide could change the orientation of the seventh α-helix, allowing the release of the catalytic site and the activation of the enzyme. The "two carboxylate clamp" mechanism for recognition of the Hsp90 and Hsp70 C-terminal residues by TPR domains presents differences in its binding dynamics. The interaction models arising from these studies can be applied in the structural study of the interaction of other TPR-containing cochaperones with Hsp90 and Hsp70. Analyzing all the data and concepts exposed above for PP5 interactions, we find a complex scenario where conformational changes, hydrogen, electrostatic and hydrophobic bonds as well as long-range interactions are part of this dynamic variability. The small number of differences in the surface residues between Hop and PP5 produce a great variability in Hsp90 binding mechanism and demonstrate the adaptability of TPR domains to mediate protein-protein interactions.

### *PP5 Folding Biophysical Studies*

Studies published by Ladbury's laboratory [203] suggest that the stabilization of the superhelical form of TPR domains is achieved by means the interactions with a negatively charged ligand. Spectroscopic studies by means NMR and CD of the isolated TPR domain from PP5 demonstrated that this domain is basically unfolded at physiological temperatures, and becomes fully folded upon binding to the Hsp90 C-terminal peptide (MEEVD) [203]. By means ITC experiments, this coupled folding-binding mechanism was characterized. It was observed that the enthalpy of binding displayed a non-linear relationship with temperature; it means that there is strong temperature dependence to the observed binding enthalpies. A further characterization was necessary to determine separately the thermodynamic contributions of the intrinsic folding and binding events to the overall coupled process. For this purpose, the combination of data from ITC and CD experiments were analyzed following a nested Gibbs-Helmholtz model [203]. The best-fit models of the coupled process demonstrate that the overall recognition process exhibits temperature-dependent enthalpy-entropy compensation. Besides, it was observed that despite large changes in the enthalpic and entropic components, tight-binding is maintained over a wide range of temperature. Thus, the coupled folding process has small effect on the overall affinity of peptide recognition. Finally, the authors suggest that a coupled folding-binding mechanism is a common feature, but not universal in TPR domains.

## CHIP

### *Introduction*

CHIP or carboxyl-terminus of Hsc70-interacting protein, is a 35-kDa TPR-containing protein which is located in cytoplasm. CHIP is expressed particularly in adult striated muscle and brain. It exhibits a highly conserved protein sequence across species. Its N-terminal region is composed by three tandem TPR motifs, similar to those found in other cochaperones.

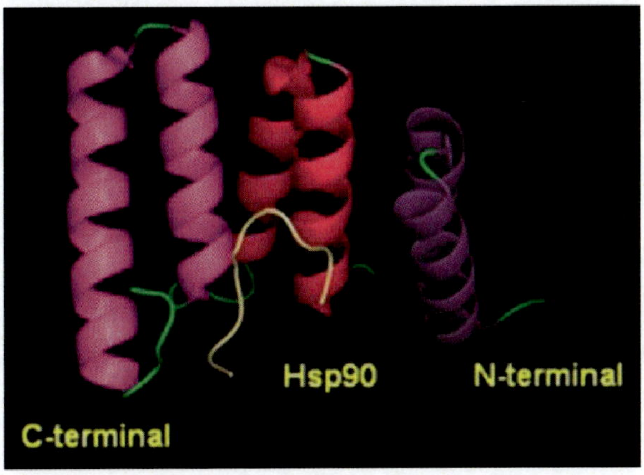


**Fig. (31).** CHIP TPR-domain. This domain is composed by three pairs of anti-parallel α-helices (residues 26–131), colored in violet, pink and magenta from N-terminus of C-terminus. Besides, the Hsp90 peptide (yellow) interacting with TPR groove is represented. Despite the fact, that the crystal structure was solved at low resolution (3.2 Å), all residues could be defined.

Using the yeast two-hybrid system, Patterson and colleagues [205] have demonstrated that CHIP binds Hsc70 and Hsp70. In addition, biochemical assays showed that recombinant CHIP negatively regulates ATPase and chaperone functions of HSPs. Therefore, CHIP must obligatorily interact with HSPs and must participate as regulator of the Hsc70-Hsp70 substrate-binding cycle. The discovery of CHIP supported the concept that Hsp90 is implicated in the ubiquitin-proteasome pathway [205]. CHIP is defined as an E3 ubiquitin ligase and is able to ubiquitinate unfolded proteins [206, 207]. This function is performed by the C-terminal U-box domain contained in CHIP structure. It must be remembered that the selection of proteins for degradation by the proteasome is mediated by E3 ubiquitin ligases. Moreover, CHIP is an example of a TPR protein which mediates the degradation of misfolded proteins and participates in protein quality control [208]. As cochaperone, it regulates both Hsp70 and Hsp90 by means the interaction with the C-terminus of these chaperones through its TPR domain [209].

### *Structure*

Pearl and coworkers [210] have resolved the crystal structure of mouse CHIP in complex with a C-terminal decapeptide of human Hsp90α. The purpose of this approach was to study the mechanism of ubiquitination coupled to the union with molecular chaperones.

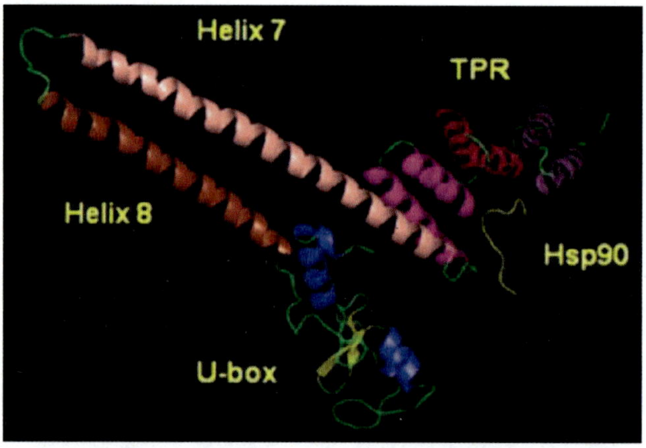


**Fig. (32).** Conformational changes in CHIP protomers. The N-terminus of the elongated seventh helix (salmon) packs against the third helical pair of TPR repeat (magenta). The α7-helix presents different conformations in the structure of the two protomers contained in CHIP dimer. In the protomer represented here, the seventh helix appears as a straight α-helix composed by residues Asp 134 to Arg 183. Beyond α8-helix, U-box domain can be found, it comprises a β-hairpin (residues 232-240) and (244-254) that runs into a short α helix (255-265) followed by a third hairpin (268-277) leading to a C-terminal α-helix (274-298). α-Helices are represented in blue, β-structures in yellow.

The structure shows how CHIP binds with Hsp90 and Hsp70 revealing an unusual asymmetric homodimer in which, the protomers adopt absolutely different conformations. In this structure [210] the N-terminus of CHIP consists of a TPR-domain composed by three pairs of anti-parallel α-helices (residues 26-131) (Fig. **31**), with an elongated α7-helix. The N-terminus of this α-helix packs against the third helical pair of the TPR repeats. The α7-helix presents different conformations in the two protomers structure of the CHIP dimer. In one protomer, the α7-helix appears as a straight α-helix composed by residues Asp 134 to Arg 183 (Fig. **32**). In the second protomer, the polypeptide seems to break into two distinct and mutually perpendicular α-helices (Fig. **33**).

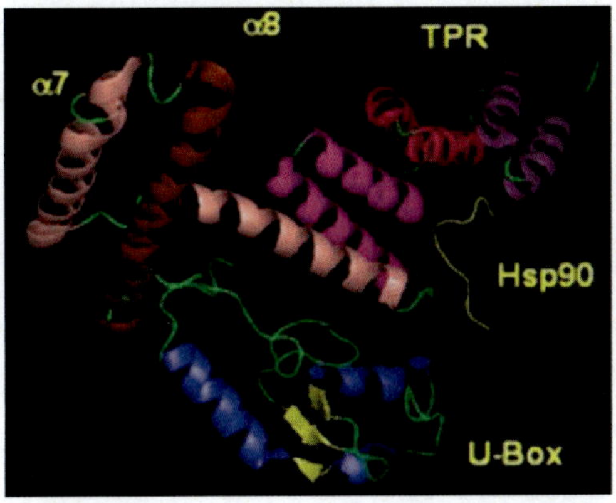


**Fig. (33).** Conformational changes in CHIP protomers. This illustration shows that the polypeptide seems to break into two separated and mutually perpendicular α helices. In both protomers, the chain changes to opposite direction via a disordered loop (Asn 184-Asp 190) shaping another α-helix, the α8-helix that is colored in chocolate and constitutes an anti-parallel hairpin with the previous helix. Beyond α8-helix, U-box domain is represented, it is composed by a β- hairpins (residues 232-240) and (244-254) that runs into a short α helix (255-265) followed by a third hairpin (268-277) ending in a C-terminal α-helix (274-298), α-helices are represented in blue, β-structures in yellow.

In both protomers contained in the dimer, the chain changes to an opposite direction via a disordered loop (Asn 184-Asp 190). Thus, in the second protomer the change in polypeptide chain direction gives place to another α-helix (α-8), which constitutes an anti-parallel hairpin with the previous α-helix. Similar to the N-terminal, the C-terminal side of the α-helical hairpin shows a different conformation in both protomers. In two cases, α8-helix is a continuous helix beyond α7-helix; it starts at Gly 192 but ends at Lys 224 in the protomer with the extended α7-helix and at Gln 217 in the protomer with broken polypeptide chain. Beyond α8-helix (in both protomers) the polypeptide chain runs into a coil shaping the beginning of the C-terminal U-box domain. This domain presents a similar structure in both protomers (Figs **32** and **33**). The U-box domain is composed by a pair of β-hairpins that runs into a short α-helix followed by a third hairpin ending in a C-terminal α-helix. Beyond the end of this α-helix the two protomers also presents different conformations.

On the other hand, CHIP dimerization implicates to both the U-box domain and the distal segment of the helical hairpins (Fig. **34**). The U-box domain constitutes a parallel dimer, which buries > 900Å2 of predominantly hydrophobic molecular surface. It is notable that the isolated U-box domain exhibits some degree of dimerization in solution. The core of protomer-protomer interface is mainly constituted by the symmetrical interaction between amide side-chains of Asn 284 from each protomer. As expected, all residues involved in the promoters interacting region are strongly conserved [210]. More in detail, the core of the second dimer interface is composed by a hydrophobic patch on the surface of the helical hairpin, and by some residues from the N-terminal region of the hairpin and from the C-terminal arm; all these structures interact with their equivalents in the other protomer to organize a four-helix bundle with a ~2 -fold symmetry (Fig. **34**).

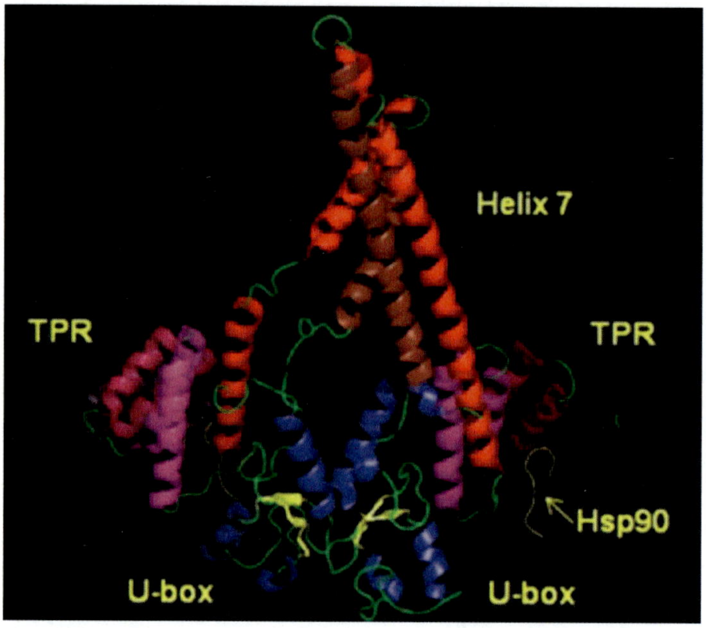


**Fig. (34).** Architecture of the CHIP homodimer. CHIP homodimer is represented in secondary structure, showing a superposition of both protomers. α7-Helix and α8-helix are depicted in red and in chocolate, respectively. The Hsp90 C-terminal decapeptide (yellow) is bound to each TPR domain located in opposite directions. The contacts between the two U-box domains and between the distal ends of the two helical hairpins are well defined resulting in a four helix bundle.

The residues shaping the second dimer interface are also highly conserved in CHIP sequences, although the length of the hairpin differs between species. The human CHIP fragment constituted by residues 128 to 229 was found to be substantially α-helical and spontaneously dimerized in solution. The deletion of those residues impedes CHIP dimerization and disrupts E3 ligase activity [211]. Additionally, Pearl and coworkers characterized CHIP as a binding partner for Ubc13-Uev1a, the Lys63-specific ubiquitin-conjugating enzyme. The resolution of the heterotrimeric complex composed by Ubc13-Uev1a and the CHIP U-box domain by Pearl and colleagues confirmed that CHIP collaborates as an E3 ubiquitin ligase [210].

## *Hsp90 C-Terminal Binding to the CHIP TPR Domain*

Previous works demonstrated that CHIP can interact with the C-terminus of either Hsp70 or Hsp90 [205, 209] by means its single TPR domain, whereas Hop/Sti1 is able to bind both chaperones but, utilizing different TPR domains, what it means a selective binding [29]. In order to understand the promiscuity of the TPR-domain in CHIP, Pearl and coworkers [210] cocrystallized CHIP with a ten residues peptide ($NH3^+$-DDTSRMEEVD-$CO2^-$) from the C-terminal of Hsp90α. Binding constants for such ten residues peptide and for another seven residue peptide ($NH3^+$-SRMEEVD-$CO2^-$) were calculated by ITC, giving the following values: 2.8 and 2.2 μM, respectively ($K_A$=3.6 x $10^5$ and 4.5 x $10^5$ $M^{-1}$ respectively). Comparing these $K_D$ values with those obtained for the C-terminal pentapeptide with MEEVD sequence ($K_D$~90 μM or $K_A$~1.1 x $10^4$ $M^{-1}$) and for the whole Hsp90 molecule ($K_D$=4.9 μM or $K_A$=2.0 x $10^5$ $M^{-1}$), it can be observed that the most tightly binding corresponds to the seven residues peptide from C-terminal of Hsp90, suggesting that those residues are essential to CHIP-TPR-Hsp-0-Cterminal recognition. The TPR domain in CHIP (as other TPR cochaperones) shapes a concave surface that contains the binding groove necessary to interact with the C-terminal peptide from Hsp90 and Hsp70. Thus, homodimeric CHIP can bind simultaneously two peptides according to the structure published by Pearl and coworkers. Carboxyl group of the C-terminal Asp 731 from Hsp90α makes polar interactions with Lys 31, Asn 35, Asn 66, and Lys 96 side chains contained in the TPR domain of CHIP. Those polar bonds are similar to interactions between carboxyl group of Asp731 and side chains from Lys 229, Asn 233, Asn 264, and Lys 301 contained in TPR domain of Hop [29]. Likewise, the hydrophobic interaction between Val 730 from Hsp90 with the side chains of Phe 38 and Leu 69 in CHIP resembles the interaction with Tyr 236 and Ala 267 in TPR domain of Hop. Besides, the side chain of Hsp90α Glu 729 makes a polar bond with Lys 96 which corresponds with the weaker interaction with Lys 301 in Hop. In accordance with described so far, similarly the terminal EVD tripeptide fragment from Hsp90 showed an analogous main chain conformation in CHIP and Hop TPR-domain complexes, respectively [210]. Another important aspect to be considered from the work of Pearl's group structure is that CHIP can be characterized as a singular example of a homodimeric protein in which the protomers adopt significantly different conformations. In CHIP, the homodimer asymmetry (confirmed experimentally) is an inherent property and it is not concomitant with ligand binding. The main consequence of the homodimer asymmetry consists of the binding of the U box domains in the CHIP dimer to a one of two binding sites present in Ubc enzymes. Such interaction produces the blockage of one site in those enzymes leading to display "half of sites" activity. This is the reason why the two U box E3 ligase domains might cooperate in development of a uniformed polyubiquitin chain. In addition, the TPR domains are positioned differently in the CHIP dimer however, they are available to bind the C-terminal peptide from Hsp90. Consequently, CHIP and Hsp90 configure a dimer-dimer complex (Fig. **34**). In summary, these studies explain the ability of CHIP to interact with Hsp90 or Hsp70, establish the structural basis for selective cooperation with specific ubiquitin conjugating enzymes, and demonstrate how the generation of an asymmetric homodimer is efficient to couple a dimeric chaperone to a single ubiquitination system.

### *Folding and Degradation Balance Mechanism*

Kundrat and Regan [140] focusing in Hsp70, Hsp90, Hop and CHIP, questioned whether the protein folding and degradation mechanism can manifest simultaneously or they compete and are totally exclusive. Their concern was whether either Hsp70 or Hsp90 can generate a complex with Hop and CHIP at the same time. Both CHIP and Hop interacts with the C-terminal of Hsp70 and Hsp90, accordingly, only oligomerization of Hsp70 or Hsp90 could achieve the generation of a complex that contains both CHIP and Hop. Hsp70 is a monomer, then, a Hop-Hsp70-CHIP complex cannot be conceived but as Hsp90 is a dimer, in theory, a Hop-Hsp90 dimer-CHIP complex could be formed. Moreover, Kundrat and Regan [140] studied the balance between protein folding and degradation for mammalian client proteins which are Hsp90-dependent. They investigated the interplay between the molecular chaperones Hsp70 and Hsp90, the cochaperone Hop, and ubiquitin ligase CHIP and their influence in the folding/degradation balance. Hsp70 and Hsp90 bind to Hop to make up a ternary folding complex while, the interaction of CHIP with these chaperones conduct to ubiquitination and finally to degradation of the client proteins as well as the chaperones. In order to understand the folding/degradation balance in more detail, Kundrat and Regan [140] measured and calculated the $K_D$ and the stoichiometry of binding for the CHIP-Hsp70 and CHIP-Hsp90 complexes. $K_D$ between CHIP and C-terminal peptides of Hsp70 and Hsp90 was measured using surface plasmon resonance (SPR), $K_D$ values for CHIP-Hsp70 peptide was ~2 μM ($K_A$=5 x $10^5$ $M^{-1}$) and for CHIP-Hsp90 peptide was ~5 μM ($K_A$=2 x $10^5$ $M^{-1}$). In order to confirm these $K_D$ values and to calculate the stoichiometry of the interactions they performed ITC experiments, $K_D$ values for CHIP-Hsp70 was ~1μM and stoichiometry of the interaction ~0.8 C-terminal Hsp70 peptide per CHIP monomer, whereas the $K_D$ value for CHIP-Hsp90 was ~4.4 μM and the stoichiometry of the interaction ~0.9 Hsp90 C-terminal peptide per CHIP monomer. Besides, by size-exclusion chromatography they could analyze the nature of the complexes composed by the whole Hsp70 molecule or entire Hsp90 molecule and CHIP. The stoichiometry of the CHIP-Hsp70 complex (~0.8 Hsp70 monomer per CHIP monomer) shows that a CHIP dimer interacts with two Hsp70 monomers. However, the stoichiometry of the CHIP-Hsp90 complex (~2.1 Hsp90 monomers per CHIP monomer) demonstrates that a CHIP dimer binds to two Hsp90 dimers. All these results are in concordance with those previously reported by Pearl's group [210]. As a conclusion, Kundrat and Regan [140] found out that the C-terminal residues of Hsp70 or Hsp90 are essential and at the same time enough to ensure the interaction with CHIP. In addition the authors determined that CHIP ubiquitinates Hsp70 and Hsp90 at comparable rates. The rate of ubiquitination of Hsp70 by CHIP is not dependent of the conformational state of the chaperone. At the same time, the ubiquitination of Hsp70 by CHIP is a robust process even when Hsp70 is bound to a client protein. Another conclusion from that work was that ubiquitination of client proteins takes place mainly in the client protein-Hsp70-CHIP complex. The authors came to that conclusion after relating the binding affinities of Hsp70 and Hsp90 for CHIP with the intracellular concentrations of chaperones. Ubiquitination in the client protein-Hsp90-CHIP complex represents a secondary pathway, since the binding affinity of CHIP for Hsp70 is higher than for Hsp90 and the *in vivo* concentration of Hsp70 is greater than that of Hsp90. Therefore, it can be considered that CHIP is generally in complex with Hsp70. Nevertheless, the main conclusion from the Kundrat and Regan work [140] was that dimeric Hsp90 cannot bind to both Hop and CHIP at the same time, which implies that, the folding and degradation processes cannot co-exist in a single complex and they are competitive mechanisms. Since under basal conditions the concentration of the Hsp70-Hop-Hsp90 complex is significantly higher than that of the Hsp70-CHIP complex, client proteins are more likely to be folded than are ubiquitinated and therefore degraded. Nevertheless, even under basal conditions a low background of client proteins that follows the ubiquitination and degrading pathway can be detected. The authors focused on Hsp90-dependent client proteins, but several client proteins only need Hsp70 for proper folding. They argued that the proposed model of balance between protein folding and degradation can be also valid for those Hsp70-dependent client proteins, since a minuscule quantity of these proteins will steadily be turned-over by CHIP. In addition, Kundrat and Regan postulated that the rates of both client protein folding and ubiquitination mechanisms are practically the same therefore, the pathway followed by a protein is determined by whether it is bound to an active Hsp70-HOP-Hsp90 complex or to an Hsp70-CHIP complex. Finally, the authors concluded that Hop and CHIP compete for interacting with Hsp90 since CHIP does not pull-down Hop through the chaperone. According to the experiments and structural analysis carried out by Kundrat and Regan, folding and degradation machineries are mutually exclusive and like Hsp70, Hsp90 can constitute complexes with either Hop or CHIP.

## TARGETING HSP90 INTERACTIONS FOR BIOTECHNOLOGICAL APPLICATIONS

### Introduction

The ultimate goal of cellular mechanisms research at the molecular level involves the study of the functioning of these mechanisms under normal and under pathological conditions, respectively. After the mechanism was determined to be exacerbated in a particular disease or in related diseases, inhibitors are designed to disrupt that mechanism but preventing such inhibitors from interfering with the functioning of healthy cells. The nature and origin of the inhibitors is varied, they can be natural or synthetic, they can be of protein origin or organic compounds, both generated through synthesis in the laboratory. In the case of protein design, a number of strategies have been used for generating DNA or protein-binding proteins based on computational methods, combinatorial libraries or rational design [212].

As it was seen in all the extent of this chapter, the TPR domain-Hsp90 interaction is crucial to perform the Hsp90 stabilization function to client proteins. This TPR-Hsp90 interaction constitutes the target of new biotechnology approach to design Hsp90 inhibitors to attack malignant cancer cells. However, not a lot of publications can be found about inhibitors which interfere specifically with the cochaperone TPR-Hsp90 C-terminal peptides interaction, perhaps because is a very complicated strategy or (as was sketched in this Chapter) that particular interaction involves more than a small peptide interacting with a TPR domain. This approach is based on the fact that Hsp90 is a molecular chaperone that is critical for the stability and function of many essential proteins for cell survival but it is also involved in the maturation and stabilization of multiple client proteins that are fundamental for oncogenesis and malignant progression. A large number of Hsp90 client proteins as transmembrane tyrosine kinases (*e.g.* Her-2), metastable signaling proteins (*e.g.* Akt), transcription factor (*e.g.* p53), chimeric signaling proteins (*e.g.* Bcr-Abl), cell cycle regulators (*e.g.* Cdk4, Cdk6), and SHRs (AR, ER, and PR) among others have been found to be inappropriately modulated in cancer. Inhibition of Hsp90 provokes client protein degradation via the ubiquitin-proteasome pathway, and this inhibition might (in theory) concurrently downregulates several redundant pathways essential for cell viability and tumor development. Hsp90 inhibitors have been and are currently being developed as anticancer agents. They have exhibited early promising results in the treatment of particular subgroups of solid tumors (*e.g.* qALK-rearranged non-small-cell lung cancer and HER2-amplified breast cancer) and some haematological malignancies (*e.g.* multiple myeloma).

As was described above, HSPs are found in almost all living organisms, and their expression is increased in response to various cellular insults, including raised temperature, presence of heavy metals, and oxidative stress. Cellular stress causes protein denaturation; denatured and aggregated proteins cannot play their function and must be rescued or eliminated with the assistance of chaperones. Heat shock factors (HSFs) transcriptionally regulate the increase of HSPs expression in response to stress. This regulation is part of the heat shock response that conducts to cellular protection from an aggression that would otherwise cause lethal damage. Defective chaperone function is noted in cellular senescence and in several diseases [213]. If the effects of cellular stress go unnoticed by the mechanisms of protein-refolding performed by chaperones, intracellular proteins become denatured and insoluble, form aggregates, and precipitates. As an example, in neurodegenerative disorders such as Parkinson's, Alzheimer's, Huntington's, and prion-related diseases, inclusion bodies formation can be found as a common pathological process, even in the absence of cellular stress. Drugs that can induce the heat shock response and promote HSP expression might have therapeutic value in these diseases [213]. In neoplastic cells, by contrast, HSPs (particularly Hsp90) are often overexpressed, they are found mostly as a multichaperone complex whereas in non-cancerous cells, in basal conditions, HSPs are in an uncomplex state. Hsp90, due to its ubiquitous nature and importance for cellular viability, has turned into an obligatory target for cancer therapeutics.

Functionally, as was shown in this chapter, Hsp90 does not work alone. It appears dependent upon a group of cochaperones in the client protein maturation in an ATP-dependent manner. Nevertheless, the mechanism by which Hsp90 selectively binds to various client proteins has not been fully elucidated. Thus, disrupting the complex formed by Hsp90 and its specific client proteins in cancer cells has been considered to be a potential therapeutic approach [213], although it has not yet been too successful. The oncogenic potential of cells is highly dependent on their ability to survive despite endogenous (hypoxia, pH changes, nutrient deprivation, dysregulated signalling pathways) and exogenous (radiation or chemotherapy) insults. Additionally, one of the main aspects to consider is that increased HSP expression can stabilize oncogenic proteins that are key drivers of the malignant phenotype. Accordingly, HSPs promote independence of growth factors, tumor-cell survival, proliferation, immortalization, neovascularization, and metastasis, and indirectly modulate response to DNA damage and cell metabolism, in other words HSPs modulate most of the hallmarks of cancer. Furthermore, cancer cells seem to be more dependent than normal cells on HSPs, and could therefore be more sensitive to HSP inhibition.

## Hsp90 and Cancer

Cancer cells possess genetic instability [214] moreover, genetic alterations appear necessary for the generation of the malignant phenotype. Genetic instability seems to be the reason why cancer cells acquire the following characteristics defined by Hanahan and Weinberg [215, 216]. They comprise: (i) self-sufficiency in growth signals; (ii) insensitivity to anti-growth signals; (iii) evasion of the apoptosis; (iv) sustained angiogenesis; (v) tissue invasion and metastasis; (vi) unlimited replicative potential; (vii) reprogramming of energy metabolism; and (viii) evasion of the killing by the immune system. This feature set reflects genetic alterations in multiple protection genes responsible for the regulation and strong coordination of diverse processes, such as cell survival, proliferation, growth, differentiation, and motility [217]. Remarkably, this genetic flexibility allows cancer cells evading a fine molecular regulation of a particular signaling node or pathway, making them often insensitive to molecular targets used as therapeutic strategy [214]. In this regard, Hsp90 consists of a central node in signaling networks and plays a central role in the acquirement and preservation of each of these capabilities [214]. Therefore, inhibition of Hsp90 conducts to the degradation of the oncogenic clients and abrogates the most of hallmarks of a cancer cell simultaneously. Consequently, targeting Hsp90 appears to be a reasonable anticancer strategy. Hsp90 inhibitors tested to date inhibit Hsp90 functions mostly by competing with ATP binding. In this way, the strategy consists of freezing the chaperone cycle, which in turn leads to a decrease in the affinity for client proteins ending in the client protein degradation by proteasome pathway. This strategy was postulated as successful because it has been observed that Hsp90 inhibitors kill cancer cells selectively compared to normal cells [218]. This therapeutic selectivity results from the activated, high affinity chaperone of Hsp90 in tumors [218]. In the past decades, Hsp90 has arisen as promising new target for the development of new anti-neoplastic drugs for the treatment of a variety of human cancers. Moreover, studies combining proteasome and Hsp90 inhibitors showed a synergistic cytotoxic effect, with accumulation of ubiquitinated, misfolded oncogenic proteins [219]. The inhibition of Hsp90 chaperone function might concurrently down regulate several pathways essential for cell viability, this methodology has resulted in marked antitumor effects in preclinical models and could possibly preclude the development of tumor drug resistance. The degree, reach and extent of protein degradation via Hsp90 inhibition are not the same for all client proteins. In addition to the ATP regulatory cycle, some studies have suggested that in tumors, Hsp90 constitutes multichaperone, biochemically distinct complexes that stabilize oncogenic proteins and show a higher affinity to specific inhibitors compared to those Hsp90 complexes found in normal cells. Consequently, Hsp90 inhibitors can be considered as selective disruptors of tumor cells. The development of Hsp90 inhibitors is by far most advanced, in the literature there is a vast amount of Hsp90 inhibitors from different chemical origin, natural and synthetic and targeting different Hsp90 structural regions. N-terminal and especially C-terminal inhibitors will be described in the following sections. In addition to the inhibitors of the different functional regions of the Hsp90 structure it can added that Hsp90 can be regulated post-translationally by phosphorylation, nitrosylation, and acetylation [220]. Importantly, Hsp90 inhibitors can be used as potentiators of the effects of cytotoxic drugs or radiotherapy for the treatment of various tumor types. Different drug combination strategies that inhibit different mechanisms as well as the combination of Hsp90 inhibitors with radiotherapy have been very successful in some clinical trials but in others showed high toxicity.

## Inhibitors of the N-terminal Domain of Hsp90

Hsp90 inhibitors can interact in many regions on its molecular surface, but the first type that was explored and the most studied was those that perform inhibition of the N-terminal region. This type of inhibitors competes with ATP and perturbs ATPase activity, thereby disrupting client protein maturation. In general, N-terminal inhibitors when bind to ATP-binding pocket adopt a bent conformation and are highly selective for Hsp90 over other ATP-binding proteins. The Hsp90 N-terminal inhibitors under clinical investigation comprise derivatives of GA, RD, purine, dihydroindazolone derivatives and other small molecules inhibitors not included in mentioned chemical groups. GA is a benzoquinone ansamycin (ansamycin antibiotic) first isolated from a fermentation broth of the actinomycete *Streptomyces hygroscopicus var geldanus* in 1970 [221], being the first natural product inhibitor of Hsp90. Initially GA was identified by possessing anti-parasitic and antineoplastic activity, but from 1990, after *in vitro* assays, it was studied as an inhibitor of Hsp90 [222]. Because of its interaction with the nucleotide-binding site of Hsp90 with higher affinity than ATP, GA impedes ATP binding and hydrolysis. This inhibition conducts to the disruption of the Hsp90-client protein complex and to the increasing of depletion of oncogenic client proteins. In the attempt to get the more potent Hsp90 N-terminal inhibitors, binding affinities of different compounds were measured by ITC and other techniques; however, it was found no obvious correlation between the binding affinity to the isolated N-terminal domain of human Hsp90 and cytostatic activity. It suggests that binding to the isolated N-terminal domain is necessary but not sufficient for inhibition activity [223]. GA presents a potent antitumor activity in more than 50 cell lines. The unchaperoned client proteins are subsequently degraded by proteasome-E3 ligase resulting in inhibition of cell proliferation and induction of apoptosis [224]. However, GA was never evaluated in clinical trials because of its poor solubility, limited *in vivo* stability and significant hepatotoxicity in animals. In the last decade, many research projects have been carried out to discover new inhibitors of Hsp90, advancing from first-generation derivatives of natural products to second and third-generation fully synthetic small molecules. The structure of GA comprises a quinone ring moiety that is the responsible for the observed hepatotoxicity, along with a pendant macrocycle that contains a carbamate group necessary for binding. The substitution of the methoxy group on C-17 of the quinone ring by amines has been a very useful approach to generate less toxic compounds [225]. In this sense, less toxic agents than GA, namely, 17-AAG (17-allylamino-17-demethoxy-geldanamycin) and 17-DMAG (17-dimethylaminoethylamino-17-demethoxy-geldanamycin) have been synthesized and preceded to clinical trials. 17-AAG (Tanespimycin) was the first Hsp90 inhibitor to be tested in clinical trials. GA and its derivatives have been reported to possess various pharmacological properties as anti-tumoral properties, inhibition of angiogenesis and metastasis of diseases such as multiple myeloma, as well as breast or prostate cancer [225]. 17-AAG showed important clinical limitations due to its modest bioavailability, instability and toxicity, in addition to the bone loss by increasing osteoclast formation [226]. In order to avoid toxicity some nanoparticle formulations were developed as nanoparticle albumin-bound, a β-cyclodextrin-17AAG complex, polymeric micelles carriers, PEGylated nanostructured lipid carriers, poly (lactic-co-glycolic acid) coated, 17AAG and Fe3O4 loaded magnetic nanoparticle formulations, among others. All these nanoformulations were tested with limited application in treatment of diverse type of tumors. Then, given apparent toxicity of 17-AAG, 17-DMAG (Alvespimycin) and IPI-504 (Retaspimycin) were developed as water-soluble analogues. 17-DMAG is the product of the substitution of the C-17 methoxy group (of GA) with N, N-dimethylethylamine. This substitution produced a compound with higher water solubility and improved oral bioavailability than 17-AAG but with conserved or superior anti-tumor activity [227]. Although 17-DMAG showed longer plasma half-life, better distribution to tissues and less extensive metabolism than 17-AAG, Phase I clinical trials in 2005 demonstrated that patients suffered from peripheral neuropathy, renal dysfunction, fatigue, cardiac problems, ocular adverse events comprised of blurred vision, keratitis, dry eyes. Other symptoms as pneumonitis with dyspnea and thrombocytopenia, anorexia, proteinuria and peripheral edema have also been observed so that the clinical development of 17-DMAG was discontinued in 2008. IPI-504 represents the highly water-soluble hydroquinone hydrochloride salt derivative of 17-AAG [228, 229]. Nevertheless, *in vivo* IPI-504 and 17-AAG coexist in redox equilibrium between the hydroquinone (IPI-504) and quinone (17-AAG) forms because of the action of oxidoreductases. The hydroquinone form in IPI-504 is a more potent inhibitor of Hsp90 and is likely the more relevant form *in vivo,* and due to this chemical group it is expectable that presents less hepatotoxicity. IPI-504 has been found to be a potent destabilizing agent of its client proteins along with other key signaling proteins essentially involved in the progression and development of tumor growth [230]. Various clinical trials have been conducted in patients with multiple myeloma, non-small cell lung cancer, castrate resistant prostate cancer, soft tissue sarcomas, breast cancer and gastrointestinal stromal tumor, among others. However, the IPI-504 administration caused toxic effects as renal failure, liver failure, metabolic acidosis, and cardiopulmonary arrest. This fact, in addition to its capacity to interconvert with the known agent 17-AAG via oxidation-reduction equilibrium as shown *in vitro* and *in vivo* experiments [230], the medical interest in this compound was lost. The toxicity and adverse effects found in GA and it derivatives led to the subsequent development of synthetic Hsp90 inhibitors devoid of some of those unfavorable properties, such as liver toxicity, attributed to quinone metabolism, or other toxic effects that have severely limited the ability to administer a tolerable dose with therapeutic benefits. Resorcinol derivatives have been on the basis of RD that is a resorcyclic acid lactone which was first isolated in 1953 from the fungus *Monosporium bonorden*. Resorcinol derivatives are not structurally related to GA; however, they also inhibit Hsp90 by binding to the ATP-binding pocket of its NTD. Resorcinol derivatives are used in clinical trials but not RD because it does not exhibit *in vivo* activity, probably because of the high instability of its reactive epoxide moiety. But, its resorcinol core, which is crucial for binding, provided the chemical basis for several drugs that subsequently entered to clinical development, including STA-9090, NVP-AUY922, AT-13387, and KW-2478. STA-9090 is a resorcinol-containing triazole that has shown greater potency, improved tumor penetration, and a more favorable toxicity profile than 17-AAG in preclinical models. Overall, STA-9090 seems to be well tolerated, with no reports of severe ocular, cardiac, liver, or renal toxic effects. Positive responses were observed in patients with rectal carcinoma, melanoma, and other types of carcinomas. NVP-AUY922 (VER52296) is an isoxazole derivative, a second generation resorcinol derivative, initially developed by optimization of a lead compound (CCT018159). It was selected by high-throughput screening for its potent inhibition of Hsp90. The first clinical trials comprised the treatment of solid tumors via intravenous administration. The agent was well tolerated with some adverse effects as atrial flutter, diarrhea, fatigue, darkening of vision and anorexia. Other common adverse events included nausea, vomiting, and night blindness [231]. Further studies are planned to confirm high activity showed in treated patients. At the same time, AUY922 is being studied in combination with trastuzumab in HER2+ breast and gastric cancer, with cetuximab in KRAS-wild-type colorectal cancer, and with capecitabine in solid tumors. A Phase I-II study of AUY922 in combination with trastuzumab in patients with HER2-positive breast cancer concluded that full doses of both drugs were tolerable. AT-13387 was discovered after optimization of a resorcinol-derivative synthesized through a fragment-based drug discovery approach.

Common toxic effects and a durable partial response were noted in the patients in Phase I clinical trials. In combination with imatinib, AT-13387 showed good tolerance, it exhibited limited antitumor activity in the treatment of gastrointestinal stromal tumor. KW-2478 is another novel resorcinol derivative that was discovered through a singular lead-optimization strategy that included microbial screening, X-ray crystallography, cell-based screening, and *in-vivo* models [232]. It exhibits a real high affinity value to Hsp90 ($K_D$=3.8 nM). A Phase I study in patients with relapsed/refractory multiple myeloma, chronic lymphocytic leukemia or B-cell non-Hodgkin's lymphoma showed that this drug presents low toxicity. Side effects are similar to those described by other resorcinol derivatives as NVP-AUY922, though there are currently no clinical trials in process or recruitment stage. New development in resorcinol derivatives as the Schiff bases derived from 2,4-dihydroxybenzaldehyde and 5-chloro-2-4-dihydroxybenzaldehyde [233], Mannich bases derived from resacetophenone and 4-chloro resacetophenone compounds [234] among others, were also reported but are not yet being tested in clinical trials. The availability of X-ray crystal structures of Hsp90 bound to ATP, ADP, and inhibitors such as GA and RD enabled the development of purine or purine-like synthetic analogues with Hsp90 inhibition properties. Since the structure of ATP includes adenine, a purine base, Hsp90 inhibitors based on a purine structure showed a possibility to block ATP binding to Hsp0, and inhibit Hsp90 function. In order to synthesize compounds specific to Hsp90, Chiosis *et al.* [235] carefully examined the Hsp90 ATP binding pocket, and realized that an inhibitor would have to satisfy several hydrophilic and lipophilic interactions at this site, while retaining higher affinity for the pocket than ADP [235]. The comprehension about the differences between the ATP pocket of Hsp90 and homologues pockets in other proteins led to the authors to design the first reported synthetic Hsp90 inhibitor, PU3 by a strategy based on structure-activity relationship. X-ray structure of the Hsp90-PU3 complex showed that the ligand lies in the ADP binding site with essentially no change in conformation across most of the protein structure. Besides, PU3 was more soluble than 17-AAG but less potent than GA against tumor cell lines *in vitro*. This compound was in consequence optimized to impact selective molecules with higher potency and enhanced pharmacological properties. This optimization resulted in the discovery of PU-H58, which demonstrated higher potency than the original compound, and PU24F-Cl that showed an affinity 30 times greater than PU3 for the N-terminus of Hsp90 [236]. Purine derivatives tested in advanced clinical trials comprise CNF2024/BIIB021, MPC-3100, and PU-H71 in conjunction with the purine-like Debio 0932 (CUDC-305). BIIB021 was synthesized by the attachment of an aromatic group to the 9-position of the purine [237] being exceptional amongst this class of inhibitors. But, in order to conserve the critical distance of 5 Å, the -NH2 group was shifted to the 2-position. BIIB021 is the first synthetic Hsp90 inhibitor to start clinical trials. It was tested in patients with chronic lymphocytic leukemia, with advanced solid tumors and with gastrointestinal stromal tumor. Adverse effects as syncope and dizziness in patients with advanced solid tumors and other toxicities as fatigue, hyponatremia and hypoglycemia, and abnormal liver enzymes was reported in patients with chronic lymphocytic leukemia. An apparently more potent version of BIIB028 was synthesized and evaluated in clinical trials for refractory metastatic or locally advanced solid tumors. Although, BIIB028 was a well-tolerated molecule that showed a relative success in some patients, no more publications have been reported about this drug. MPC-3100 is the product of a conducted optimization studies centered on the purine-based Hsp90 inhibitor 28a that contains a piperidine moiety at the purine N9 position. This drug exhibited good *in vitro* profiles as well as a characteristic molecular biomarker signature of Hsp90 inhibition and the Hsp70 mRNA induction, however, due to the poor solubility and bioavailability of MPC-3100, researchers proceeded to introduce a pro-drug, MPC-0767. It is a novel L-alanine ester pro-drug of MPC-3100, designed to have improved aqueous solubility compared to MPC-3100, but no clinical trials were reported [238]. Moreover, Debio 0932 was generated replacing the N3 of the purine on PU3 with a carbon atom. Debio 0932 displays more favorable pharmacologic properties, including high oral bioavailability, sustained tumor retention, blood-brain barrier penetration, and potentially a better therapeutic window, compared with other Hsp90 inhibitors. It exhibits potent antitumor activity against a variety of tumor types *in vitro* and *in vivo*, including epidermal growth factor receptor inhibitor-resistant non-small cell lung cancer, glioblastoma, triple-negative breast cancer, and acute myelogenous leukemia. Debio 0932 displays some adverse effects as febrile neutropenia, diarrhea, asthenia gastrointestinal events, although no ocular toxicities were observed, due to these effects clinical trials were discontinued. PU-H7, another purine like Hsp90 inhibitors, possess a iodide functional group, the positron emission tomography (PET) radionuclide $^{124}$I can be inserted to produce the imaging agent $^{124}$I-PU-H71. This radioactive PU-H71 derivative allows serial imaging monitoring tumor PU-H71 concentrations for multiple days [223]. PU-H71 was evaluated in a Phase I clinical trial in patients with low-grade non Hodgkin's lymphoma and solid tumors. In combination with radiotherapy it has been tested *in vitro* for human lung cancer cell lines showing synergic effects. Results from different approaches indicate that PU-H71 as Hsp90 inhibitor may be successfully utilized in the treatment of radioresistant carcinomas [240]. Currently and in the last decades, the use of inhibitors of the N-terminal region of Hsp90 has grown in interest both academically and patent literature. Second and third generation inhibitors are constantly being discovered and developed.

## Inhibitors of the C-terminal Domain of Hsp90

As was described above, the molecule of Hsp90 is composed by different domains. In addition to the NTD other areas of the molecule were intended as target inhibition, such is the case of the dimerization domain of Hsp90 that is located within the CTD. In 2000, Neckers and coworkers [241] discovered that novobiocin (NB), a coumarin antibiotic, bound to a previously unknown Hsp90 C-terminal nucleotide binding site induced the degradation of Hsp90-dependent client proteins. Due to the poor affinity constant of NB to Hsp90 (approximately 700 μM), researchers paid very little attention toward the synthesis of NB analogues to perform Hsp90 inhibition [242]. Blagg and colleagues prepared a parallel library of 20 NB derivatives then, they assessed the biological activity of each compound by Western blot analysis of Hsp90 client proteins. In this approach the A4 agent was discovered to be a strong inhibitor of Hsp90. This property was inferred due to its ability to cause the degradation of several Hsp90 client proteins in both breast and prostate cancer cell lines. As an example, in the presence of 1 μM A4, several Hsp90 client proteins were degraded, including Akt, HER2, Hif-1α, and AR [242]. Interestingly, it has been demonstrated that the C-terminal nucleotide binding pocket in Hsp90 structure not only binds ATP, but cisplatin, NB, epilgallocatechin-3-gallate (EGCG) and taxol, among other Hsp90 inhibitors. The coumarin antibiotics NB, clorobiocin, and coumermycin A1, isolated from several *Streptomyces* strains, show potent activity against Gram-positive bacteria. Another function performed by all these compounds consist to inhibit ATPases and to interact with type II topoisomerases, including the DNA gyrase. Nevertheless, it is remarkable that structural modification of NB produced an improvement in anti-proliferative activity in about 1000-fold. In this way, new more powerful derivatives can be developed.

At the same time, another compound as cisplatin, a platinum-containing chemotherapeutic utilized to treat various cancer types, including testicular, ovarian, bladder, and small cell lung cancer was studied. Notably, cisplatin, due to the Pt element in its structure, coordinates to DNA bases, resulting in cross-linked DNA. This chemical property prevents cells division because interferes in DNA duplication during mitosis. Further, in 1999 Itoh and coworkers reported that cisplatin reduces the chaperone activity of Hsp90 [243]. These researchers, making several types of experiments, pointed out that cisplatin exhibits high affinity for Hsp90. Moreover, Csermely and co-workers demonstrated that the cisplatin binding site is located proximal to the C-terminal ATP binding site [244]. Nevertheless, it has also been shown that due to its chemical reactivity, cisplatin interacts with various proteins, phospholipids, and RNA [245]. Rosenhagen and colleagues [246] by use of a heat-shock factor (HSF)-dependent luciferase reporter assay, showed that cisplatin does not induce the heat-shock response. These results suggest that unlike compounds that bind Hsp90, cisplatin selectively inhibits some Hsp90 functions and thus, could be useful to elucidate novel ways to regulate its chaperone activity. On the other hand, LA-12 obtained as an optimized derivative of cisplatin, exhibits higher affinity for Hsp90 than cisplatin with a more benign pharmacokinetic and intensified cytotoxicity against various cancer cell lines, including those that are cisplatin resistant. These effects are caused as a result of the degradation of additional Hsp90 client proteins, such as mutant p53, Cyclin D1, and ER by LA-12 induction [247]. Another C-terminal Hsp90 inhibitor, epilgallocatechin-3-gallate (EGCG) does not appear to have the ability to prevent Hsp90 from forming multiprotein complexes. EGCG is a polyphenolic compound, the major component of green tea, it is well-recognized for its antioxidant, antimicrobial, and anticancer properties [248]. Gasiewicz and coworkers demonstrated that EGCG binds the same amino acids (538–728) as NB [249]. But, contrary to other inhibitors, it stabilizes the interaction of cochaperones Hsp70, Cyp40, and XAP-2 to Hsp90 [249]. Besides, it was demonstrated that EGCG induce the degradation of several Hsp90-dependen- -oncoprotein as ErbB2, Raf-1, and pAkt as well as a slightly increase in Hsp70 concentration. Some efforts to improve the EGCG inhibitory activity have been done in the last years, but further research about the interaction between ECGC and Hsp90 must be carried out.

Taxol, is a compound that is frequently used for the treatment of various cancers, this property has been attributed to the inhibition of mitosis via stabilization of microtubules by means binding to tubulin [250]. Previously, taxol was characterized as inductor of the activation of kinases and transcription factors, besides of mimetizing the effect of bacterial lipopolysaccharide (LPS). Byrd and coworkers, performing affinity purification experiments with biotinylated taxol, have identified Hsp90 and Hsp70, as potential mediators of its LPS-mimicking activity [251]. Nevertheless, taxol seems to stimulate Hsp90 functions rather than to inhibit them. Currently, taxol is defined as an Hsp90 inhibitor that interacts with the C-terminal region of the chaperone though, no structural study describing the complex between taxol and Hsp90 has been reported. Again, further investigations about structural basis of the Hsp90 C-terminal inhibition are needed to have a more real and complete idea about the effect of the inhibitors described above.

## *Inhibitors of the Hsp90-TPR Interactions*

Therefore, it is important the detailed study and the understanding about the Hsp90 machinery, since every day more diseases associated with such machinery are finding. In this way, it is necessary to design new drugs to treat such diseases following new strategies. The deep knowledge of the Hsp90-TPR proteins interaction is a requisite to design an inhibitor that interferes with this interaction in cancer cells dependent of steroid hormones. As it was expressed above, the CTD (11-15 kDa), is the dimerization site for Hsp90 and contains a MEEVD sequence. Of the twenty six client proteins and cochaperones recognized to interact with the C-terminus of Hsp90, 19 contain a TPR region in their structure, where this TPR domain binds to the MEEVD region of Hsp90. The TPR-containing cochaperones, including Hop/Sti1, immunophilins (FKBP51, FKBP52, and Cyp40), PP5/Ppt1, and CHIP have specific functional domains involved in the regulation of Hsp90 client proteins or of the Hsp90 chaperone machinery itself. Particularly, Hsp90-CyP40 and Hsp90-FKBP51/FKBP52 interactions modulate hormone receptor activity; Hsp90-Hop regulates protein folding; and Hsp90-CHIP controls protein degradation. Some examples about the importance of interaction Hsp90-TPR proteins can be considered. FKBP38 and FKBP52 are immunophilins that both interact with the CTD of Hsp90 via the MEEVD-TPR interaction. When free in the cytosol, FKBP38 prevents Bcl-2 from inducing anti-apoptotic effects. Consequently, inhibiting the interaction of FKBP38 with Hsp90 presumably apoptosis via the Bcl-2 pathway is induced [252]. As was described before FKBP52 is an important immunosuppressant, it is involved in the intracellular trafficking of SHRs and also in its regulation via a TPR-C-terminal Hsp90 interaction. Another example to mention is Hop that possesses 3 TPR binding domains in its structure containing one binding site to the MEEVD region of Hsp90. At the same time, Hop is also responsible for docking Hsp70 to Hsp90 and affording the transfer of the newly formed polypeptides from Hsp70 to Hsp90. The inhibition of the Hsp90 C-terminal region has been reveled as a promising strategy for regulating Hsp90 activity to replace the old strategies of inhibition. A recent work has shown that molecules that inhibit the C-terminus of Hsp90 do not produce the heat shock response, and act via a distinct mechanism than those employed for the N-terminal inhibitors [253]. Some evidence shows that inhibiting cochaperone binding at C-terminus of Hsp90 actively decreases the cell protection mechanisms, e.g. the heat shock response, opposed to the classical inhibitors that stimulate the heat shock response.

In this sense, Kawakami and coworkers [254], focusing on the interaction between Hsp90 and its cochaperone p60/Hop, designed a hybrid Antp-TPR peptide that consist of a cell permeable peptidomimetic and was created from the modeling of the interaction between Hsp90-peptide and TPR2A (223K-352L) domain of Hop. This protein contains three TPR domains (see above), where TPR2A interacts with the MEEVD region on Hsp90. The designed "TPR peptide" contains a 12- amino acid sequence and was based on the α-helix 3 of TPR2A what it means the fragment from Lys-301 to Lys-312. The named "TPR peptide" was further attached by its N-terminus to α-helix 3 of the *Antennapedia* homeodomain protein to design a cell-permeable hybrid Antp-TPR peptide. By means biosensor SPR assay the authors demonstrated that both Hsp90 and Hsp70 associate to the immobilized TPR peptide with similar $K_D$ values: 1.42 x $10^{-6}$ M and 0.68 x $10^{-6}$ M at increasing ligand concentrations, respectively. Furthermore, the $K_D$ value of the interaction of Hsp90 with Hop was also similar (4.43 x $10^{-6}$ M). TPR peptide inhibited the interaction between Hsp90 and Hop protein, but neither the association of Hsp70 with Hop nor that of Hsp90 with FKBP5 or PP5 proteins. It is interesting to note that Hop, FKBP5 and PP5 are all TPR proteins provided with a binding site for the C-terminal region of Hsp90 as described previously [29]. Kawakami and colleagues demonstrated that the designed hybrid Antp-TPR peptide was effective in provoking cell death of breast, pancreatic, renal, lung, prostate, and gastric cancer cell lines *in vitro*. On the contrary, Antp-TPR peptide did not produce any change on the viability of normal cells. Moreover, analysis *in vivo* demonstrated that Antp-TPR peptide exhibited an appreciable antitumor activity in a xenograft model of human pancreatic cancer in mice [254]. Thus, Antp-TPR hybrid peptide had selective cytotoxic activity discriminating between normal and cancer cells to produce cancer cell death. Further studies with this peptide revealed that Antp-TPR peptide provoked acute myeloid leukemia cell death in cell lines such as U937, K562, THP-1, and HL-60 via activation of caspases 3 and 7, and interference of mitochondrial membrane potential. Experiments with peripheral blood mononuclear cells (PBMCs) demonstrated that the Antp-TPR peptide does not decrease the viability of these cells, whereas both GA and 17-AAG do so at low concentrations. In addition, mutation analysis of TPR peptide revealed that the highly conserved amino acids Lys 301 and Arg 305 in the α-helix 3 were especially essential to the cytotoxic activity due to their respective binding by hydrogen bonding with the side chains of the Asp and Glu amino acids from the Hsp90 C-terminal peptide [29]. These results showed that Antp-TPR hybrid peptide would provide an effective and selective therapeutic option for the treatment of acute myeloid leukemia [255]. In addition, the authors evaluated the inhibitory power of Antp-TPR hybrid peptide in glioblastoma cell lines (U251, A172, and SN19) because it is assumed that glioblastoma is one of the most malignant cancers with a worse prognosis. They found that the addition of Antp-TPR peptide to glioblastoma cells resulted in concentration-dependent cytotoxicity, and at 50 μM all the cells tested lost their viability. That lethal action on glioblastoma cells was through the loss of Hsp90 client proteins such as p53, Akt, CDK4, and cRaf. It should be noted that Antp-TPR did not induce the up-regulation of Hsp70 and Hsp90 proteins, the contrary effect of 17-AAG, the Hsp90 N-terminal inhibitor.

In 2014 Kawakami and colleagues to further improve the cytotoxic activity of Antp-TPR toward cancer cells, investigated the effect of an Hsp70-targeted peptide (R11-Hsp70). The cell-permeability property was given by the addition of the polyarginine with a linker sequence. Then, the action of this construct over the cytotoxic activity of Antp-TPR in breast cancer cell lines was investigated. Antp-TPR in the presence of Hsp70-targeted peptide provoked effective cytotoxic activity toward breast cancer cells due to the decrease of Hsp90 client proteins such as p53, Akt, and cRaf. Again, the combination of Antp-TPR peptide with Hsp70-targeted peptide did not produce the up-regulation of Hsp70 protein as was revealed by western blotting [256].

It is well-known that in the cells, Hsp70 interacts with a great variety of hydrophobic polypeptide sequences in recently synthesized protein and partially folded substrates. Hsp70 plays important roles to collaborate for protein folding of client proteins which, as was revised in previous paragraphs, those roles are vital for cancer cell growth, especially in the context of Hsp70/Hsp90 multi-chaperon system. The researchers hypothesized that the inhibition of Hsp70 might contribute to the increase to cytotoxic activity via the inhibition of Hsp90 even in Hsp70-low expressing cancer cells. Then, they suggested that the cellular uptake of Antp-TPR would be enhanced by R11-Hsp70 and the combination of both strategies would be successful due to the effective decrease of Hsp90 client proteins such as Akt, p53, and cRaf in the cytosol.

Besides, it is believed that the combinatory treatment of Antp-TPR with R11-Hsp70 may create a change of environment and protein homeostasis in cancer cells. On the other hand, Kawakami and co-workers also found that combinatory treatment or with Antp-TPR and Hsp70-targeted peptide separately, did not produce an increase in the glutathione concentrations in the cancer cells. The strategy of targeting both Hsp90 and Hsp70 with Antp-TPR and Hsp70-targeted peptide may be very promising as a selective strategy to increase cytotoxicity in cancer cells [256].

Another type of inhibitors also called C-terminal inhibitors as San A derivatives [253] act allosterically by binding to the ND-MD of Hsp90 and impeding the access of cochaperones to C-terminus of Hsp90. This structural strategy, in designing molecules that produce steric hindrance is highly challenging since it is not easy to predict structure-activity relationship as was demonstrated in the work using San A derivatives. But, at this time, in McAlpine laboratory, molecules that directly block access to the C-terminus of Hsp90 (MEEVD region) were developed [257]. In order to improve Kawakami's laboratory development, McAlpine and coworkers designed and synthesized molecules that mimic the TPR2A domain to interfere in the binding between the C-terminus of Hsp90 and TPR-containing cochaperones. They synthesized short amino acid variants from the 12-amino acid peptide previously designed by Kawakami *et al.* Then, the authors assessed the capability of these compounds to hinder the interaction of Hsp90 with CyP40 by means a binding assay. Also, the binding to the MEEVD region was visualized by NMR technique and the repercussion in Hsp90-mediated protein folding was evaluated by luciferase renaturation assay. Inspired by C-terminus of the 12-amino acid fragment McAlpine and colleagues constructed four linear molecules of five, six, seven or eight amino acids in length. Protein-binding assays were performed to assess the capability of 17 potential inhibitors of the Hsp90-CyP40 binding, the screening was made using both Hsp90α and Hsp90β and the totality of the compounds was compared to NB (C-terminal Hsp90 inhibitor). It is important to remember that TPR-containing cochaperones do not discriminate between the two Hsp90 isoforms and bind to Hsp90α and Hsp90β with the same affinity. On the contrary, compounds with binding ability to Hsp90 generally discriminate between Hsp90α and Hsp90β due to their difference in protein sequence. It should be remembered that the sequence alignment of the C-terminal region from Hsp90α and Hsp90β (used in these binding assays) displays 85% similarity.

Thus, if the compounds interact with CyP40, they should possess the same IC50 values when assayed with Hsp90α or Hsp90β but, if they bind to Hsp90 it is likely they present different IC50 values for each isoform. NB, as was published by other authors, presents a distinct IC50 value for each isoform, what it means that can discriminate between isoforms; even more NB displays a 10 fold lower IC50 for Hsp90α than for Hsp90β. On the other hand, the compounds 7.1 CYC and 8.1 CYC also exhibit isoform selectivity, considering that both by preference bind to Hsp90β. Furthermore, the authors have shown that 5.1 CYC, a small molecule containing only five amino acids exhibited remarkable inhibition (IC50 binding ~4 mM) and was adequately effective to inhibit the Cyp40-Hsp90β interaction. The NMR structure of 5.1 CYC provides the basis for the design of new potent compounds to inhibit the C-terminus of Hsp90.

## FINAL CONSIDERATIONS

Hsp90 and Hsp70 chaperones constitute a complex machinery to carry out the maintenance of folding and integrity of both the newly synthesized peptides and the mature and functional proteins. The great quantity of papers about this subject have been published in last decades, however a lot of work have to be done to elucidate several structural aspects, still unknown, to explain differential behavior of chaperones and cochaperones in the stabilization of client proteins. At the same time, great effort has been made to find the "right" inhibitor that may preferentially inhibit the folding machine of the tumor cells. Many researchers have followed the "easy way" focusing mainly on the ATP binding pocket at the N-terminal of Hsp90 however; their attempts have not been successful so far. The contents presented in this chapter aim to invite scientists to explore different strategies to design inhibitors used in the treatment of cancer and other diseases.

Those new strategies should be elaborated taking into account all the structural complexity of the Hsp90/Hsp70 folding machine. In this sense, new technologies as well as improved old techniques should be applied to follow a valid approach in the inhibition of cancer cells without disturbing the homeostasis of normal cells.

## CONSENT FOR PUBLICATION

Not applicable.

## CONFLICT OF INTEREST

The author declares no conflict of interest, financial or otherwise.

## ACKNOWLEDGEMENTS

The authors acknowledge financial support of The University of Buenos Aires (UBACYT 2014-2017), The Argentine Agency for Science and Technology (PICT 2014-3433), and The Argentine Institute of Cancer (INC2016/17).

## REFERENCES

[1] Brocard C, Hartig A. Peroxisome targeting signal 1: is it really a simple tripeptide? Biochim Biophys Acta 2006; 1763: 1565-73.
[http://dx.doi.org/10.1016/j.bbamcr.2006.08.022] [PMID: 17007944]

[2] Zeytuni N, Ozyamak E, Ben-Harush K, *et al.* Self-recognition mechanism of MamA, a magnetosome-associated TPR-containing protein, promotes complex assembly. Proc Natl Acad Sci USA 2011; 108(33): E480-7.
[http://dx.doi.org/10.1073/pnas.1103367108] [PMID: 21784982]

[3] Zeytuni N, Zarivach R. Structural and functional discussion of the tetra-trico-peptide repeat, a protein interaction module. Structure 2012; 20(3): 397-405.
[http://dx.doi.org/10.1016/j.str.2012.01.006] [PMID: 22404999]

[4] D'Andrea LD, Regan L. TPR proteins: the versatile helix. Trends Biochem Sci 2003; 28(12): 655-62.
[http://dx.doi.org/10.1016/j.tibs.2003.10.007] [PMID: 14659697]

[5] Pratt WB, Toft DO. Regulation of signaling protein function and trafficking by the hsp90/hsp70-based chaperone machinery. Exp Biol Med (Maywood) 2003; 228(2): 111-33.
[http://dx.doi.org/10.1177/153537020322800201] [PMID: 12563018]

[6] Hirano T, Kinoshita N, Morikawa K, Yanagida M. Snap helix with knob and hole: essential repeats in S. pombe nuclear protein nuc2+. Cell 1990; 60(2): 319-28.
[http://dx.doi.org/10.1016/0092-8674(90)90746-2] [PMID: 2297790]

[7] Sikorski RS, Boguski MS, Goebl M, Hieter P. A repeating amino acid motif in CDC23 defines a family of proteins and a new relationship among genes required for mitosis and RNA synthesis. Cell 1990; 60(2): 307-17.
[http://dx.doi.org/10.1016/0092-8674(90)90745-Z] [PMID: 2404612]

[8] Goebl M, Yanagida M. The TPR snap helix: a novel protein repeat motif from mitosis to transcription. Trends Biochem Sci 1991; 16(5): 173-7.
[http://dx.doi.org/10.1016/0968-0004(91)90070-C] [PMID: 1882418]

[9] Lamb JR, Tugendreich S, Hieter P. Tetratrico peptide repeat interactions: to TPR or not to TPR? Trends Biochem Sci 1995; 20(7): 257-9. [http://dx.doi.org/10.1016/S0968-0004(00)89037-4] [PMID: 7667876]

[10] Ponting CP. Novel domains in NADPH oxidase subunits, sorting nexins, and PtdIns 3-kinases: binding partners of SH3 domains? Protein Sci 1996; 5(11): 2353-7. [http://dx.doi.org/10.1002/pro.5560051122] [PMID: 8931154]

[11] Callahan MA, Handley MA, Lee Y-H, Talbot KJ, Harper JW, Panganiban AT. Functional interaction of human immunodeficiency virus type 1 Vpu and Gag with a novel member of the tetratricopeptide repeat protein family. J Virol 1998; 72(6): 5189-97. [PMID: 9573291]

[12] Cziepluch C, Kordes E, Poirey R, Grewenig A, Rommelaere J, Jauniaux JC. Identification of a novel cellular TPR-containing protein, SGT, that interacts with the nonstructural protein NS1 of parvovirus H-1. J Virol 1998; 72(5): 4149-56. [PMID: 9557704]

[13] Mamane Y, Sharma S, Petropoulos L, Lin R, Hiscott J. Posttranslational regulation of IRF-4 activity by the immunophilin FKBP52. Immunity 2000; 12(2): 129-40. [http://dx.doi.org/10.1016/S1074-7613(00)80166-1] [PMID: 10714679]

[14] Blatch GL, Lässle M. The tetratricopeptide repeat: a structural motif mediating protein-protein interactions. BioEssays 1999; 21(11): 932-9. [http://dx.doi.org/10.1002/(SICI)1521-1878(199911)21:11<932::AID-BIES5>3.0.CO;2-N] [PMID: 10517866]

[15] Gatto GJ Jr, Geisbrecht BV, Gould SJ, Berg JM. A proposed model for the PEX5-peroxisomal targeting signal-1 recognition complex. Proteins 2000; 38(3): 241-6. [http://dx.doi.org/10.1002/(SICI)1097-0134(20000215)38:3<241::AID-PROT1>3.0.CO;2-1] [PMID: 10713985]

[16] Lapouge K, Smith SJ, Walker PA, Gamblin SJ, Smerdon SJ, Rittinger K. Structure of the TPR domain of p67phox in complex with Rac.GTP. Mol Cell 2000; 6(4): 899-907. [http://dx.doi.org/10.1016/S1097-2765(05)00091-2] [PMID: 11090627]

[17] Main ER, Jackson SE, Regan L. The folding and design of repeat proteins: reaching a consensus. Curr Opin Struct Biol 2003; 13(4): 482-9. [http://dx.doi.org/10.1016/S0959-440X(03)00105-2] [PMID: 12948778]

[18] Main ER, Xiong Y, Cocco MJ, D'Andrea L, Regan L. Design of stable alpha-helical arrays from an idealized TPR motif. Structure 2003; 11(5): 497-508. [http://dx.doi.org/10.1016/S0969-2126(03)00076-5] [PMID: 12737816]

[19] Das AK, Cohen PW, Barford D. The structure of the tetratricopeptide repeats of protein phosphatase 5: implications for TPR-mediated protein-protein interactions. EMBO J 1998; 17(5): 1192-9. [http://dx.doi.org/10.1093/emboj/17.5.1192] [PMID: 9482716]

[20] Schultz J, Milpetz F, Bork P, Ponting CP. SMART, a simple modular architecture research tool: identification of signaling domains. Proc Natl Acad Sci USA 1998; 95(11): 5857-64. [http://dx.doi.org/10.1073/pnas.95.11.5857] [PMID: 9600884]

[21] Letunic I, Doerks T, Bork P. SMART 7: recent updates to the protein domain annotation resource. Nucleic Acids Res 2012; 40(Database issue): D302-5. [http://dx.doi.org/10.1093/nar/gkr931] [PMID: 22053084]

[22] Gasteiger E, Gattiker A, Hoogland C, Ivanyi I, Appel RD, Bairoch A. ExPASy: The proteomics server for in-depth protein knowledge and analysis. Nucleic Acids Res 2003; 31(13): 3784-8. [http://dx.doi.org/10.1093/nar/gkg563] [PMID: 12824418]

[23] Biegert A, Mayer C, Remmert M, Söding J, Lupas AN. The MPI Bioinformatics Toolkit for protein sequence analysis. Nucleic Acids Res 2006; 34(Web Server issue): W335-9.

[http://dx.doi.org/10.1093/nar/gkl217]

[24] Andrade MA, Perez-Iratxeta C, Ponting CP. Protein repeats: structures, functions, and evolution. J Struct Biol 2001; 134(2-3): 117-31. [http://dx.doi.org/10.1006/jsbi.2001.4392] [PMID: 11551174]

[25] Main ER, Lowe AR, Mochrie SG, Jackson SE, Regan L. A recurring theme in protein engineering: the design, stability and folding of repeat proteins. Curr Opin Struct Biol 2005; 15(4): 464-71. [http://dx.doi.org/10.1016/j.sbi.2005.07.003] [PMID: 16043339]

[26] Liu D, Bienkowska J, Petosa C, Collier RJ, Fu H, Liddington R. Crystal structure of the zeta isoform of the 14-3-3 protein. Nature 1995; 376(6536): 191-4. [http://dx.doi.org/10.1038/376191a0] [PMID: 7603574]

[27] Xiao B, Smerdon SJ, Jones DH, *et al.* Structure of a 14-3-3 protein and implications for coordination of multiple signalling pathways. Nature 1995; 376(6536): 188-91. [http://dx.doi.org/10.1038/376188a0] [PMID: 7603573]

[28] Itzhaki LS, Lowe AR. From artificial antibodies to nanosprings: the biophysical properties of repeat proteins. Adv Exp Med Biol 2012; 747: 153-66. [http://dx.doi.org/10.1007/978-1-4614-3229-6_10] [PMID: 22949117]

[29] Scheufler C, Brinker A, Bourenkov G, *et al.* Structure of TPR domain-peptide complexes: critical elements in the assembly of the Hsp70-Hsp90 multichaperone machine. Cell 2000; 101(2): 199-210. [http://dx.doi.org/10.1016/S0092-8674(00)80830-2] [PMID: 10786835]

[30] Bahadur RP, Zacharias M. The interface of protein-protein complexes: analysis of contacts and prediction of interactions. Cell Mol Life Sci 2008; 65(7-8): 1059-72. [http://dx.doi.org/10.1007/s00018-007-7451-x] [PMID: 18080088]

[31] Kim K, Oh J, Han D, Kim EE, Lee B, Kim Y. Crystal structure of PilF: functional implication in the type 4 pilus biogenesis in Pseudomonas aeruginosa. Biochem Biophys Res Commun 2006; 340(4): 1028-38. [http://dx.doi.org/10.1016/j.bbrc.2005.12.108] [PMID: 16403447]

[32] Kloss E, Courtemanche N, Barrick D. Repeat-protein folding: new insights into origins of cooperativity, stability, and topology. Arch Biochem Biophys 2008; 469(1): 83-99. [http://dx.doi.org/10.1016/j.abb.2007.08.034] [PMID: 17963718]

[33] Jackson SE. How do small single-domain proteins fold? Fold Des 1998; 3(4): R81-91. [http://dx.doi.org/10.1016/S1359-0278(98)00033-9] [PMID: 9710577]

[34] Lane TJ, Schwantes CR, Beauchamp KA, Pande VS. Probing the origins of two-state folding. J Chem Phys 2013; 139(14): 145104. [http://dx.doi.org/10.1063/1.4823502] [PMID: 24116650]

[35] Zweifel ME, Barrick D. Studies of the ankyrin repeats of the Drosophila melanogaster Notch receptor. 2. Solution stability and cooperativity of unfolding. Biochemistry 2001; 40(48): 14357-67. [http://dx.doi.org/10.1021/bi011436+] [PMID: 11724547]

[36] Cortajarena AL, Regan L. Calorimetric study of a series of designed repeat proteins: modular structure and modular folding. Protein Sci 2011; 20(2): 336-40. [http://dx.doi.org/10.1002/pro.564] [PMID: 21280125]

[37] Kajander T, Cortajarena AL, Main ER, Mochrie SG, Regan L. A new folding paradigm for repeat proteins. J Am Chem Soc 2005; 127(29): 10188-90. [http://dx.doi.org/10.1021/ja0524494] [PMID: 16028928]

[38] Faccin M, Bruscolini P, Pelizzola A. Analysis of the equilibrium and kinetics of the ankyrin repeat protein myotrophin. J Chem Phys 2011; 134(7): 075102. [http://dx.doi.org/10.1063/1.3535562] [PMID: 21341874]

[39] Plaxco KW, Simons KT, Baker D. Contact order, transition state placement and the refolding rates of

single domain proteins. J Mol Biol 1998; 277(4): 985-94. [http://dx.doi.org/10.1006/jmbi.1998.1645] [PMID: 9545386]

[40] Binz HK, Stumpp MT, Forrer P, Amstutz P, Plückthun A. Designing repeat proteins: well-expressed, soluble and stable proteins from combinatorial libraries of consensus ankyrin repeat proteins. J Mol Biol 2003; 332(2): 489-503. [http://dx.doi.org/10.1016/S0022-2836(03)00896-9] [PMID: 12948497]

[41] Phillips JJ, Javadi Y, Millership C, Main ER. Modulation of the multistate folding of designed TPR proteins through intrinsic and extrinsic factors. Protein Sci 2012; 21(3): 327-38. [http://dx.doi.org/10.1002/pro.2018] [PMID: 22170589]

[42] Hagai T, Levy Y. Folding of elongated proteins: conventional or anomalous? J Am Chem Soc 2008; 130(43): 14253-62. [http://dx.doi.org/10.1021/ja804280p] [PMID: 18834131]

[43] Hagai T, Azia A, Trizac E, Levy Y. Modulation of folding kinetics of repeat proteins: interplay between intra- and interdomain interactions. Biophys J 2012; 103(7): 1555-65. [http://dx.doi.org/10.1016/j.bpj.2012.08.018] [PMID: 23062348]

[44] Berman HM, Westbrook J, Feng Z, *et al.* The protein data bank. Nucleic Acids Res 2000; 28(1): 235-42. [http://dx.doi.org/10.1093/nar/28.1.235] [PMID: 10592235]

[45] Zimm BH, Bragg JK. Theory of the phase transition between helix and random coil in polypeptide chains. J Chem Phys 1959; 31: 526-35. [http://dx.doi.org/10.1063/1.1730390]

[46] Kajander T, Cortajarena AL, Mochrie S, Regan L. Structure and stability of designed TPR protein superhelices: unusual crystal packing and implications for natural TPR proteins. Acta Crystallogr D Biol Crystallogr 2007; 63(Pt 7): 800-11. [http://dx.doi.org/10.1107/S0907444907024353] [PMID: 17582171]

[47] Ferreiro DU, Walczak AM, Komives EA, Wolynes PG. The energy landscapes of repeat-containing proteins: topology, cooperativity, and the folding funnels of one-dimensional architectures. PLOS Comput Biol 2008; 4(5): e1000070. [http://dx.doi.org/10.1371/journal.pcbi.1000070] [PMID: 18483553]

[48] González-Charro V, Rey A. Intermediates in the folding equilibrium of repeat proteins from the TPR family. Eur Biophys J 2014; 43(8-9): 433-43. [http://dx.doi.org/10.1007/s00249-014-0975-8] [PMID: 25048829]

[49] Sampathkumar P, Roach C, Michels PA, Hol WG. Structural insights into the recognition of peroxisomal targeting signal 1 by Trypanosoma brucei peroxin 5. J Mol Biol 2008; 381(4): 867-80. [http://dx.doi.org/10.1016/j.jmb.2008.05.089] [PMID: 18598704]

[50] Lunelli M, Lokareddy RK, Zychlinsky A, Kolbe M. IpaB-IpgC interaction defines binding motif for type III secretion translocator. Proc Natl Acad Sci USA 2009; 106(24): 9661-6. [http://dx.doi.org/10.1073/pnas.0812900106] [PMID: 19478065]

[51] Zhang Z, Kulkarni K, Hanrahan SJ, Thompson AJ, Barford D. The APC/C subunit Cdc16/Cut9 is a contiguous tetratricopeptide repeat superhelix with a homo-dimer interface similar to Cdc27. EMBO J 2010; 29(21): 3733-44. [http://dx.doi.org/10.1038/emboj.2010.247] [PMID: 20924356]

[52] Ramarao MK, Bianchetta MJ, Lanken J, Cohen JB. Role of rapsyn tetratricopeptide repeat and coiled-coil domains in self-association and nicotinic acetylcholine receptor clustering. J Biol Chem 2001; 276(10): 7475-83. [http://dx.doi.org/10.1074/jbc.M009888200] [PMID: 11087759]

[53] Eckler SA, Kuehn R, Gautam M. Deletion of N-terminal rapsyn domains disrupts clustering and has dominant negative effects on clustering of full-length rapsyn. Neuroscience 2005; 131(3): 661-70.

[http://dx.doi.org/10.1016/j.neuroscience.2004.11.035] [PMID: 15730871]

[54] Bansal PK, Nourse A, Abdulle R, Kitagawa K. Sgt1 dimerization is required for yeast kinetochore assembly. J Biol Chem 2009; 284(6): 3586-92.
[http://dx.doi.org/10.1074/jbc.M806281200] [PMID: 19073600]

[55] Serasinghe MN, Yoon Y. The mitochondrial outer membrane protein hFis1 regulates mitochondrial morphology and fission through self-interaction. Exp Cell Res 2008; 314(19): 3494-507.
[http://dx.doi.org/10.1016/j.yexcr.2008.09.009] [PMID: 18845145]

[56] Krissinel E, Henrick K. Inference of macromolecular assemblies from crystalline state. J Mol Biol 2007; 372(3): 774-97.
[http://dx.doi.org/10.1016/j.jmb.2007.05.022] [PMID: 17681537]

[57] Krachler AM, Sharma A, Kleanthous C. Self-association of TPR domains: Lessons learned from a designed, consensus-based TPR oligomer. Proteins 2010; 78(9): 2131-43.
[PMID: 20455268]

[58] Javadi Y, Main ER. Exploring the folding energy landscape of a series of designed consensus tetratricopeptide repeat proteins. Proc Natl Acad Sci USA 2009; 106(41): 17383-8.
[http://dx.doi.org/10.1073/pnas.0907455106] [PMID: 19805120]

[59] Dutta R, Inouye M. GHKL, an emergent ATPase/kinase superfamily. Trends Biochem Sci 2000; 25(1): 24-8.
[http://dx.doi.org/10.1016/S0968-0004(99)01503-0] [PMID: 10637609]

[60] Obermann WM, Sondermann H, Russo AA, Pavletich NP, Hartl FU. In vivo function of Hsp90 is dependent on ATP binding and ATP hydrolysis. J Cell Biol 1998; 143(4): 901-10.
[http://dx.doi.org/10.1083/jcb.143.4.901] [PMID: 9817749]

[61] Panaretou B, Siligardi G, Meyer P, *et al.* Activation of the ATPase activity of hsp90 by the stress-regulated cochaperone aha1. Mol Cell 2002; 10(6): 1307-18.
[http://dx.doi.org/10.1016/S1097-2765(02)00785-2] [PMID: 12504007]

[62] Bohen SP, Kralli A, Yamamoto KR. Hold 'em and fold 'em: chaperones and signal transduction. Science 1995; 268(5215): 1303-4. [http://dx.doi.org/10.1126/science.7761850]. [PMID: 7761850].
[http://dx.doi.org/10.1126/science.7761850] [PMID: 7761850]

[63] Jakob U, Buchner J. Assisting spontaneity: the role of Hsp90 and small Hsps as molecular chaperones. Trends Biochem Sci 1994; 19(5): 205-11.
[http://dx.doi.org/10.1016/0968-0004(94)90023-X] [PMID: 7914036]

[64] Freeman BC, Morimoto RI. The human cytosolic molecular chaperones hsp90, hsp70 (hsc70) and hdj-1 have distinct roles in recognition of a non-native protein and protein refolding. EMBO J 1996; 15(12): 2969-79.
[PMID: 8670798]

[65] Wiech H, Buchner J, Zimmermann R, Jakob U. Hsp90 chaperones protein folding in vitro. Nature 1992; 358(6382): 169-70.
[http://dx.doi.org/10.1038/358169a0] [PMID: 1614549]

[66] Nathan DF, Vos MH, Lindquist S. In vivo functions of the Saccharomyces cerevisiae Hsp90 chaperone. Proc Natl Acad Sci USA 1997; 94(24): 12949-56.
[http://dx.doi.org/10.1073/pnas.94.24.12949] [PMID: 9371781]

[67] Pratt WB, Toft DO. Regulation of signaling protein function and trafficking by the hsp90/hsp70-based chaperone machinery. Exp Biol Med (Maywood) 2003; 228(2): 111-33.
[http://dx.doi.org/10.1177/153537020322800201] [PMID: 12563018]

[68] Li W, Sahu D, Tsen F. Secreted heat shock protein-90 (Hsp90) in wound healing and cancer. Biochim Biophys Acta 2012; 1823(3): 730-41.
[http://dx.doi.org/10.1016/j.bbamcr.2011.09.009] [PMID: 21982864]

[69] Lai BT, Chin NW, Stanek AE, Keh W, Lanks KW. Quantitation and intracellular localization of the 85K heat shock protein by using monoclonal and polyclonal antibodies. Mol Cell Biol 1984; 4(12): 2802-10.
[http://dx.doi.org/10.1128/MCB.4.12.2802] [PMID: 6396506]

[70] Sreedhar AS, Kalmár E, Csermely P, Shen YF. Hsp90 isoforms: functions, expression and clinical importance. FEBS Lett 2004; 562(1-3): 11-5.
[http://dx.doi.org/10.1016/S0014-5793(04)00229-7] [PMID: 15069952]

[71] Grammatikakis N, Vultur A, Ramana CV, *et al.* The role of Hsp90N, a new member of the Hsp90 family, in signal transduction and neoplastic transformation. J Biol Chem 2002; 277(10): 8312-20.
[http://dx.doi.org/10.1074/jbc.M109200200] [PMID: 11751906]

[72] Csermely P, Schnaider T, Soti C, Prohászka Z, Nardai G. The 90-kDa molecular chaperone family: structure, function, and clinical applications. A comprehensive review. Pharmacol Ther 1998; 79(2): 129-68.
[http://dx.doi.org/10.1016/S0163-7258(98)00013-8] [PMID: 9749880]

[73] Felts SJ, Owen BA, Nguyen P, Trepel J, Donner DB, Toft DO. The hsp90-related protein TRAP1 is a mitochondrial protein with distinct functional properties. J Biol Chem 2000; 275(5): 3305-12.
[http://dx.doi.org/10.1074/jbc.275.5.3305] [PMID: 10652318]

[74] Kim SH, Kim D, Jung GS, Um JH, Chung BS, Kang CD. Involvement of c-Jun NH(2)-terminal kinase pathway in differential regulation of heat shock proteins by anticancer drugs. Biochem Biophys Res Commun 1999; 262(2): 516-22.
[http://dx.doi.org/10.1006/bbrc.1999.1229] [PMID: 10462506]

[75] Hickey E, Brandon SE, Smale G, Lloyd D, Weber LA. Sequence and regulation of a gene encoding a human 89-kilodalton heat shock protein. Mol Cell Biol 1989; 9(6): 2615-26.
[http://dx.doi.org/10.1128/MCB.9.6.2615] [PMID: 2527334]

[76] Gupta RS. Phylogenetic analysis of the 90 kD heat shock family of protein sequences and an examination of the relationship among animals, plants, and fungi species. Mol Biol Evol 1995; 12(6): 1063-73.
[PMID: 8524040]

[77] Stechmann A, Cavalier-Smith T. Phylogenetic analysis of eukaryotes using heat-shock protein Hsp90. J Mol Evol 2003; 57(4): 408-19.
[http://dx.doi.org/10.1007/s00239-003-2490-x] [PMID: 14708574]

[78] Pepin K, Momose F, Ishida N, Nagata K. Molecular cloning of horse Hsp90 cDNA and its comparative analysis with other vertebrate Hsp90 sequences. J Vet Med Sci 2001; 63(2): 115-24.
[http://dx.doi.org/10.1292/jvms.63.115] [PMID: 11258446]

[79] Chen B, Zhong D, Monteiro A. Comparative genomics and evolution of the HSP90 family of genes across all kingdoms of organisms. BMC Genomics 2006; 7: 156-74.
[http://dx.doi.org/10.1186/1471-2164-7-156] [PMID: 16780600]

[80] Eustace BK, Sakurai T, Stewart JK, *et al.* Functional proteomic screens reveal an essential extracellular role for hsp90 alpha in cancer cell invasiveness. Nat Cell Biol 2004; 6(6): 507-14.
[http://dx.doi.org/10.1038/ncb1131] [PMID: 15146192]

[81] HUGO Gene Nomenclature Committee http://www.gene.ucl.ac.uk/nomenclature

[82] Schroda M. The *Chlamydomonas* genome reveals its secrets: chaperone genes and the potential roles of their gene products in the chloroplast. Photosynth Res 2004; 82(3): 221-40.
[http://dx.doi.org/10.1007/s11120-004-2216-y] [PMID: 16143837]

[83] Goodsell D. The molecule of the month: Hsp90. http://pdb101.rcsb.org/motm/108

[84] Pearl LH, Prodromou C. Structure and in vivo function of Hsp90. Curr Opin Struct Biol 2000; 10(1): 46-51.

[http://dx.doi.org/10.1016/S0959-440X(99)00047-0] [PMID: 10679459]

[85] Pearl LH, Prodromou C. Structure, function, and mechanism of the Hsp90 molecular chaperone. Adv Protein Chem 2001; 59: 157-86.
[http://dx.doi.org/10.1016/S0065-3233(01)59005-1] [PMID: 11868271]

[86] Prodromou C, Pearl LH. Structure and functional relationships of Hsp90. Curr Cancer Drug Targets 2003; 3(5): 301-23.
[http://dx.doi.org/10.2174/1568009033481877] [PMID: 14529383]

[87] Stankiewicz M, Mayer MP. The universe of Hsp90. Biomol Concepts 2012; 3(1): 79-97.
[http://dx.doi.org/10.1515/bmc.2011.054] [PMID: 25436526]

[88] Obermann WM, Sondermann H, Russo AA, Pavletich NP, Hartl FU. In vivo function of Hsp90 is dependent on ATP binding and ATP hydrolysis. J Cell Biol 1998; 143(4): 901-10.
[http://dx.doi.org/10.1083/jcb.143.4.901] [PMID: 9817749]

[89] Stebbins CE, Russo AA, Schneider C, Rosen N, Hartl FU, Pavletich NP. Crystal structure of an Hsp90-geldanamycin complex: targeting of a protein chaperone by an antitumor agent. Cell 1997; 89(2): 239-50.
[http://dx.doi.org/10.1016/S0092-8674(00)80203-2] [PMID: 9108479]

[90] Pearl LH, Prodromou C. Structure and mechanism of the Hsp90 molecular chaperone machinery. Annu Rev Biochem 2006; 75: 271-94.
[http://dx.doi.org/10.1146/annurev.biochem.75.103004.142738] [PMID: 16756493]

[91] Harris SF, Shiau AK, Agard DA. The crystal structure of the carboxy-terminal dimerization domain of htpG, the Escherichia coli Hsp90, reveals a potential substrate binding site. Structure 2004; 12(6): 1087-97.
[http://dx.doi.org/10.1016/j.str.2004.03.020] [PMID: 15274928]

[92] Ratzke C, Mickler M, Hellenkamp B, Buchner J, Hugel T. Dynamics of heat shock protein 90 C-terminal dimerization is an important part of its conformational cycle. Proc Natl Acad Sci USA 2010; 107(37): 16101-6.
[http://dx.doi.org/10.1073/pnas.1000916107] [PMID: 20736353]

[93] Hainzl O, Lapina MC, Buchner J, Richter K. The charged linker region is an important regulator of Hsp90 function. J Biol Chem 2009; 284(34): 22559-67.
[http://dx.doi.org/10.1074/jbc.M109.031658] [PMID: 19553666]

[94] Li J, Buchner J. Structure, function and regulation of the hsp90 machinery. Biomed J 2013; 36(3): 106-17.
[http://dx.doi.org/10.4103/2319-4170.113230] [PMID: 23806880]

[95] Prodromou C, Roe SM, Piper PW, Pearl LH. A molecular clamp in the crystal structure of the N-terminal domain of the yeast Hsp90 chaperone. Nat Struct Biol 1997; 4(6): 477-82.
[http://dx.doi.org/10.1038/nsb0697-477] [PMID: 9187656]

[96] Ali MM, Roe SM, Vaughan CK, *et al.* Crystal structure of an Hsp90-nucleotide-p23/Sba1 closed chaperone complex. Nature 2006; 440(7087): 1013-7.
[http://dx.doi.org/10.1038/nature04716] [PMID: 16625188]

[97] Shiau AK, Harris SF, Southworth DR, Agard DA. Structural Analysis of E. coli hsp90 reveals dramatic nucleotide-dependent conformational rearrangements. Cell 2006; 127(2): 329-40.
[http://dx.doi.org/10.1016/j.cell.2006.09.027] [PMID: 17055434]

[98] Dollins DE, Warren JJ, Immormino RM, Gewirth DT. Structures of GRP94-nucleotide complexes reveal mechanistic differences between the hsp90 chaperones. Mol Cell 2007; 28(1): 41-56.
[http://dx.doi.org/10.1016/j.molcel.2007.08.024] [PMID: 17936703]

[99] Krukenberg KA, Forster F, Rice LM, Sali A, Agard DA. Multiple conformations of E. coli Hsp90 in solution: insights into the conformational dynamics of Hsp90. Structure 2008; 16(5): 755-65.
[http://dx.doi.org/10.1016/j.str.2008.01.021] [PMID: 18462680]

[100] Mickler M, Hessling M, Ratzke C, Buchner J, Hugel T. The large conformational changes of Hsp90 are only weakly coupled to ATP hydrolysis. Nat Struct Mol Biol 2009; 16(3): 281-6. [http://dx.doi.org/10.1038/nsmb.1557] [PMID: 19234469]

[101] Street TO, Lavery LA, Agard DA. Substrate binding drives large-scale conformational changes in the Hsp90 molecular chaperone. Mol Cell 2011; 42(1): 96-105. [http://dx.doi.org/10.1016/j.molcel.2011.01.029] [PMID: 21474071]

[102] Krukenberg KA, Böttcher UM, Southworth DR, Agard DA. Grp94, the endoplasmic reticulum Hsp90, has a similar solution conformation to cytosolic Hsp90 in the absence of nucleotide. Protein Sci 2009; 18(9): 1815-27. [http://dx.doi.org/10.1002/pro.191] [PMID: 19554567]

[103] Krukenberg KA, Street TO, Lavery LA, Agard DA. Conformational dynamics of the molecular chaperone Hsp90. Q Rev Biophys 2011; 44(2): 229-55. [http://dx.doi.org/10.1017/S0033583510000314] [PMID: 21414251]

[104] Southworth DR, Agard DA. Species-dependent ensembles of conserved conformational states define the Hsp90 chaperone ATPase cycle. Mol Cell 2008; 32(5): 631-40. [http://dx.doi.org/10.1016/j.molcel.2008.10.024] [PMID: 19061638]

[105] Prodromou C, Roe SM, O'Brien R, Ladbury JE, Piper PW, Pearl LH. Identification and structural characterization of the ATP/ADP-binding site in the Hsp90 molecular chaperone. Cell 1997; 90(1): 65-75. [http://dx.doi.org/10.1016/S0092-8674(00)80314-1] [PMID: 9230303]

[106] Hessling M, Richter K, Buchner J. Dissection of the ATP-induced conformational cycle of the molecular chaperone Hsp90. Nat Struct Mol Biol 2009; 16(3): 287-93. [http://dx.doi.org/10.1038/nsmb.1565] [PMID: 19234467]

[107] Richter K, Moser S, Hagn F, *et al.* Intrinsic inhibition of the Hsp90 ATPase activity. J Biol Chem 2006; 281(16): 11301-11. [http://dx.doi.org/10.1074/jbc.M510142200] [PMID: 16461354]

[108] Cunningham CN, Krukenberg KA, Agard DA. Intra- and intermonomer interactions are required to synergistically facilitate ATP hydrolysis in Hsp90. J Biol Chem 2008; 283(30): 21170-8. [http://dx.doi.org/10.1074/jbc.M800046200] [PMID: 18492664]

[109] Vaughan CK, Piper PW, Pearl LH, Prodromou C. A common conformationally coupled ATPase mechanism for yeast and human cytoplasmic HSP90s. FEBS J 2009; 276(1): 199-209. [http://dx.doi.org/10.1111/j.1742-4658.2008.06773.x] [PMID: 19032597]

[110] Richter K, Muschler P, Hainzl O, Buchner J. Coordinated ATP hydrolysis by the Hsp90 dimer. J Biol Chem 2001; 276(36): 33689-96. [http://dx.doi.org/10.1074/jbc.M103832200] [PMID: 11441008]

[111] Graf C, Stankiewicz M, Kramer G, Mayer MP. Spatially and kinetically resolved changes in the conformational dynamics of the Hsp90 chaperone machine. EMBO J 2009; 28(5): 602-13. [http://dx.doi.org/10.1038/emboj.2008.306] [PMID: 19165152]

[112] Wegele H, Muschler P, Bunck M, Reinstein J, Buchner J. Dissection of the contribution of individual domains to the ATPase mechanism of Hsp90. J Biol Chem 2003; 278(41): 39303-10. [http://dx.doi.org/10.1074/jbc.M305751200] [PMID: 12890674]

[113] Morra G, Verkhivker G, Colombo G. Modeling signal propagation mechanisms and ligand-based conformational dynamics of the Hsp90 molecular chaperone full-length dimer. PLOS Comput Biol 2009; 5(3): e1000323. [http://dx.doi.org/10.1371/journal.pcbi.1000323] [PMID: 19300478]

[114] Retzlaff M, Hagn F, Mitschke L, *et al.* Asymmetric activation of the hsp90 dimer by its cochaperone aha1. Mol Cell 2010; 37(3): 344-54. [http://dx.doi.org/10.1016/j.molcel.2010.01.006] [PMID: 20159554]

[115] Pratt WB, Morishima Y, Murphy M, Harrell M. Chaperoning of glucocorticoid receptors. Handb Exp Pharmacol 2006; 172(172): 111-38.
[http://dx.doi.org/10.1007/3-540-29717-0_5] [PMID: 16610357]

[116] Pratt WB, Morishima Y, Peng HM, Osawa Y. Proposal for a role of the Hsp90/Hsp70-based chaperone machinery in making triage decisions when proteins undergo oxidative and toxic damage. Exp Biol Med (Maywood) 2010; 235(3): 278-89.
[http://dx.doi.org/10.1258/ebm.2009.009250] [PMID: 20404045]

[117] Gamerdinger M, Hajieva P, Kaya AM, Wolfrum U, Hartl FU, Behl C. Protein quality control during aging involves recruitment of the macroautophagy pathway by BAG3. EMBO J 2009; 28(7): 889-901.
[http://dx.doi.org/10.1038/emboj.2009.29] [PMID: 19229298]

[118] Taipale M, Jarosz DF, Lindquist S. HSP90 at the hub of protein homeostasis: emerging mechanistic insights. Nat Rev Mol Cell Biol 2010; 11(7): 515-28.
[http://dx.doi.org/10.1038/nrm2918] [PMID: 20531426]

[119] Li J, Soroka J, Buchner J. The Hsp90 chaperone machinery: Conformational dynamics and regulation by co-chaperones. Biochim Biophys Acta 2012; 1823: 624-35.

[120] Pearl LH, Prodromou C, Workman P. The Hsp90 molecular chaperone: an open and shut case for treatment. Biochem J 2008; 410(3): 439-53.
[http://dx.doi.org/10.1042/BJ20071640] [PMID: 18290764]

[121] Escriva H, Delaunay F, Laudet V. Ligand binding and nuclear receptor evolution. BioEssays 2000; 22(8): 717-27.
[http://dx.doi.org/10.1002/1521-1878(200008)22:8<717::AID-BIES5>3.0.CO;2-I] [PMID: 10918302]

[122] McKenna NJ, O'Malley BW. Combinatorial control of gene expression by nuclear receptors and coregulators. Cell 2002; 108(4): 465-74.
[http://dx.doi.org/10.1016/S0092-8674(02)00641-4] [PMID: 11909518]

[123] Smith DF. Tetratricopeptide repeat cochaperones in steroid receptor complexes. Cell Stress Chaperones 2004; 9(2): 109-21.
[http://dx.doi.org/10.1379/CSC-31.1] [PMID: 15497498]

[124] Smith DF. Dynamics of heat shock protein 90-progesterone receptor binding and the disactivation loop model for steroid receptor complexes. Mol Endocrinol 1993; 7(11): 1418-29.
[PMID: 7906860]

[125] Scherrer LC, Hutchison KA, Sanchez ER, Randall SK, Pratt WB. A heat shock protein complex isolated from rabbit reticulocyte lysate can reconstitute a functional glucocorticoid receptor-Hsp90 complex. Biochemistry 1992; 31(32): 7325-9.
[http://dx.doi.org/10.1021/bi00147a017] [PMID: 1510923]

[126] Hernández MP, Chadli A, Toft DO. HSP40 binding is the first step in the HSP90 chaperoning pathway for the progesterone receptor. J Biol Chem 2002; 277(14): 11873-81.
[http://dx.doi.org/10.1074/jbc.M111445200] [PMID: 11809754]

[127] Chen S, Smith DF. Hop as an adaptor in the heat shock protein 70 (Hsp70) and hsp90 chaperone machinery. J Biol Chem 1998; 273(52): 35194-200.
[http://dx.doi.org/10.1074/jbc.273.52.35194] [PMID: 9857057]

[128] Johnson BD, Schumacher RJ, Ross ED, Toft DO. Hop modulates Hsp70/Hsp90 interactions in protein folding. J Biol Chem 1998; 273(6): 3679-86.
[http://dx.doi.org/10.1074/jbc.273.6.3679] [PMID: 9452498]

[129] Song Y, Masison DC. Independent regulation of Hsp70 and Hsp90 chaperones by Hsp70/Hsp90-organizing protein Sti1 (Hop1). J Biol Chem 2005; 280(40): 34178-85.
[http://dx.doi.org/10.1074/jbc.M505420200] [PMID: 16100115]

[130] Freeman BC, Felts SJ, Toft DO, Yamamoto KR. The p23 molecular chaperones act at a late step in

intracellular receptor action to differentially affect ligand efficacies. Genes Dev 2000; 14(4): 422-34. [PMID: 10691735]

[131] Li J, Richter K, Buchner J. Mixed Hsp90-cochaperone complexes are important for the progression of the reaction cycle. Nat Struct Mol Biol 2011; 18(1): 61-6. [http://dx.doi.org/10.1038/nsmb.1965] [PMID: 21170051]

[132] Chen S, Sullivan WP, Toft DO, Smith DF. Differential interactions of p23 and the TPR-containing proteins Hop, Cyp40, FKBP52 and FKBP51 with Hsp90 mutants. Cell Stress Chaperones 1998; 3(2): 118-29. [http://dx.doi.org/10.1379/1466-1268(1998)003<0118:DIOPAT>2.3.CO;2] [PMID: 9672247]

[133] Ratajczak T, Allan RK, Cluning C, Ward BK. Functional Protein Interactions in Steroid Receptor-Chaperone Complexes.Protein Interactions. Rijeka, Croatia: Intech Open 2012; pp. 71-102. [http://dx.doi.org/10.5772/38339]

[134] Prodromou C, Siligardi G, O'Brien R, *et al.* Regulation of Hsp90 ATPase activity by tetratricopeptide repeat (TPR)-domain co-chaperones. EMBO J 1999; 18(3): 754-62. [http://dx.doi.org/10.1093/emboj/18.3.754] [PMID: 9927435]

[135] Honoré B, Leffers H, Madsen P, Rasmussen HH, Vandekerckhove J, Celis JE. Molecular cloning and expression of a transformation-sensitive human protein containing the TPR motif and sharing identity to the stress-inducible yeast protein STI1. J Biol Chem 1992; 267(12): 8485-91. [PMID: 1569099]

[136] Smith DF, Sullivan WP, Marion TN, *et al.* Identification of a 60-kilodalton stress-related protein, p60, which interacts with hsp90 and hsp70. Mol Cell Biol 1993; 13(2): 869-76. [http://dx.doi.org/10.1128/MCB.13.2.869] [PMID: 8423808]

[137] Bose S, Weikl T, Bügl H, Buchner J. Chaperone function of Hsp90-associated proteins. Science 1996; 274(5293): 1715-7. [http://dx.doi.org/10.1126/science.274.5293.1715] [PMID: 8939863]

[138] Freeman BC, Toft DO, Morimoto RI. Molecular chaperone machines: chaperone activities of the cyclophilin Cyp-40 and the steroid aporeceptor-associated protein p23. Science 1996; 274(5293): 1718-20. [http://dx.doi.org/10.1126/science.274.5293.1718] [PMID: 8939864]

[139] Schmid AB, Lagleder S, Gräwert MA, *et al.* The architecture of functional modules in the Hsp90 co-chaperone Sti1/Hop. EMBO J 2012; 31(6): 1506-17. [http://dx.doi.org/10.1038/emboj.2011.472] [PMID: 22227520]

[140] Kundrat L, Regan L. Balance between folding and degradation for Hsp90-dependent client proteins: a key role for CHIP. Biochemistry 2010; 49(35): 7428-38. [http://dx.doi.org/10.1021/bi100386w] [PMID: 20704274]

[141] Prodromou C, Siligardi G, O'Brien R, *et al.* Regulation of Hsp90 ATPase activity by tetratricopeptide repeat (TPR)-domain co-chaperones. EMBO J 1999; 18(3): 754-62. [http://dx.doi.org/10.1093/emboj/18.3.754] [PMID: 9927435]

[142] Yi F, Doudevski I, Regan L. HOP is a monomer: investigation of the oligomeric state of the co-chaperone HOP. Protein Sci 2010; 19(1): 19-25. [PMID: 19866486]

[143] Young JC, Obermann WM, Hartl FU. Specific binding of tetratricopeptide repeat proteins to the C-terminal 12-kDa domain of hsp90. J Biol Chem 1998; 273(29): 18007-10. [http://dx.doi.org/10.1074/jbc.273.29.18007] [PMID: 9660753]

[144] Owens-Grillo JK, Czar MJ, Hutchison KA, Hoffmann K, Perdew GH, Pratt WB. A model of protein targeting mediated by immunophilins and other proteins that bind to hsp90 via tetratricopeptide repeat domains. J Biol Chem 1996; 271(23): 13468-75. [http://dx.doi.org/10.1074/jbc.271.23.13468] [PMID: 8662874]

[145] Kuriyan J, Cowburn D. Modular peptide recognition domains in eukaryotic signaling. Annu Rev Biophys Biomol Struct 1997; 26: 259-88.
[http://dx.doi.org/10.1146/annurev.biophys.26.1.259] [PMID: 9241420]

[146] Richter K, Muschler P, Hainzl O, Reinstein J, Buchner J. Sti1 is a non-competitive inhibitor of the Hsp90 ATPase. Binding prevents the N-terminal dimerization reaction during the ATPase cycle. J Biol Chem 2003; 278(12): 10328-33.
[http://dx.doi.org/10.1074/jbc.M213094200] [PMID: 12525481]

[147] Cheung-Flynn J, Roberts PJ, Riggs DL, Smith DF. C-terminal sequences outside the tetratricopeptide repeat domain of FKBP51 and FKBP52 cause differential binding to Hsp90. J Biol Chem 2003; 278(19): 17388-94.
[http://dx.doi.org/10.1074/jbc.M300955200] [PMID: 12611898]

[148] Hatherley R, Clitheroe CL, Faya N, Tastan Bishop Ö. Plasmodium falciparum Hop: detailed analysis on complex formation with Hsp70 and Hsp90. Biochem Biophys Res Commun 2015; 456(1): 440-5.
[http://dx.doi.org/10.1016/j.bbrc.2014.11.103] [PMID: 25482441]

[149] Pratt WB, Toft DO. Steroid receptor interactions with heat shock protein and immunophilin chaperones. Endocr Rev 1997; 18(3): 306-60.
[PMID: 9183567]

[150] Dolinski KJ, Cardenas ME, Heitman J. CNS1 encodes an essential p60/Sti1 homolog in Saccharomyces cerevisiae that suppresses cyclophilin 40 mutations and interacts with Hsp90. Mol Cell Biol 1998; 18(12): 7344-52.
[http://dx.doi.org/10.1128/MCB.18.12.7344] [PMID: 9819421]

[151] Buchner J. Hsp90 & Co. - a holding for folding. Trends Biochem Sci 1999; 24(4): 136-41.
[http://dx.doi.org/10.1016/S0968-0004(99)01373-0] [PMID: 10322418]

[152] Taylor P, Dornan J, Carrello A, Minchin RF, Ratajczak T, Walkinshaw MD. Two structures of cyclophilin 40: folding and fidelity in the TPR domains. Structure 2001; 9(5): 431-8.
[http://dx.doi.org/10.1016/S0969-2126(01)00603-7] [PMID: 11377203]

[153] Sinars CR, Cheung-Flynn J, Rimerman RA, Scammell JG, Smith DF, Clardy J. Structure of the large FK506-binding protein FKBP51, an Hsp90-binding protein and a component of steroid receptor complexes. Proc Natl Acad Sci USA 2003; 100(3): 868-73.
[http://dx.doi.org/10.1073/pnas.0231020100] [PMID: 12538866]

[154] Wu B, Li P, Liu Y, *et al.* 3D structure of human FK506-binding protein 52: implications for the assembly of the glucocorticoid receptor/Hsp90/immunophilin heterocomplex. Proc Natl Acad Sci USA 2004; 101(22): 8348-53.
[http://dx.doi.org/10.1073/pnas.0305969101] [PMID: 15159550]

[155] Nair SC, Rimerman RA, Toran EJ, *et al.* Molecular cloning of human FKBP51 and comparisons of immunophilin interactions with Hsp90 and progesterone receptor. Mol Cell Biol 1997; 17(2): 594-603.
[http://dx.doi.org/10.1128/MCB.17.2.594] [PMID: 9001212]

[156] Barent RL, Nair SC, Carr DC, *et al.* Analysis of FKBP51/FKBP52 chimeras and mutants for Hsp90 binding and association with progesterone receptor complexes. Mol Endocrinol 1998; 12(3): 342-54.
[http://dx.doi.org/10.1210/mend.12.3.0075] [PMID: 9514152]

[157] Ratajczak T, Carrello A. Cyclophilin 40 (CyP-40), mapping of its hsp90 binding domain and evidence that FKBP52 competes with CyP-40 for hsp90 binding. J Biol Chem 1996; 271(6): 2961-5.
[http://dx.doi.org/10.1074/jbc.271.6.2961] [PMID: 8621687]

[158] Pirkl F, Buchner J. Functional analysis of the Hsp90-associated human peptidyl prolyl cis/trans isomerases FKBP51, FKBP52 and Cyp40. J Mol Biol 2001; 308(4): 795-806.
[http://dx.doi.org/10.1006/jmbi.2001.4595] [PMID: 11350175]

[159] Sivils JC, Storer CL, Galigniana MD, Cox MB. Regulation of steroid hormone receptor function by the 52-kDa FK506-binding protein (FKBP52). Curr Opin Pharmacol 2011; 11(4): 314-9.

[http://dx.doi.org/10.1016/j.coph.2011.03.010] [PMID: 21511531]

[160] Storer CL, Dickey CA, Galigniana MD, Rein T, Cox MB. FKBP51 and FKBP52 in signaling and disease. Trends Endocrinol Metab 2011; 22(12): 481-90.
[http://dx.doi.org/10.1016/j.tem.2011.08.001] [PMID: 21889356]

[161] Braun W, Kallen J, Mikol V, Walkinshaw MD, Wüthrich K. Three-dimensional structure and actions of immunosuppressants and their immunophilins. FASEB J 1995; 9(1): 63-72.
[http://dx.doi.org/10.1096/fasebj.9.1.7529736] [PMID: 7529736]

[162] Kallen J, Mikol V, Taylor P, Walkinshaw MD. X-ray structures and analysis of 11 cyclosporin derivatives complexed with cyclophilin A. J Mol Biol 1998; 283(2): 435-49.
[http://dx.doi.org/10.1006/jmbi.1998.2108] [PMID: 9769216]

[163] Allan RK, Ratajczak T. Versatile TPR domains accommodate different modes of target protein recognition and function. Cell Stress Chaperones 2011; 16(4): 353-67.
[http://dx.doi.org/10.1007/s12192-010-0248-0] [PMID: 21153002]

[164] Kieffer LJ, Thalhammer T, Handschumacher RE. Isolation and characterization of a 40-kDa cyclophilin-related protein. J Biol Chem 1992; 267(8): 5503-7.
[PMID: 1544925]

[165] Liu J, Albers MW, Chen CM, Schreiber SL, Walsh CT. Cloning, expression, and purification of human cyclophilin in Escherichia coli and assessment of the catalytic role of cysteines by site-directed mutagenesis. Proc Natl Acad Sci USA 1990; 87(6): 2304-8.
[http://dx.doi.org/10.1073/pnas.87.6.2304] [PMID: 2179953]

[166] Russell LC, Whitt SR, Chen MS, Chinkers M. Identification of conserved residues required for the binding of a tetratricopeptide repeat domain to heat shock protein 90. J Biol Chem 1999; 274(29): 20060-3.
[http://dx.doi.org/10.1074/jbc.274.29.20060] [PMID: 10400612]

[167] Reynolds PD, Ruan Y, Smith DF, Scammell JG. Glucocorticoid resistance in the squirrel monkey is associated with overexpression of the immunophilin FKBP51. J Clin Endocrinol Metab 1999; 84(2): 663-9.
[PMID: 10022435]

[168] Van Duyne GD, Standaert RF, Karplus PA, Schreiber SL, Clardy J. Atomic structure of FKBP-FK506, an immunophilin-immunosuppressant complex. Science 1991; 252(5007): 839-42.
[http://dx.doi.org/10.1126/science.1709302] [PMID: 1709302]

[169] Wilson KP, Yamashita MM, Sintchak MD, *et al.* Comparative X-ray structures of the major binding protein for the immunosuppressant FK506 (tacrolimus) in unliganded form and in complex with FK506 and rapamycin. Acta Crystallogr D Biol Crystallogr 1995; 51(Pt 4): 511-21.
[http://dx.doi.org/10.1107/S0907444994014514] [PMID: 15299838]

[170] Ward BK, Allan RK, Mok D, *et al.* A structure-based mutational analysis of cyclophilin 40 identifies key residues in the core tetratricopeptide repeat domain that mediate binding to Hsp90. J Biol Chem 2002; 277(43): 40799-809.
[http://dx.doi.org/10.1074/jbc.M207097200] [PMID: 12145316]

[171] Bracher A, Kozany C, Hähle A, Wild P, Zacharias M, Hausch F. Crystal structures of the free and ligand-bound FK1-FK2 domain segment of FKBP52 reveal a flexible inter-domain hinge. J Mol Biol 2013; 425(22): 4134-44.
[http://dx.doi.org/10.1016/j.jmb.2013.07.041] [PMID: 23933011]

[172] Riboldi-Tunnicliffe A, König B, Jessen S, *et al.* Crystal structure of Mip, a prolylisomerase from Legionella pneumophila. Nat Struct Biol 2001; 8(9): 779-83.
[http://dx.doi.org/10.1038/nsb0901-779] [PMID: 11524681]

[173] Craescu CT, Rouvière N, Popescu A, *et al.* Three-dimensional structure of the immunophilin-like domain of FKBP59 in solution. Biochemistry 1996; 35(34): 11045-52.

[http://dx.doi.org/10.1021/bi960975p] [PMID: 8780506]

[174] Riggs DL, Cox MB, Tardif HL, Hessling M, Buchner J, Smith DF. Noncatalytic role of the FKBP52 peptidyl-prolyl isomerase domain in the regulation of steroid hormone signaling. Mol Cell Biol 2007; 27(24): 8658-69.
[http://dx.doi.org/10.1128/MCB.00985-07] [PMID: 17938211]

[175] Hartl FU, Bracher A, Hayer-Hartl M. Molecular chaperones in protein folding and proteostasis. Nature 2011; 475(7356): 324-32.
[http://dx.doi.org/10.1038/nature10317] [PMID: 21776078]

[176] Craescu CT, Rouvière N, Popescu A, *et al.* Three-dimensional structure of the immunophilin-like domain of FKBP59 in solution. Biochemistry 1996; 35(34): 11045-52.
[http://dx.doi.org/10.1021/bi960975p] [PMID: 8780506]

[177] Galigniana MD, Radanyi C, Renoir JM, Housley PR, Pratt WB. Evidence that the peptidylprolyl isomerase domain of the hsp90-binding immunophilin FKBP52 is involved in both dynein interaction and glucocorticoid receptor movement to the nucleus. J Biol Chem 2001; 276(18): 14884-9.
[http://dx.doi.org/10.1074/jbc.M010809200] [PMID: 11278753]

[178] Galigniana MD, Harrell JM, Murphy PJ, *et al.* Binding of hsp90-associated immunophilins to cytoplasmic dynein: direct binding and in vivo evidence that the peptidylprolyl isomerase domain is a dynein interaction domain. Biochemistry 2002; 41(46): 13602-10.
[http://dx.doi.org/10.1021/bi020399z] [PMID: 12427021]

[179] Galigniana MD, Echeverría PC, Erlejman AG, Piwien-Pilipuk G. Role of molecular chaperones and TPR-domain proteins in the cytoplasmic transport of steroid receptors and their passage through the nuclear pore. Nucleus 2010; 1(4): 299-308.
[http://dx.doi.org/10.4161/nucl.1.4.11743] [PMID: 21113270]

[180] Bracher A, Kozany C, Thost AK, Hausch F. Structural characterization of the PPIase domain of FKBP51, a cochaperone of human Hsp90. Acta Crystallogr D Biol Crystallogr 2011; 67(Pt 6): 549-59.
[http://dx.doi.org/10.1107/S0907444911013862] [PMID: 21636895]

[181] Wang Y, Kirschner A, Fabian AK, *et al.* Increasing the efficiency of ligands for FK506-binding protein 51 by conformational control. J Med Chem 2013; 56(10): 3922-35.
[http://dx.doi.org/10.1021/jm400087k] [PMID: 23647266]

[182] Silverstein AM, Galigniana MD, Kanelakis KC, Radanyi C, Renoir JM, Pratt WB. Different regions of the immunophilin FKBP52 determine its association with the glucocorticoid receptor, hsp90, and cytoplasmic dynein. J Biol Chem 1999; 274(52): 36980-6.
[http://dx.doi.org/10.1074/jbc.274.52.36980] [PMID: 10601253]

[183] Marsh JA, Kalton HM, Gaber RF. Cns1 is an essential protein associated with the hsp90 chaperone complex in Saccharomyces cerevisiae that can restore cyclophilin 40-dependent functions in cpr7Delta cells. Mol Cell Biol 1998; 18(12): 7353-9.
[http://dx.doi.org/10.1128/MCB.18.12.7353] [PMID: 9819422]

[184] Owens-Grillo JK, Hoffmann K, Hutchison KA, *et al.* The cyclosporin A-binding immunophilin CyP-40 and the FK506-binding immunophilin hsp56 bind to a common site on hsp90 and exist in independent cytosolic heterocomplexes with the untransformed glucocorticoid receptor. J Biol Chem 1995; 270(35): 20479-84.
[http://dx.doi.org/10.1074/jbc.270.35.20479] [PMID: 7657624]

[185] Mayr C, Richter K, Lilie H, Buchner J. Cpr6 and Cpr7, two closely related Hsp90-associated immunophilins from Saccharomyces cerevisiae, differ in their functional properties. J Biol Chem 2000; 275(44): 34140-6.
[http://dx.doi.org/10.1074/jbc.M005251200] [PMID: 10942767]

[186] Cohen PT. Novel protein serine/threonine phosphatases: variety is the spice of life. Trends Biochem Sci 1997; 22(7): 245-51.
[http://dx.doi.org/10.1016/S0968-0004(97)01060-8] [PMID: 9255065]

[187] Ali A, Zhang J, Bao S, *et al.* Requirement of protein phosphatase 5 in DNA-damage-induced ATM activation. Genes Dev 2004; 18(3): 249-54. [http://dx.doi.org/10.1101/gad.1176004] [PMID: 14871926]

[188] Yang J, Roe SM, Cliff MJ, *et al.* Molecular basis for TPR domain-mediated regulation of protein phosphatase 5. EMBO J 2005; 24(1): 1-10. [http://dx.doi.org/10.1038/sj.emboj.7600496] [PMID: 15577939]

[189] Zeke T, Morrice N, Vázquez-Martin C, Cohen PT. Human protein phosphatase 5 dissociates from heat-shock proteins and is proteolytically activated in response to arachidonic acid and the microtubule-depolymerizing drug nocodazole. Biochem J 2005; 385(Pt 1): 45-56. [http://dx.doi.org/10.1042/BJ20040690] [PMID: 15383005]

[190] Borthwick EB, Zeke T, Prescott AR, Cohen PT. Nuclear localization of protein phosphatase 5 is dependent on the carboxy-terminal region. FEBS Lett 2001; 491(3): 279-84. [http://dx.doi.org/10.1016/S0014-5793(01)02177-9] [PMID: 11240142]

[191] Chinkers M. Protein phosphatase 5 in signal transduction. Trends Endocrinol Metab 2001; 12(1): 28-32. [http://dx.doi.org/10.1016/S1043-2760(00)00335-0] [PMID: 11137038]

[192] Chen MS, Silverstein AM, Pratt WB, Chinkers M. The tetratricopeptide repeat domain of protein phosphatase 5 mediates binding to glucocorticoid receptor heterocomplexes and acts as a dominant negative mutant. J Biol Chem 1996; 271(50): 32315-20. [http://dx.doi.org/10.1074/jbc.271.50.32315] [PMID: 8943293]

[193] Silverstein AM, Galigniana MD, Chen MS, Owens-Grillo JK, Chinkers M, Pratt WB. Protein phosphatase 5 is a major component of glucocorticoid receptor.hsp90 complexes with properties of an FK506-binding immunophilin. J Biol Chem 1997; 272(26): 16224-30. [http://dx.doi.org/10.1074/jbc.272.26.16224] [PMID: 9195923]

[194] Ramsey AJ, Chinkers M. Identification of potential physiological activators of protein phosphatase 5. Biochemistry 2002; 41(17): 5625-32. [http://dx.doi.org/10.1021/bi016090h] [PMID: 11969423]

[195] Haslbeck V, Drazic A, Eckl JM, *et al.* Selective activators of protein phosphatase 5 target the auto-inhibitory mechanism. Biosci Rep 2015; 35(3): 1-11. [PMID: 26182372]

[196] Ramsey AJ, Russell LC, Whitt SR, Chinkers M. Overlapping sites of tetratricopeptide repeat protein binding and chaperone activity in heat shock protein 90. J Biol Chem 2000; 275(23): 17857-62. [http://dx.doi.org/10.1074/jbc.M001625200] [PMID: 10751404]

[197] Swingle MR, Honkanen RE, Ciszak EM. Structural basis for the catalytic activity of human serine/threonine protein phosphatase-5. J Biol Chem 2004; 279(32): 33992-9. [http://dx.doi.org/10.1074/jbc.M402855200] [PMID: 15155720]

[198] Egloff MP, Cohen PT, Reinemer P, Barford D. Crystal structure of the catalytic subunit of human protein phosphatase 1 and its complex with tungstate. J Mol Biol 1995; 254(5): 942-59. [http://dx.doi.org/10.1006/jmbi.1995.0667] [PMID: 7500362]

[199] Griffith JP, Kim JL, Kim EE, *et al.* X-ray structure of calcineurin inhibited by the immunophilin-immunosuppressant FKBP12-FK506 complex. Cell 1995; 82(3): 507-22. [http://dx.doi.org/10.1016/0092-8674(95)90439-5] [PMID: 7543369]

[200] Chen MX, Cohen PT. Activation of protein phosphatase 5 by limited proteolysis or the binding of polyunsaturated fatty acids to the TPR domain. FEBS Lett 1997; 400(1): 136-40. [http://dx.doi.org/10.1016/S0014-5793(96)01427-5] [PMID: 9000529]

[201] Cliff MJ, Harris R, Barford D, Ladbury JE, Williams MA. Conformational diversity in the TPR domain-mediated interaction of protein phosphatase 5 with Hsp90. Structure 2006; 14(3): 415-26. [http://dx.doi.org/10.1016/j.str.2005.12.009] [PMID: 16531226]

[202] Kang H, Sayner SL, Gross KL, Russell LC, Chinkers M. Identification of amino acids in the tetratricopeptide repeat and C-terminal domains of protein phosphatase 5 involved in autoinhibition and lipid activation. Biochemistry 2001; 40(35): 10485-90. [http://dx.doi.org/10.1021/bi010999i] [PMID: 11523989]

[203] Cliff MJ, Williams MA, Brooke-Smith J, Barford D, Ladbury JE. Molecular recognition via coupled folding and binding in a TPR domain. J Mol Biol 2005; 346(3): 717-32. [http://dx.doi.org/10.1016/j.jmb.2004.12.017] [PMID: 15713458]

[204] Hof P, Pluskey S, Dhe-Paganon S, Eck MJ, Shoelson SE. Crystal structure of the tyrosine phosphatase SHP-2. Cell 1998; 92(4): 441-50. [http://dx.doi.org/10.1016/S0092-8674(00)80938-1] [PMID: 9491886]

[205] Ballinger CA, Connell P, Wu Y, *et al.* Identification of CHIP, a novel tetratricopeptide repeat-containing protein that interacts with heat shock proteins and negatively regulates chaperone functions. Mol Cell Biol 1999; 19(6): 4535-45. [http://dx.doi.org/10.1128/MCB.19.6.4535] [PMID: 10330192]

[206] Jiang J, Ballinger CA, Wu Y, *et al.* CHIP is a U-box-dependent E3 ubiquitin ligase: identification of Hsc70 as a target for ubiquitylation. J Biol Chem 2001; 276(46): 42938-44. [http://dx.doi.org/10.1074/jbc.M101968200] [PMID: 11557750]

[207] Murata S, Minami Y, Minami M, Chiba T, Tanaka K. CHIP is a chaperone-dependent E3 ligase that ubiquitylates unfolded protein. EMBO 2001; Rep 2: 1133-38. [http://dx.doi.org/10.1093/embo-reports/kve246]

[208] Cyr DM, Höhfeld J, Patterson C. Protein quality control: U-box-containing E3 ubiquitin ligases join the fold. Trends Biochem Sci 2002; 27(7): 368-75. [http://dx.doi.org/10.1016/S0968-0004(02)02125-4] [PMID: 12114026]

[209] Connell P, Ballinger CA, Jiang J, *et al.* The co-chaperone CHIP regulates protein triage decisions mediated by heat-shock proteins. Nat Cell Biol 2001; 3(1): 93-6. [http://dx.doi.org/10.1038/35050618] [PMID: 11146632]

[210] Zhang M, Windheim M, Roe SM, *et al.* Chaperoned ubiquitylation--crystal structures of the CHIP U box E3 ubiquitin ligase and a CHIP-Ubc13-Uev1a complex. Mol Cell 2005; 20(4): 525-38. [http://dx.doi.org/10.1016/j.molcel.2005.09.023] [PMID: 16307917]

[211] Nikolay R, Wiederkehr T, Rist W, Kramer G, Mayer MP, Bukau B. Dimerization of the human E3 ligase CHIP via a coiled-coil domain is essential for its activity. J Biol Chem 2004; 279(4): 2673-8. [http://dx.doi.org/10.1074/jbc.M311112200] [PMID: 14610072]

[212] Kohl A, Binz HK, Forrer P, Stumpp MT, Plückthun A, Grütter MG. Designed to be stable: crystal structure of a consensus ankyrin repeat protein. Proc Natl Acad Sci USA 2003; 100(4): 1700-5. [http://dx.doi.org/10.1073/pnas.0337680100] [PMID: 12566564]

[213] Garcia-Carbonero R, Carnero A, Paz-Ares L. Inhibition of HSP90 molecular chaperones: moving into the clinic. Lancet Oncol 2013; 14(9): e358-69. [http://dx.doi.org/10.1016/S1470-2045(13)70169-4] [PMID: 23896275]

[214] Neckers L. Heat shock protein 90: the cancer chaperone. J Biosci 2007; 32(3): 517-30. [http://dx.doi.org/10.1007/s12038-007-0051-y] [PMID: 17536171]

[215] Hanahan D, Weinberg RA. The hallmarks of cancer. Cell 2000; 100(1): 57-70. [http://dx.doi.org/10.1016/S0092-8674(00)81683-9] [PMID: 10647931]

[216] Hanahan D, Weinberg RA. Hallmarks of cancer: the next generation. Cell 2011; 144(5): 646-74. [http://dx.doi.org/10.1016/j.cell.2011.02.013] [PMID: 21376230]

[217] Sidera K, Patsavoudi E. HSP90 inhibitors: current development and potential in cancer therapy. Recent Patents Anticancer Drug Discov 2014; 9(1): 1-20. [http://dx.doi.org/10.2174/15748928113089990031] [PMID: 23312026]

[218] Kamal A, Thao L, Sensintaffar J, *et al.* A high-affinity conformation of Hsp90 confers tumour selectivity on Hsp90 inhibitors. Nature 2003; 425(6956): 407-10.
[http://dx.doi.org/10.1038/nature01913] [PMID: 14508491]

[219] Mimnaugh EG, Xu W, Vos M, *et al.* Simultaneous inhibition of hsp 90 and the proteasome promotes protein ubiquitination, causes endoplasmic reticulum-derived cytosolic vacuolization, and enhances antitumor activity. Mol Cancer Ther 2004; 3(5): 551-66.
[PMID: 15141013]

[220] Murphy PJ, Morishima Y, Kovacs JJ, Yao TP, Pratt WB. Regulation of the dynamics of hsp90 action on the glucocorticoid receptor by acetylation/deacetylation of the chaperone. J Biol Chem 2005; 280(40): 33792-9.
[http://dx.doi.org/10.1074/jbc.M506997200] [PMID: 16087666]

[221] DeBoer C, Meulman PA, Wnuk RJ, Peterson DH. Geldanamycin, a new antibiotic. J Antibiot 1970; 23(9): 442-7.
[http://dx.doi.org/10.7164/antibiotics.23.442] [PMID: 5459626]

[222] Whitesell L, Mimnaugh EG, De Costa B, Myers CE, Neckers LM. Inhibition of heat shock protein HSP90-pp60v-src heteroprotein complex formation by benzoquinone ansamycins: essential role for stress proteins in oncogenic transformation. Proc Natl Acad Sci USA 1994; 91(18): 8324-8.
[http://dx.doi.org/10.1073/pnas.91.18.8324] [PMID: 8078881]

[223] Menzella HG, Tran TT, Carney JR, *et al.* Potent non-benzoquinone ansamycin heat shock protein 90 inhibitors from genetic engineering of Streptomyces hygroscopicus. J Med Chem 2009; 52(6): 1518-21.
[http://dx.doi.org/10.1021/jm900012a] [PMID: 19231864]

[224] Stebbins CE, Russo AA, Schneider C, Rosen N, Hartl FU, Pavletich NP. Crystal structure of an Hsp90-geldanamycin complex: targeting of a protein chaperone by an antitumor agent. Cell 1997; 89(2): 239-50.
[http://dx.doi.org/10.1016/S0092-8674(00)80203-2] [PMID: 9108479]

[225] Jhaveri K, Ochiana SO, Dunphy MP, *et al.* Heat shock protein 90 inhibitors in the treatment of cancer: current status and future directions. Expert Opin Investig Drugs 2014; 23(5): 611-28.
[http://dx.doi.org/10.1517/13543784.2014.902442] [PMID: 24669860]

[226] Workman P, Burrows F, Neckers L, Rosen N. Drugging the cancer chaperone HSP90: combinatorial therapeutic exploitation of oncogene addiction and tumor stress. Ann N Y Acad Sci 2007; 1113: 202-16.
[http://dx.doi.org/10.1196/annals.1391.012] [PMID: 17513464]

[227] Messaoudi S, Peyrat JF, Brion JD, Alami M. Recent advances in Hsp90 inhibitors as antitumor agents. Anticancer Agents Med Chem 2008; 8(7): 761-82.
[http://dx.doi.org/10.2174/187152008785914824] [PMID: 18855578]

[228] Sydor JR, Normant E, Pien CS, *et al.* Development of 17-allylamino-17-demethoxygeldanamycin hydroquinone hydrochloride (IPI-504), an anti-cancer agent directed against Hsp90. Proc Natl Acad Sci USA 2006; 103(46): 17408-13.
[http://dx.doi.org/10.1073/pnas.0608372103] [PMID: 17090671]

[229] Hanson BE, Vesole DH. Retaspimycin hydrochloride (IPI-504): a novel heat shock protein inhibitor as an anticancer agent. Expert Opin Investig Drugs 2009; 18(9): 1375-83.
[http://dx.doi.org/10.1517/13543780903158934] [PMID: 19642950]

[230] Patterson J, Palombella VJ, Fritz C, Normant E. IPI-504, a novel and soluble HSP-90 inhibitor, blocks the unfolded protein response in multiple myeloma cells. Cancer Chemother Pharmacol 2008; 61(6): 923-32.
[http://dx.doi.org/10.1007/s00280-007-0546-0] [PMID: 17624530]

[231] Samuel TA, Sessa C, Britten C, *et al.* AUY922, a novel HSP90 inhibitor: final results of a first-i-

-human study in patients with advanced solid malignancies. J Clin Oncol 2010; 28. [Abstract 2528].

[232] Nakashima T, Ishii T, Tagaya H, *et al.* New molecular and biological mechanism of antitumor activities of KW-2478, a novel nonansamycin heat shock protein 90 inhibitor, in multiple myeloma cells. Clin Cancer Res 2010; 16(10): 2792-802.
[http://dx.doi.org/10.1158/1078-0432.CCR-09-3112] [PMID: 20406843]

[233] Dutta Gupta S, Snigdha D, Mazaira GI, *et al.* Molecular docking study, synthesis and biological evaluation of Schiff bases as Hsp90 inhibitors. Biomed Pharmacother 2014; 68(3): 369-76.
[http://dx.doi.org/10.1016/j.biopha.2014.01.003] [PMID: 24486109]

[234] Dutta Gupta S, Revathi B, Mazaira GI, *et al.* 2,4-dihydroxy benzaldehyde derived Schiff bases as small molecule Hsp90 inhibitors: rational identification of a new anticancer lead. Bioorg Chem 2015; 59: 97-105.
[http://dx.doi.org/10.1016/j.bioorg.2015.02.003] [PMID: 25727264]

[235] Chiosis G, Timaul MN, Lucas B, *et al.* A small molecule designed to bind to the adenine nucleotide pocket of Hsp90 causes Her2 degradation and the growth arrest and differentiation of breast cancer cells. Chem Biol 2001; 8(3): 289-99.
[http://dx.doi.org/10.1016/S1074-5521(01)00015-1] [PMID: 11306353]

[236] Chiosis G. Discovery and development of purine-scaffold Hsp90 inhibitors. Curr Top Med Chem 2006; 6(11): 1183-91.
[http://dx.doi.org/10.2174/156802606777812013] [PMID: 16842155]

[237] Trendowski M. PU-H71: An improvement on nature's solutions to oncogenic Hsp90 addiction. Pharmacol Res 2015; 99: 202-16.
[http://dx.doi.org/10.1016/j.phrs.2015.06.007] [PMID: 26117427]

[238] Kim SH, Tangallapally R, Kim IC, *et al.* Discovery of an L-alanine ester prodrug of the Hsp90 inhibitor, MPC-3100. Bioorg Med Chem Lett 2015; 25(22): 5254-7.
[http://dx.doi.org/10.1016/j.bmcl.2015.09.053] [PMID: 26483201]

[239] Williams JC, Weijland A, Gonfloni S, *et al.* The 2.35 A crystal structure of the inactivated form of chicken Src: a dynamic molecule with multiple regulatory interactions. J Mol Biol 1997; 274(5): 757-75.
[http://dx.doi.org/10.1006/jmbi.1997.1426] [PMID: 9405157]

[240] Segawa T, Fujii Y, Tanaka A, *et al.* Radiosensitization of human lung cancer cells by the novel purine-scaffold Hsp90 inhibitor, PU-H71. Int J Mol Med 2014; 33(3): 559-64.
[http://dx.doi.org/10.3892/ijmm.2013.1594] [PMID: 24366006]

[241] Marcu MG, Schulte TW, Neckers L. Novobiocin and related coumarins and depletion of heat shock protein 90-dependent signaling proteins. J Natl Cancer Inst 2000; 92(3): 242-8.
[http://dx.doi.org/10.1093/jnci/92.3.242] [PMID: 10655441]

[242] Yu XM, Shen G, Neckers L, *et al.* Hsp90 inhibitors identified from a library of novobiocin analogues. J Am Chem Soc 2005; 127(37): 12778-9.
[http://dx.doi.org/10.1021/ja0535864] [PMID: 16159253]

[243] Itoh H, Ogura M, Komatsuda A, Wakui H, Miura AB, Tashima Y. A novel chaperone-activit- -reducing mechanism of the 90-kDa molecular chaperone HSP90. Biochem J 1999; 343(Pt 3): 697-703.
[http://dx.doi.org/10.1042/bj3430697] [PMID: 10527951]

[244] Söti C, Rácz A, Csermely P. A Nucleotide-dependent molecular switch controls ATP binding at the C-terminal domain of Hsp90. N-terminal nucleotide binding unmasks a C-terminal binding pocket. J Biol Chem 2002; 277(9): 7066-75.
[http://dx.doi.org/10.1074/jbc.M105568200] [PMID: 11751878]

[245] Sreedhar AS, Soti C, Csermely P. Inhibition of Hsp90: a new strategy for inhibiting protein kinases. Biochim Biophys Acta 2004; 1697(1-2): 233-42.

[http://dx.doi.org/10.1016/j.bbapap.2003.11.027] [PMID: 15023364]

[246] Rosenhagen MC, Sōti C, Schmidt U, *et al.* The heat shock protein 90-targeting drug cisplatin selectively inhibits steroid receptor activation. Mol Endocrinol 2003; 17(10): 1991-2001. [http://dx.doi.org/10.1210/me.2003-0141] [PMID: 12869591]

[247] Kvardova V, Hrstka R, Walerych D, *et al.* The new platinum(IV) derivative LA-12 shows stronger inhibitory effect on Hsp90 function compared to cisplatin. Mol Cancer 2010; 9: 147-55. [http://dx.doi.org/10.1186/1476-4598-9-147] [PMID: 20550649]

[248] Zaveri NT. Green tea and its polyphenolic catechins: medicinal uses in cancer and noncancer applications. Life Sci 2006; 78(18): 2073-80. [http://dx.doi.org/10.1016/j.lfs.2005.12.006] [PMID: 16445946]

[249] Yin Z, Henry EC, Gasiewicz TA. (-)-Epigallocatechin-3-gallate is a novel Hsp90 inhibitor. Biochemistry 2009; 48(2): 336-45. [http://dx.doi.org/10.1021/bi801637q] [PMID: 19113837]

[250] Wani MC, Taylor HL, Wall ME, Coggon P, McPhail AT. Plant antitumor agents. VI. The isolation and structure of taxol, a novel antileukemic and antitumor agent from Taxus brevifolia. J Am Chem Soc 1971; 93(9): 2325-7. [http://dx.doi.org/10.1021/ja00738a045] [PMID: 5553076]

[251] Byrd CA, Bornmann W, Erdjument-Bromage H, *et al.* Heat shock protein 90 mediates macrophage activation by Taxol and bacterial lipopolysaccharide. Proc Natl Acad Sci USA 1999; 96(10): 5645-50. [http://dx.doi.org/10.1073/pnas.96.10.5645] [PMID: 10318938]

[252] Edlich F, Erdmann F, Jarczowski F, Moutty MC, Weiwad M, Fischer G. The Bcl-2 regulator FKBP38-calmodulin-$Ca^{2+}$ is inhibited by Hsp90. J Biol Chem 2007; 282(21): 15341-8. [http://dx.doi.org/10.1074/jbc.M611594200] [PMID: 17379601]

[253] Ardi VC, Alexander LD, Johnson VA, McAlpine SR. Macrocycles that inhibit the binding between heat shock protein 90 and TPR-containing proteins. ACS Chem Biol 2011; 6(12): 1357-66. [http://dx.doi.org/10.1021/cb200203m] [PMID: 21950602]

[254] Horibe T, Kohno M, Haramoto M, Ohara K, Kawakami K. Designed hybrid TPR peptide targeting Hsp90 as a novel anticancer agent. J Transl Med 2011; 9: 8. [http://dx.doi.org/10.1186/1479-5876-9-8] [PMID: 21235734]

[255] Horibe T, Torisawa A, Kohno M, Kawakami K. Molecular mechanism of cytotoxicity induced by Hsp90-targeted Antp-TPR hybrid peptide in glioblastoma cells. Mol Cancer 2012; 11(59): 59. [http://dx.doi.org/10.1186/1476-4598-11-59] [PMID: 22913813]

[256] Horibe T, Torisawa A, Kohno M, Kawakami K. Synergetic cytotoxic activity toward breast cancer cells enhanced by the combination of Antp-TPR hybrid peptide targeting Hsp90 and Hsp70-targeted peptide. BMC Cancer 2014; 14(615): 615. [http://dx.doi.org/10.1186/1471-2407-14-615] [PMID: 25159299]

CHAPTER 5

# GroEL Chaperonin: Interaction with Polypeptides Lacking a Rigid Tertiary Structure

**Victor V. Marchenkov, Natalia Yu Marchenko** and **Gennady V. Semisotnov***

*Institute of Protein Research, Russian Academy of Sciences, 4 Institutskaya St., 142290 Pushchino, Moscow Region, Russia*

**Abstract:** Molecular chaperones and especially chaperonins are primarily known as special members of the family of heat shock proteins which participate in folding, assembly, transmembrane transport and degradation of a wide variety of cellular proteins both in prokaryotes and eukaryotes. The multifunctional character of chaperones underlies their involvement in many diseases. One of their functions is binding to polypeptides lacking a rigid tertiary structure, *i.e.*, to those with many exposed hydrophobic amino acids. The current review is focused on interaction of polypeptides of this type with GroEL, the best-studied *Escherichia coli* chaperonin. The literature data on driving forces of their interaction, localization of substrate polypeptides on the GroEL surface, and the effect of GroEL ligands on its interaction with substrate polypeptides are considered. Some biotechnological applications of this event are also discussed.

**Keywords:** Protein folding, Chaperones, GroEL chaperonin, Protein-protein interaction.

## INTRODUCTION

The majority of structural and functional properties of GroEL are presently well-studied [1 - 12]. Its spatial structure, as established by electron microscopy [8, 12] and X-ray crystallography [9, 11], comprises 14 identical subunits arranged into two stacking seven-subunit ring-like structures possessing an extensive inner cavity. In its turn, each subunit consists of three well distinguished domains playing an important role in GroEL functioning. The apical domain contains hydrophobic clusters on its top, thus providing multiple interactions of GroEL with exposed hydrophobic clusters of substrate polypeptides [9, 10, 13, 14]. The equatorial domain mainly provides intersubunit and interring contacts and contains the binding site of adenine nucleotides (ADP or ATP) [9, 11, 13, 15].

* **Corresponding author Gennady V. Semisotnov:** Institute of Protein Research, Russian Academy of Sciences, 4 Institutskaya St., 142290 Pushchino, Moscow Region, Russia; Tel/Fax: +7-495-632-7871; Email: nina@vega.protres.ru

**Mario D. Galigniana (Ed.)**

The intermediate domain is some hinge between the apical and equatorial domains providing large-scale conformational changes of the GroEL particle upon binding of its ligands and substrate polypeptides [8, 11, 12, 14 - 16,]. In the presence of Mg-ADP or Mg-ATP, GroEL interacts with its co-chaperonin GroES, which results in large scale turning of the apical and intermediate domains and an essential increase of the GroEL inner cavity under the GroES lid [8, 11]. GroES, another representative of the heat shock protein family (Hsp10), has a dome-like quaternary structure combining 7 identical subunits (10 kDa each) [11, 17]. GroEL displays a weak ATPase activity [3, 4, 6, 7], but the principal event in its functioning is binding to polypeptides of various size that have no rigid tertiary structure [16, 18, 19]. Many studies have been devoted to various aspects of GroEL interaction with substrate polypeptides, and the current review is an attempt to distinguish between argued and unreasoned reports.

## Driving Forces of GroEL Interaction with Substrate Polypeptides

Taking into account that GroEL interacts mainly with proteins lacking a rigid tertiary structure (denatured or non-folded) [18 - 21] and hence, having many exposed hydrophobic groups [22, 23], it was proposed that the general force providing recognition of target polypeptides by GroEL is hydrophobicity [19, 22, 24].

Later, this assumption was strongly confirmed by X-ray crystallography [9, 25] and mutations [10, 26]. The X-ray crystallographic structure of GroEL [9, 13, 25] reveals that the top of the apical domain of each GroEL subunit contains hydrophobic clusters exposed within its inner cavity and involved in GroEL binding to GroES [11] and substrate polypeptides [10, 25]. These hydrophobic clusters are partially mobile (not resolved by X-ray crystallography) [9, 13], but their mutations lead to the loss of GroEL ability to bind substrate polypeptides [10, 14]. It was suggested that high mobility of GroEL hydrophobic clusters allows its binding to a wide variety of substrate polypeptides without a pronounced specificity [18, 19, 27], thus forming one of GroEL peculiarities.

The hydrophobic interactions can influence GroEL functioning. Specifically, GroEL displays increasing ATPase activity in the presence of hydrophobic amino acids, which is inhibited by its interaction with GroES [27]. An unstructured solution polypeptide comprising 13 N-terminal rhodanese residues forms an alpha-helix upon binding to GroEL [28]. Another more hydrophobic polypeptide also generates an alpha-helix upon binding to GroEL, although in solution it is only partially helical [29]. A contribution of hydrophobic interactions to stabilization of the GroEL complex with a protein target was evaluated using isothermal calorimetry [24]. It was shown that an expanded version of subtilisin interacts with GroEL with a positive enthalpy change (+19.9 kcal/mol) and a negative change of the heat capacity ($\Delta Cp = -0.85$ kcal/mol × grad$^{-1}$). Binding to alpha-casein occurs with a similar jump in heat capacity, suggesting that hydrophobic interactions are predominant in this process [24]. An estimate for the GroEL-interacting surface of subtilisin is ~3Å$^2$ [24]. Note that the hydrophobic surface of the GroEL apical domain is ~4.5Å$^2$ [13]. Binding of GroEL ligands (ATP, ADP, and GroES) changes hydrophobicity of the GroEL inner cavity [14], probably to give a decreased residence time of polypeptides associated with this chaperonin [30, 31]. Thus, it can be concluded that GroEL shows low specificity in respect to amino acid residues in target sites of the interacting polypeptides. The degree of exposure of hydrophobic residues in the target polypeptides can determine the relative stability of the GroEL-polypeptide complex. Protein chains capable of binding to GroEL are usually in the "molten globule" state [22] with typically collapsed and poorly packed secondary structure elements [32, 33]. This intermediate conformation is characterized by much more amply exposed hydrophobic groups, as compared with the native (rigidly packed) state [23, 32].

In recent years, due attention was paid to electrostatic interactions, which, similar to hydrophobic interactions, may play an important role in GroEL interaction with substrate polypeptides [34 - 37]. The apical domain of each GroEL subunit (through which the interaction with substrate polypeptides is performed) is negatively charged at neutral pH (pI 4.7). There are data showing that this event may influence GroEL interaction with charged substrates. Thus, the polylysine chain (positively charged at neutral pH and containing no hydrophobic groups) binds to GroEL, forming a stable complex that inhibits interaction with non-native malate dehydrogenase (natural GroEL substrate target) [38]. Moreover, GroEL affinity for negatively charged protein substrates varies depending on ionic conditions of the bulk solution [34 - 37]. For example, positively charged barnase (pI 8.8) during its refolding interacts with GroEL at a low ionic strength, while at a high ionic strength (more than 600 mM) this interaction is negligible [36]. Note that the refolding rate of barnase in the absence of GroEL is independent of ionic strength. For positively charged apocytochrome (pI 10.1), its affinity for GroEL was also shown to decrease with increasing salt concentration of the solution [35]. In contrast, GroEL affinity for negatively charged non-native alpha-lactalbumin increased upon salt addition to the solution [35]. The same results were obtained in our studies on interaction of a number of negatively and positively charged denatured proteins with GroEL [34]. The above data suggest a conclusion that GroEL affinity for substrate polypeptides is determined by a combination of hydrophobic and electrostatic interactions controlled by both GroEL ligands and environmental conditions (temperature, ionic strength, pH, *etc.*) [31, 34, 35].

## Location of Substrate Polypeptides within a GroEL Particle and Stoichiometry of this Complex

The presence of multiple hydrophobic and charged clusters on the apical domain tops forming the GroEL inner channel is the main reason to believe that it is the region which accommodates polypeptide targets [9, 13, 14]. Indeed, a crystal structure of the GroEL complex with a small (2 kDa) polypeptide demonstrates binding of this polypeptide to each of 14 subunits at the apical domain tops [25, 39]. The presence of substrate proteins within the inner cavity of GroEL was confirmed by cryo-electron microscopy [8, 12, 16] and mass spectrometry [40].

Phage display allowed selecting a 12 a.a. polypeptide binding to the GroEL apical domain with high affinity [25]. The complex of this peptide with GroEL was crystallized [25], and the interaction site was identified. The bound polypeptide had a β-hairpin conformation and interacted with the cavity-facing surface of the apical domain tops through a number of hydrophobic contacts with GroEL residues which had been previously found important for GroEL interaction with substrate proteins [10]. These residues are placed on two mobile helices and the extended part of GroEL apical domains forming a hydrophobic tier of the inner cavity [25, 41]. Interestingly, these contacts are very much similar to those of the GroES mobile loop when it is bound to GroEL apical domains [11, 14, 17]. Substituting Ser for Val263 (the principal residue in GroEL polypeptide binding site), Farr and co-workers showed that 3-4 apical domains of the GroEL ring are able to efficiently bind to a substrate protein [26]. This was confirmed by using substrate protein Rubisco possessive cysteines which showed an ability to form disulfide bonds with cysteines introduced in the GroEL apical domain binding site [26]. Multivalent binding of substrate proteins to GroEL was also directly demonstrated by cryo-electron microscopy [12, 42]. In these studies MDH [42] and T4 phage capsid protein gp23 (Mw 54 kDa) [12] were used as substrate proteins to show that their electron density covered the three or four and five or seven of GroEL apical domains, respectively. Thus, it may be concluded that substrate proteins are located within the GroEL inner cavity and interact simultaneously with several apical domains of GroEL. Our scanning calorimetry data also demonstrate that GroEL is temperature stabilized when bound to substrate proteins (unpublished). Moreover, the higher molecular weight of the substrate protein, the higher melting temperature of the complex. Apparently, multiple interactions of substrate proteins with GroEL apical domains result in increasing rigidity of the GroEL particle.

An important question arises as to how many substrate protein molecules can be accommodated by one GroEL particle. The GroEL-polypeptide complex stoichiometry is still poorly studied, and experimental data are rather contradictory. Small polypeptides (~2 kDa) can be bound to each of 14 apical domains of GroEL [25]. In the case of large polypeptides (such as proteins), the available data point to a role of the nature of substrate proteins and experimental conditions. Titration calorimetry demonstrated a stoichiometry of 1:1 for pepsin at pH 7 and reduced α-lactalbumin [37]. The same stoichiometry was reported for a mutant form of subtilisin [24]. In contrast, a 1:2 stoichiometry of the GroEL-substrate complex was found for maltose-binding protein [43] and for staphylococcal nuclease [44]. GroEL is able to accommodate up to 4-5 molecules of non-native barnase [45] or mutant forms of DHFR [46]. Being saturated with non-native α-ketoacid dehydrogenase up to a 1:1 stoichiometry, GroEL is able also to accommodate a stoichiometric amount of denatured lysozyme [47]. Solubilization of the membrane protein λ-holin by GroEL revealed six λ-holin molecules per GroEL particle [48]. Note that error stoichiometry measurements may result from incomplete GroEL purification [49]. After additional GroEL purification, stoichiometry of its complexes with a number of denatured proteins was measured using fluorescence anisotropy titration and size-exclusion chromatography. Two molecules of a substrate protein were found to be accommodated by one GroEL particle in the cases of denatured pepsin, reduced α-lactalbumin, lysozyme, and bovine serum albumin (our unpublished data). Interestingly, under certain conditions, a GroEL particle could bind two molecules of GroES, forming a symmetrical complex [50, 51].

## The Role of Ligands in GroEL Functioning

GroEL functioning as a molecular chaperonin requires a number of ligands, including $Mg^{2+}$ and $K^+$, ADP, ATP, and GroES (Hsp10) [14, 21, 52]. $Mg^{2+}$ is known to stimulate binding to adenine nucleotides (ADP or ATP) [11, 15, 21, 52] and negatively charged substrate proteins [34]. It may be believed that GroEL surface-bound $Mg^{2+}$ decreases a strong negative charge of this chaperonin and promotes its interaction with negatively charged partners. Monovalent $K^+$ was reported to be a regulator of GroEL ATPase activity [52]. GroEL-bound adenine nucleotides (ADP and ATP) induce large-scale conformational changes of the GroEL particle [15, 53 - 56] that result in GroES binding [8, 11, 14] and an impaired interaction with substrate targets [31, 34, 57 - 59]. The GroEL-ATP$_\gamma$S complex structurally differs from the unliganded one [15]: its apical domain is slightly elevated and turned by 20°. Some minor changes in the GroEL intermediate domain were also observed [15]. With bound Mg-ADP, GroES binding induces additional elevation of the GroEL apical domain by 50° and a turn by 120°, thus preventing hydrophobic binding sites from facing the cavity [11, 14]. ATP hydrolysis occurs solely on subunits of one heptameric ring to give a tight complex of GroEL with 7ADP ($K_{diss}$ = 2-7 μM) [59]. The presence of a GroEL ring-bound ADP leads to a dramatic (by three orders of magnitude) decrease in ADP specificity of another (the opposite) GroEL ring ($K_{diss}$ = 2.3 mM) upon the GroEL-$ADP_{14}$ complex formation [59], probably due to allosteric changes of the opposite ring, thereby inhibiting its interaction with another group of seven ADP molecules [60]. ATP hydrolysis in the $GroEL_{14}$-$ATP_7$-$GroES_7$ complex leads to formation of a most stable complex $GroEL_{14}$-$ADP_7$-$GroES_7$ ($K_{diss}$ = 0.3 nM), which allows binding of another group of seven ATP molecules to the opposite (*trans*) ring. Subsequent hydrolysis of these ATP molecules results in dissociation of GroES from the opposite (*cis*) ring [59]. Thus, in the presence of GroES, adenine nucleotides regulate interaction of GroEL with the GroES. However, even in the GroES absence, GroEL remains asymmetric about adenine nucleotide binding [61]. Note that adenine nucleotides display different affinity for GroEL. So, Mg-ADP readily interacts with the GroEL-substrate protein complex, while Mg-ATP binds mainly to free GroEL [62]. At the same time, GroEL-assisted folding of many proteins requires no full set of the ligands and can occur in the presence of solely Mg-ATP (or its unhydrolyzed analog AMP-PNP) or Mg-ADP, and even in the absence of the oligomeric structure and the ligands [39, 63 - 69]. Nevertheless, there are two basic mechanisms (active and passive) proposed for $GroEL_{14}$-assisted protein folding mediated by GroEL ligands. The active mechanism offers a direct repairing action of GroEL that implies unfolding of a misfolded protein chain through its multiple interactions with the mobile GroEL apical domains [70, 71]. The passive mechanism implies that GroEL interacts with polypeptide chains lacking the rigid tertiary structure within its inner cavity and prevents their fatal aggregation. However, in this case substrate protein folding occurs spontaneously either within the GroEL inner cavity [14, 72, 73] or in the bulk solution [74, 75]. These models of GroEL functioning as a molecular chaperone initiated intensive studies of GroEL interaction with substrate proteins. The most popular and supported by numerous experimental data mechanism is isolation of a folding protein within the so-called "Anfinsen cage" formed by GroEL under the GroES lid [14, 41, 72]. The main idea of this model is separation of the substrate protein from unfavorable intermolecular contacts in surrounding medium (including non-specific intermolecular association) and providing its proper spontaneous folding inside the cage. Experimental data supporting this mechanism were reviewed in detail by Horwich and Fenton [14, 41]. The main steps of this mechanism are as follows [14]. First, the substrate protein and seven ATP molecules rapidly bind to the preexisting asymmetrical complex GroEL/7ADP/GroES in the inner cavity at the level of apical domains of the GroEL toroid (*trans* ring) opposite to the one bound to GroES. The interaction of the substrate protein and ATP with the GroEL *trans* toroid induces movements of intermediate and apical domains resulting in dissociation of GroES from the opposite toroid and its interaction with the toroid containing the substrate protein and ATP. The time of this step was evaluated as

~0.2 s [14]. Second, the substrate protein becomes enclosed within the GroES-GroEL inner chamber ("Anfinsen cage") where it makes an attempt to fold [14]. This enclosure occurs due to a large scale elevation (60°) and turning (120°) of the GroEL apical domains induced by the interaction with GroES. These large-scale conformational changes make the inner cavity face hydrophilic rather than hydrophobic [11]. The time of this step (forming folding-active *cis* complex) was evaluated as 0.8 s [14]. Third, ATP hydrolysis occurs with a half-time of ~10 s and results in a decrease of GroEL affinity for GroES [14]. Fourth, GroES and hence the substrate protein dissociate from the GroEL *cis* ring, with simultaneous binding of ATP and another substrate protein to the *trans* ring of GroEL [14]. Because the life lime of the folding active *cis* chamber is only 10 s, which may be insufficient for substrate protein folding, this yet non-folded protein must be rebound to other GroEL-GroES complexes for next attempt to be folded during subsequent binding-release rounds. Thus, according to this mechanism, GroEL-GroES machinery uses ATP binding and hydrolysis as means to enclose the substrate protein within the inner chamber ("Anfinsen cage") and after some delay (~10 s) to release it, folded or non-folded, from the complex. However, despite its apparent attraction, this mechanism cannot be recognized as universal, since it is inconsistent with the following experimental data. First, GroEL assists folding of very large proteins whose size far exceeds the GroEL inner cavity even in the complex with GroES [18, 47, 76 - 79]. Second, GroEL is able to assist protein folding in the absence of GroES and even when being monomeric without nucleotides (ADP or ATP), as mentioned above [43, 63 - 69, 80, 81]. Third, the GroEL/GroES complex fails to retain denatured proteins which cannot acquire a rigid tertiary structure [82 - 84].

The other (passive) model of GroEL functioning as a molecular chaperone implies ligand-controlled binding-release of substrate proteins and their folding outside the GroEL inner cavity but in the bulk solution [75, 77 - 79]. This model implies that GroEL binding with polypeptides prone to non-specific aggregation essentially decreases their concentration in the bulk solution and hence, the probability of their fatal aggregation [75]. In this case it is not important to restrict the size or number of substrate polypeptides bound for assisting. Because GroEL affinity for substrate polypeptides is conditioned by the balance of hydrophobic and electrostatic forces, various polypeptides have various binding constants depending on their nature. Therefore, with a low binding constant, a substrate polypeptide very often dissociates from GroEL and most probably acquires its native (rigid) structure in the bulk solution without aggregation even in the absence of GroEL ligands. A high binding constant can be decreased by GroEL ligands through inducing large-scale conformational changes of GroEL (see above) that result in a decreased life time of the complex and hence, increased probability of substrate polypeptide folding or its passing to other cellular factors in the bulk solution [75]. Note that this model seems to be appropriate to other chaperones (such as Hsp100, Hsp90, Hsp70 and small heat shock proteins) which assist protein folding without using the inner cavity and sometimes in a ligand-independent manner [1, 2, 5, 85, 86].

## Biotechnological Applications of GroEL Interaction with Substrate Polypeptides

An ability of GroEL and its homologs possessing hydrophobic clusters exposed to bind to polypeptides lacking a rigid tertiary structure has been used for at least two biotechnological applications. One is a large-scale production of recombinant proteins in *Escherichia coli* cells frequently complicated by aggregation of newly synthesized proteins and formation of inclusion bodies [87]. One way to solubilize such proteins is co-expression of their genes with those encoding GroEL and GroES or their homologs (cpn60 and cpn10) [88, 89]. These chaperone systems assist folding of proteins in question and prevent or restrict formation of inclusion bodies [90]. The other way is fast and simple purification and express-analysis of the chaperones in various cells using affinity chromatography on the basis of denatured proteins [91 - 93]. This approach implies covalent immobilization of some proteins on CNBr-activated Sepharose [94] and their subsequent irreversible denaturation.

Using alcohol oxidase after its irreversible denaturation with 8 M urea as substrate protein, a crude extract of the yeast *Hansenula polymorpha* cells was analyzed [92]. It was shown that mainly hsp60 and hsp70 chaperones interacted with this substrate protein. GroEL interaction with various non-native polypeptides was thoroughly studied by Marchenko and colleagues [34, 91]. Several cheap commercial products, such as cytochrome *c*, lysozyme and ribonuclease A served as substrate proteins positively charged at neutral pH; negatively charged products were α-lactalbumin, β-casein and pepsin. These proteins were immobilized on CNBr-activated Sepharose and denatured in conditions under which GroEL retains its native structure. Cytochrome *c* was denatured by removing its heme [95], while β-casein and pepsin were initially denatured at neutral pH [96, 97]. To denature lysozyme, ribonuclease A and α-lactalbumin stabilized by intramolecular disulfide bonds, a reducing agent such as DTT (20 mM) was added. As found, GroEL interacted with all affinity carriers, although the reaction conditions were different. Negatively charged denatured proteins interacted with GroEL only in the presence of 10 mM $MgCl_2$ or 600 mM NaCl, thereby suggesting that in this case the interaction probably required suppressed electrostatic repulsion between negatively charged GroEL and substrate proteins. In contrast, GroEL interaction with positively charged denatured proteins was practically independent of moderate ionic conditions of the solution [34, 91]. GroEL elution conditions were also different for different affinity carriers, thus suggesting their different affinity for GroEL. To elute native GroEL from positively charged apocytochrome *c* and reduced ribonuclease A, as well as from negatively charged pepsin, β-casein and reduced α-lactalbumin, it was sufficient to add 20 mM Mg-ADP in elution buffer or change its ionic composition [34, 91]. In the case of positively charged reduced lysozyme, its complex with GroEL proved to be most stable and could not be disrupted by added Mg-ADP [34] or Mg-ATP or even GroES (our unpublished data). Note that complete elution of GroEL from all affinity carries can be achieved only in denaturing conditions (6M urea or pH 11, our unpublished data). Affinity chromatography on the basis of denatured pepsin was reported to be used for one-step purification of native GroEL from *E. coli* crude extract [98]. Unfortunately, there are only a few publications in this field.

## CONCLUSION

An analysis of the above literature data allows the following conclusion. There are a number of unquestionable accomplishments. Firstly, GroEL affinity for substrate proteins is conditioned by hydrophobic and electrostatic interactions, which are under control of GroEL ligands. Secondly, substrate polypeptides are bound within the GroEL inner cavity by means of multivalent interactions with hydrophobic and, probably, charged clusters localized to the tops of GroEL apical domains. Apparently, a ring-shaped arrangement of GroEL subunits allows tight binding of numerous substrate polypeptides of various size, in contrast to other chaperones (such as Hsp90, Hsp70, *etc.*). Thirdly, GroEL ligands (Mg-ADP, Mg-ATP, and GroES) decrease GroEL affinity for substrate polypeptides. At the same time, it is still unclear whether substrate proteins fold within the GroEL inner cavity under GroES lid or in the bulk solution.

## CONSENT FOR PUBLICATION

Not applicable.

## CONFLICT OF INTEREST

The author declares no conflict of interest, financial or otherwise.

## ACKNOWLEDGEMENTS

The authors thank the Russian Science Foundation (grant N° 14-24-00157) for financial support.

## REFERENCES

[1] Ellis J. Proteins as molecular chaperones. Nature 1987; 328(6129): 378-9. [http://dx.doi.org/10.1038/328378a0] [PMID: 3112578]

[2] Hartl FU, Bracher A, Hayer-Hartl M. Molecular chaperones in protein folding and proteostasis. Nature 2011; 475(7356): 324-32.
[http://dx.doi.org/10.1038/nature10317] [PMID: 21776078]

[3] Ellis RJ. Molecular chaperones: assisting assembly in addition to folding. Trends Biochem Sci 2006; 31(7): 395-401.
[http://dx.doi.org/10.1016/j.tibs.2006.05.001] [PMID: 16716593]

[4] Hemmingsen SM, Woolford C, van der Vies SM, *et al.* Homologous plant and bacterial proteins chaperone oligomeric protein assembly. Nature 1988; 333(6171): 330-4.
[http://dx.doi.org/10.1038/333330a0] [PMID: 2897629]

[5] Hartl FU, Hayer-Hartl M. Molecular chaperones in the cytosol: from nascent chain to folded protein. Science 2002; 295(5561): 1852-8.
[http://dx.doi.org/10.1126/science.1068408] [PMID: 11884745]

[6] Horwich AL, Fenton WA, Chapman E, Farr GW. Two families of chaperonin: physiology and mechanism. Annu Rev Cell Dev Biol 2007; 23: 115-45.
[http://dx.doi.org/10.1146/annurev.cellbio.23.090506.123555] [PMID: 17489689]

[7] Yébenes H, Mesa P, Muñoz IG, Montoya G, Valpuesta JM. Chaperonins: two rings for folding. Trends Biochem Sci 2011; 36(8): 424-32.
[http://dx.doi.org/10.1016/j.tibs.2011.05.003] [PMID: 21723731]

[8] Chen S, Roseman AM, Hunter AS, *et al.* Location of a folding protein and shape changes in GroEL-GroES complexes imaged by cryo-electron microscopy. Nature 1994; 371(6494): 261-4.
[http://dx.doi.org/10.1038/371261a0] [PMID: 7915827]

[9] Braig K, Otwinowski Z, Hegde R, *et al.* The crystal structure of the bacterial chaperonin GroEL at 2.8 A. Nature 1994; 371(6498): 578-86.
[http://dx.doi.org/10.1038/371578a0] [PMID: 7935790]

[10] Fenton WA, Kashi Y, Furtak K, Horwich AL. Residues in chaperonin GroEL required for polypeptide binding and release. Nature 1994; 371(6498): 614-9.
[http://dx.doi.org/10.1038/371614a0] [PMID: 7935796]

[11] Xu Z, Horwich AL, Sigler PB. The crystal structure of the asymmetric GroEL-GroES-(ADP)7 chaperonin complex. Nature 1997; 388(6644): 741-50.
[http://dx.doi.org/10.1038/41944] [PMID: 9285585]

[12] Clare DK, Bakkes PJ, van Heerikhuizen H, van der Vies SM, Saibil HR. Chaperonin complex with a newly folded protein encapsulated in the folding chamber. Nature 2009; 457(7225): 107-10.
[http://dx.doi.org/10.1038/nature07479] [PMID: 19122642]

[13] Braig K, Adams PD, Brünger AT. Conformational variability in the refined structure of the chaperonin GroEL at 2.8 A resolution. Nat Struct Biol 1995; 2(12): 1083-94.
[http://dx.doi.org/10.1038/nsb1295-1083] [PMID: 8846220]

[14] Horwich AL, Fenton WA. Chaperonin-mediated protein folding: using a central cavity to kinetically assist polypeptide chain folding. Q Rev Biophys 2009; 42(2): 83-116.
[http://dx.doi.org/10.1017/S0033583509004764] [PMID: 19638247]

[15] Boisvert DC, Wang J, Otwinowski Z, Horwich AL, Sigler PB. The 2.4 A crystal structure of the bacterial chaperonin GroEL complexed with ATP gamma S. Nat Struct Biol 1996; 3(2): 170-7.
[http://dx.doi.org/10.1038/nsb0296-170] [PMID: 8564544]

[16] Braig K, Simon M, Furuya F, Hainfeld JF, Horwich AL. A polypeptide bound by the chaperonin groEL is localized within a central cavity. Proc Natl Acad Sci USA 1993; 90(9): 3978-82.
[http://dx.doi.org/10.1073/pnas.90.9.3978] [PMID: 8097882]

[17] Hunt JF, Weaver AJ, Landry SJ, Gierasch L, Deisenhofer J. The crystal structure of the GroES co-chaperonin at 2.8 A resolution. Nature 1996; 379(6560): 37-45.

[http://dx.doi.org/10.1038/379037a0] [PMID: 8538739]

[18] Houry WA, Frishman D, Eckerskorn C, Lottspeich F, Hartl FU. Identification of *in vivo* substrates of the chaperonin GroEL. Nature 1999; 402(6758): 147-54. [http://dx.doi.org/10.1038/45977] [PMID: 10647006]

[19] Viitanen PV, Gatenby AA, Lorimer GH. Purified chaperonin 60 (groEL) interacts with the nonnative states of a multitude of *Escherichia coli* proteins. Protein Sci 1992; 1(3): 363-9. [http://dx.doi.org/10.1002/pro.5560010308] [PMID: 1363913]

[20] Mendoza JA, Rogers E, Lorimer GH, Horowitz PM. Chaperonins facilitate the in vitro folding of monomeric mitochondrial rhodanese. J Biol Chem 1991; 266(20): 13044-9. [PMID: 1677004]

[21] Goloubinoff P, Christeller JT, Gatenby AA, Lorimer GH. Reconstitution of active dimeric ribulose bisphosphate carboxylase from an unfoleded state depends on two chaperonin proteins and Mg-ATP. Nature 1989; 342(6252): 884-9. [http://dx.doi.org/10.1038/342884a0] [PMID: 10532860]

[22] Martin J, Langer T, Boteva R, Schramel A, Horwich AL, Hartl FU. Chaperonin-mediated protein folding at the surface of groEL through a 'molten globule'-like intermediate. Nature 1991; 352(6330): 36-42. [http://dx.doi.org/10.1038/352036a0] [PMID: 1676490]

[23] Semisotnov GV, Rodionova NA, Razgulyaev OI, Uversky VN, Gripas' AF, Gilmanshin RI. Study of the "molten globule" intermediate state in protein folding by a hydrophobic fluorescent probe. Biopolymers 1991; 31(1): 119-28. [http://dx.doi.org/10.1002/bip.360310111] [PMID: 2025683]

[24] Lin Z, Schwartz FP, Eisenstein E. The hydrophobic nature of GroEL-substrate binding. J Biol Chem 1995; 270(3): 1011-4. [http://dx.doi.org/10.1074/jbc.270.3.1011] [PMID: 7836352]

[25] Chen L, Sigler PB. The crystal structure of a GroEL/peptide complex: plasticity as a basis for substrate diversity. Cell 1999; 99(7): 757-68. [http://dx.doi.org/10.1016/S0092-8674(00)81673-6] [PMID: 10619429]

[26] Farr GW, Furtak K, Rowland MB, *et al.* Multivalent binding of nonnative substrate proteins by the chaperonin GroEL. Cell 2000; 100(5): 561-73. [http://dx.doi.org/10.1016/S0092-8674(00)80692-3] [PMID: 10721993]

[27] Richarme G, Kohiyama M. Amino acid specificity of the *Escherichia coli* chaperone GroEL (heat shock protein 60). J Biol Chem 1994; 269(10): 7095-8. [PMID: 7907325]

[28] Landry SJ, Gierasch LM. The chaperonin GroEL binds a polypeptide in an alpha-helical conformation. Biochemistry 1991; 30(30): 7359-62. [http://dx.doi.org/10.1021/bi00244a001] [PMID: 1677268]

[29] Landry SJ, Jordan R, McMacken R, Gierasch LM. Different conformations for the same polypeptide bound to chaperones DnaK and GroEL. Nature 1992; 355(6359): 455-7. [http://dx.doi.org/10.1038/355455a0] [PMID: 1346469]

[30] Keskin O, Bahar I, Flatow D, Covell DG, Jernigan RL. Molecular mechanisms of chaperonin GroEL-GroES function. Biochemistry 2002; 41(2): 491-501. [http://dx.doi.org/10.1021/bi011393x] [PMID: 11781087]

[31] Staniforth RA, Burston SG, Atkinson T, Clarke AR. Affinity of chaperonin-60 for a protein substrate and its modulation by nucleotides and chaperonin-10. Biochem J 1994; 300(Pt 3): 651-8. [http://dx.doi.org/10.1042/bj3000651] [PMID: 7912068]

[32] Ptitsyn OB, Semisotnov GV. The mechanism of protein folding.Conformations and Forces in Protein Folding. Washington, DC: AAAS 1991; pp. 155-68.

[33] Kuwajima K. The molten globule state of alpha-lactalbumin. FASEB J 1996; 10(1): 102-9. [PMID: 8566530]

[34] Marchenko NIu, Marchenkov VV, Kaĭsheva AL, *et al.* Affinity chromatography of GroEL chaperonin based on denatured proteins: role of electrostatic interactions in regulation of GroEL affinity for protein substrates. Biochemistry (Mosc) 2006; 71(12): 1357-64. [http://dx.doi.org/10.1134/S000629790612011X] [PMID: 17223789]

[35] Pack CG, Nishimura G, Tamura M, *et al.* Analysis of interaction between chaperonin GroEL and its substrate using fluorescence correlation spectroscopy. Cytometry 1999; 36(3): 247-53. [http://dx.doi.org/10.1002/(SICI)1097-0320(19990701)36:3<247::AID-CYTO15>3.0.CO;2-#] [PMID: 10404975]

[36] Perrett S, Zahn R, Stenberg G, Fersht AR. Importance of electrostatic interactions in the rapid binding of polypeptides to GroEL. J Mol Biol 1997; 269(5): 892-901. [http://dx.doi.org/10.1006/jmbi.1997.1081] [PMID: 9223649]

[37] Aoki K, Taguchi H, Shindo Y, *et al.* Calorimetric observation of a GroEL-protein binding reaction with little contribution of hydrophobic interaction. J Biol Chem 1997; 272(51): 32158-62. [http://dx.doi.org/10.1074/jbc.272.51.32158] [PMID: 9405415]

[38] Lau CK, Churchich JE. Binding of polylysine to GroEL. Inhibition of the refolding of mMDH. Biochim Biophys Acta 1999; 1431(2): 282-9. [http://dx.doi.org/10.1016/S0167-4838(99)00050-3] [PMID: 10350605]

[39] Zahn R, Buckle AM, Perrett S, *et al.* Chaperone activity and structure of monomeric polypeptide binding domains of GroEL. Proc Natl Acad Sci USA 1996; 93(26): 15024-9. [http://dx.doi.org/10.1073/pnas.93.26.15024] [PMID: 8986757]

[40] van Duijn E, Barendregt A, Synowsky S, Versluis C, Heck AJ. Chaperonin complexes monitored by ion mobility mass spectrometry. J Am Chem Soc 2009; 131(4): 1452-9. [http://dx.doi.org/10.1021/ja8055134] [PMID: 19138114]

[41] Fenton WA, Horwich AL. Chaperonin-mediated protein folding: fate of substrate polypeptide. Q Rev Biophys 2003; 36(2): 229-56. [http://dx.doi.org/10.1017/S0033583503003883] [PMID: 14686103]

[42] Elad N, Farr GW, Clare DK, Orlova EV, Horwich AL, Saibil HR. Topologies of a substrate protein bound to the chaperonin GroEL. Mol Cell 2007; 26(3): 415-26. [http://dx.doi.org/10.1016/j.molcel.2007.04.004] [PMID: 17499047]

[43] Sparrer H, Lilie H, Buchner J. Dynamics of the GroEL-protein complex: effects of nucleotides and folding mutants. J Mol Biol 1996; 258(1): 74-87. [http://dx.doi.org/10.1006/jmbi.1996.0235] [PMID: 8613994]

[44] Tsurupa GP, Ikura T, Makio T, Kuwajima K. Refolding kinetics of staphylococcal nuclease and its mutants in the presence of the chaperonin GroEL. J Mol Biol 1998; 277(3): 733-45. [http://dx.doi.org/10.1006/jmbi.1998.1630] [PMID: 9533891]

[45] Corrales FJ, Fersht AR. The folding of GroEL-bound barnase as a model for chaperonin-mediated protein folding. Proc Natl Acad Sci USA 1995; 92(12): 5326-30. [http://dx.doi.org/10.1073/pnas.92.12.5326] [PMID: 7777506]

[46] Clark AC, Hugo E, Frieden C. Determination of regions in the dihydrofolate reductase structure that interact with the molecular chaperonin GroEL. Biochemistry 1996; 35(18): 5893-901. [http://dx.doi.org/10.1021/bi953051v] [PMID: 8639551]

[47] Song JL, Wynn RM, Chuang DT. Interactions of GroEL/GroES with a heterodimeric intermediate during α 2β 2 assembly of mitochondrial branched-chain α-ketoacid dehydrogenase. *cis* capping of the native-like 86-kDa intermediate by GroES. J Biol Chem 2000; 275(29): 22305-12. [http://dx.doi.org/10.1074/jbc.M002038200] [PMID: 10764784]

[48] Deaton J, Savva CG, Sun J, Holzenburg A, Berry J, Young R. Solubilization and delivery by GroEL of megadalton complexes of the lambda holin. Protein Sci 2004; 13(7): 1778-86. [http://dx.doi.org/10.1110/ps.04735104] [PMID: 15215521]

[49] Clark AC, Ramanathan R, Frieden C. Purification of GroEL with low fluorescence background. Methods Enzymol 1998; 290: 100-18. [http://dx.doi.org/10.1016/S0076-6879(98)90010-6] [PMID: 9534154]

[50] Engel A, Hayer-Hartl MK, Goldie KN, *et al.* Functional significance of symmetrical *versus* asymmetrical GroEL-GroES chaperonin complexes. Science 1995; 269(5225): 832-6. [http://dx.doi.org/10.1126/science.7638600] [PMID: 7638600]

[51] Llorca O, Carrascosa JL, Valpuesta JM. Biochemical characterization of symmetric GroEL-GroES complexes. Evidence for a role in protein folding. J Biol Chem 1996; 271(1): 68-76. [http://dx.doi.org/10.1074/jbc.271.1.68] [PMID: 8550627]

[52] Viitanen PV, Lubben TH, Reed J, Goloubinoff P, O'Keefe DP, Lorimer GH. Chaperonin-facilitated refolding of ribulosebisphosphate carboxylase and ATP hydrolysis by chaperonin 60 (groEL) are $K^+$ dependent. Biochemistry 1990; 29(24): 5665-71. [http://dx.doi.org/10.1021/bi00476a003] [PMID: 1974461]

[53] Inobe T, Arai M, Nakao M, *et al.* Equilibrium and kinetics of the allosteric transition of GroEL studied by solution X-ray scattering and fluorescence spectroscopy. J Mol Biol 2003; 327(1): 183-91. [http://dx.doi.org/10.1016/S0022-2836(03)00087-1] [PMID: 12614617]

[54] Ranson NA, Farr GW, Roseman AM, *et al.* ATP-bound states of GroEL captured by cryo-electron microscopy. Cell 2001; 107(7): 869-79. [http://dx.doi.org/10.1016/S0092-8674(01)00617-1] [PMID: 11779463]

[55] Saibil HR, Zheng D, Roseman AM, *et al.* ATP induces large quaternary rearrangements in a cage-like chaperonin structure. Curr Biol 1993; 3(5): 265-73. [http://dx.doi.org/10.1016/0960-9822(93)90176-O] [PMID: 15335746]

[56] Yifrach O, Horovitz A. Transient kinetic analysis of adenosine 5′-triphosphate binding-induced conformational changes in the allosteric chaperonin GroEL. Biochemistry 1998; 37(20): 7083-8. [http://dx.doi.org/10.1021/bi980370o] [PMID: 9585518]

[57] Papo N, Kipnis Y, Haran G, Horovitz A. Concerted release of substrate domains from GroEL by ATP is demonstrated with FRET. J Mol Biol 2008; 380(4): 717-25. [http://dx.doi.org/10.1016/j.jmb.2008.05.021] [PMID: 18556021]

[58] Kipnis Y, Papo N, Haran G, Horovitz A. Concerted ATP-induced allosteric transitions in GroEL facilitate release of protein substrate domains in an all-or-none manner. Proc Natl Acad Sci USA 2007; 104(9): 3119-24. [http://dx.doi.org/10.1073/pnas.0700070104] [PMID: 17360617]

[59] Burston SG, Ranson NA, Clarke AR. The origins and consequences of asymmetry in the chaperonin reaction cycle. J Mol Biol 1995; 249(1): 138-52. [http://dx.doi.org/10.1006/jmbi.1995.0285] [PMID: 7776368]

[60] Roseman AM, Chen S, White H, Braig K, Saibil HR. The chaperonin ATPase cycle: mechanism of allosteric switching and movements of substrate-binding domains in GroEL. Cell 1996; 87(2): 241-51. [http://dx.doi.org/10.1016/S0092-8674(00)81342-2] [PMID: 8861908]

[61] Bochkareva ES, Lissin NM, Flynn GC, Rothman JE, Girshovich AS. Positive cooperativity in the functioning of molecular chaperone GroEL. J Biol Chem 1992; 267(10): 6796-800. [PMID: 1348056]

[62] Yifrach O, Horovitz A. Allosteric control by ATP of non-folded protein binding to GroEL. J Mol Biol 1996; 255(3): 356-61. [http://dx.doi.org/10.1006/jmbi.1996.0028] [PMID: 8568880]

[63] Kawata Y, Nosaka K, Hongo K, Mizobata T, Nagai J. Chaperonin GroE and ADP facilitate the folding of various proteins and protect against heat inactivation. FEBS Lett 1994; 345(2-3): 229-32. [http://dx.doi.org/10.1016/0014-5793(94)00456-0] [PMID: 7911090]

[64] Viitanen PV, Donaldson GK, Lorimer GH, Lubben TH, Gatenby AA. Complex interactions between the chaperonin 60 molecular chaperone and dihydrofolate reductase. Biochemistry 1991; 30(40): 9716-23. [http://dx.doi.org/10.1021/bi00104a021] [PMID: 1680394]

[65] Hayer-Hartl MK, Weber F, Hartl FU. Mechanism of chaperonin action: GroES binding and release can drive GroEL-mediated protein folding in the absence of ATP hydrolysis. EMBO J 1996; 15(22): 6111-21. [PMID: 8947033]

[66] Viitanen PV, Lorimer GH, Seetharam R, *et al.* Mammalian mitochondrial chaperonin 60 functions as a single toroidal ring. J Biol Chem 1992; 267(2): 695-8. [PMID: 1346131]

[67] Chatellier J, Hill F, Foster NW, Goloubinoff P, Fersht AR. From minichaperone to GroEL 3: properties of an active single-ring mutant of GroEL. J Mol Biol 2000; 304(5): 897-910. [http://dx.doi.org/10.1006/jmbi.2000.4278] [PMID: 11124035]

[68] Makino Y, Taguchi H, Yoshida M. Truncated GroEL monomer has the ability to promote folding of rhodanese without GroES and ATP. FEBS Lett 1993; 336(2): 363-7. [http://dx.doi.org/10.1016/0014-5793(93)80838-L] [PMID: 7903258]

[69] Taguchi H, Makino Y, Yoshida M. Monomeric chaperonin-60 and its 50-kDa fragment possess the ability to interact with non-native proteins, to suppress aggregation, and to promote protein folding. J Biol Chem 1994; 269(11): 8529-34. [PMID: 7907593]

[70] Lin Z, Madan D, Rye HS. GroEL stimulates protein folding through forced unfolding. Nat Struct Mol Biol 2008; 15(3): 303-11. [http://dx.doi.org/10.1038/nsmb.1394] [PMID: 18311152]

[71] Shtilerman M, Lorimer GH, Englander SW. Chaperonin function: folding by forced unfolding. Science 1999; 284(5415): 822-5. [http://dx.doi.org/10.1126/science.284.5415.822] [PMID: 10221918]

[72] Mayhew M, da Silva AC, Martin J, Erdjument-Bromage H, Tempst P, Hartl FU. Protein folding in the central cavity of the GroEL-GroES chaperonin complex. Nature 1996; 379(6564): 420-6. [http://dx.doi.org/10.1038/379420a0] [PMID: 8559246]

[73] Martin J, Mayhew M, Langer T, Hartl FU. The reaction cycle of GroEL and GroES in chaperonin-assisted protein folding. Nature 1993; 366(6452): 228-33. [http://dx.doi.org/10.1038/366228a0] [PMID: 7901770]

[74] Marchenkov VV, Sokolovskiĭ IV, Kotova NV, *et al.* [The interaction of the GroEL chaperone with early kinetic intermediates of renaturing proteins inhibits the formation of their native structure]. Biofizika 2004; 49(6): 987-94. [PMID: 15612537]

[75] Marchenkov VV, Semisotnov GV. GroEL-assisted protein folding: does it occur within the chaperonin inner cavity? Int J Mol Sci 2009; 10(5): 2066-83. [http://dx.doi.org/10.3390/ijms10052066] [PMID: 19564940]

[76] Brunschier R, Danner M, Seckler R. Interactions of phage P22 tailspike protein with GroE molecular chaperones during refolding *in vitro.* J Biol Chem 1993; 268(4): 2767-72. [PMID: 8094077]

[77] Chaudhuri TK, Farr GW, Fenton WA, Rospert S, Horwich AL. GroEL/GroES-mediated folding of a protein too large to be encapsulated. Cell 2001; 107(2): 235-46.

[http://dx.doi.org/10.1016/S0092-8674(01)00523-2] [PMID: 11672530]

[78] Paul S, Punam S, Chaudhuri TK. Chaperone-assisted refolding of *Escherichia coli* maltodextrin glucosidase. FEBS J 2007; 274(22): 6000-10.
[http://dx.doi.org/10.1111/j.1742-4658.2007.06122.x] [PMID: 17983358]

[79] Farr GW, Fenton WA, Chaudhuri TK, Clare DK, Saibil HR, Horwich AL. Folding with and without encapsulation by *cis*- and *trans*-only GroEL-GroES complexes. EMBO J 2003; 22(13): 3220-30.
[http://dx.doi.org/10.1093/emboj/cdg313] [PMID: 12839985]

[80] Jaenicke R. What does protein refolding *in vitro* tell us about protein folding in the cell? Philos Trans R Soc Lond B Biol Sci 1993; 339(1289): 287-94.
[http://dx.doi.org/10.1098/rstb.1993.0027] [PMID: 8098533]

[81] Corrales FJ, Fersht AR. Toward a mechanism for GroEL.GroES chaperone activity: an ATPase-gated and -pulsed folding and annealing cage. Proc Natl Acad Sci USA 1996; 93(9): 4509-12.
[http://dx.doi.org/10.1073/pnas.93.9.4509] [PMID: 8633099]

[82] Marchenko NIu, Marchenkov VV, Kotova NV, Semisotnov GV, Bulankina NI, Kaliman PA. [Effect of ADP and GroES on interaction of molecular chaperonin GroEL with non-native lysozyme]. Ukr Biokhim Zh (1999) 2003; 75(3): 88-94.
[PMID: 14577157]

[83] Plotnikov AN, Vasilenko KS, Kirkitadze MD, *et al.* [Biosynthesis and conformational state of 17-kDa and 27-kDa N-terminal fragments of elongation factor EF-2 in solution]. Bioorg Khim 1996; 22(7): 489-502.
[PMID: 8992954]

[84] Marchenko NYu, Marchenkov VV, Kotova NV, *et al.* Complex of GroEL chaperone with fluorescein-labeled denatured pepsin: stoichiometry and role of the ligands. J V.N. Karazin's Kharkov National Univ 2005; 1: 53-6.

[85] Calloni G, Chen T, Schermann SM, *et al.* DnaK functions as a central hub in the *E. coli* chaperone network. Cell Reports 2012; 1(3): 251-64.
[http://dx.doi.org/10.1016/j.celrep.2011.12.007] [PMID: 22832197]

[86] Haslbeck M, Vierling E. A first line of stress defense: small heat shock proteins and their function in protein homeostasis. J Mol Biol 2015; 427(7): 1537-48.
[http://dx.doi.org/10.1016/j.jmb.2015.02.002] [PMID: 25681016]

[87] Gopal GJ, Kumar A. Strategies for the production of recombinant protein in *Escherichia coli.* Protein J 2013; 32(6): 419-25.
[http://dx.doi.org/10.1007/s10930-013-9502-5] [PMID: 23897421]

[88] Levy R, Weiss R, Chen G, Iverson BL, Georgiou G. Production of correctly folded Fab antibody fragment in the cytoplasm of Escherichia coli trxB gor mutants via the coexpression of molecular chaperones. Protein Expr Purif 2001; 23(2): 338-47.
[http://dx.doi.org/10.1006/prep.2001.1520] [PMID: 11676610]

[89] de Marco A, Deuerling E, Mogk A, Tomoyasu T, Bukau B. Chaperone-based procedure to increase yields of soluble recombinant proteins produced in *E. coli.* BMC Biotechnol 2007; 7: 32.
[http://dx.doi.org/10.1186/1472-6750-7-32] [PMID: 17565681]

[90] da Silva Pinto L, Gonçales RA, Conceição FR, *et al.* Stability, oviposition attraction, and larvicidal activity of binary toxin from *Bacillus sphaericus* expressed in *Escherichia coli.* Appl Microbiol Biotechnol 2012; 95(5): 1235-41.
[http://dx.doi.org/10.1007/s00253-011-3808-z] [PMID: 22202967]

[91] Marchenko NYu, Marchenkov VV, Kaysheva AL, *et al.* A simple method for isolation, purification and express-analysis of GroEL-like chaperones. FEBS J 2006; 273: 114.

[92] Evers ME, Huhse B, Titorenko VI, *et al.* Affinity purification of molecular chaperones of the yeast *Hansenula polymorpha* using immobilized denatured alcohol oxidase. FEBS Lett 1993; 321(1): 32-6.

[http://dx.doi.org/10.1016/0014-5793(93)80615-2] [PMID: 8096822]

[93] Nam SH, Walsh MK. Affinity purification and characterization of the *Escherichia coli* molecular chaperones. Protein Expr Purif 2002; 24(2): 282-91. [http://dx.doi.org/10.1006/prep.2001.1571] [PMID: 11858724]

[94] Axén R, Porath J, Ernback S. Chemical coupling of peptides and proteins to polysaccharides by means of cyanogen halides. Nature 1967; 214(5095): 1302-4. [http://dx.doi.org/10.1038/2141302a0] [PMID: 6056841]

[95] Stellwagen E, Rysavy R, Babul G. The conformation of horse heart apocytochrome *c*. J Biol Chem 1972; 247(24): 8074-7. [PMID: 4344990]

[96] Livney YD, Schwan AL, Dalgleish DG. A study of beta-casein tertiary structure by intramolecular crosslinking and mass spectrometry. J Dairy Sci 2004; 87(11): 3638-47. [http://dx.doi.org/10.3168/jds.S0022-0302(04)73502-X] [PMID: 15483147]

[97] Kozlov LV, Meshcheriakova EA, Zavada LL, Efremov ES, Rashkovetskiĭ LG. [Dependence of pepsin conformational states on pH and temperature]. Biokhimiia 1979; 44(2): 339-49. [PMID: 35243]

[98] Marchenko NYu, Marchenkov VV, Kaysheva AL, *et al.* Single-stage purification of molecular chaperone GroEL by the affinity chromatography on a basis of denatured pepsin. Vestnik KhNU Series. Biology (Basel) 2005; 709(1-2): 48-55.

**CHAPTER 6**

# Mechanisms of Protein Folding by Type II Chaperonins

**Rebecca L. Plimpton**[1], **José M. Valpuesta**[2] and **Barry M. Willardson**[1,*]

[1] *Department of Chemistry and Biochemistry, Brigham Young University, Provo, Utah, USA*

[2] *Centro Nacional de Biotecnología, Campus de la Universidad Autónoma de Madrid, Madrid, Spain*

**Abstract:** Protein homeostasis depends on the ability of molecular chaperones to assist proteins with complex folding patterns to achieve their native state. An important molecular chaperone found in the eukaryotic cytosol is the chaperonin containing tailless complex polypeptide 1 (CCT, also called TRiC). CCT is composed of eight homologous subunits, each with ATPase activity, that form a double ring complex with protein folding cavities in the center of each ring. CCT folds primarily nascent polypeptides that have yet to achieve their native state by binding the unfolded protein in the folding cavity of the open, nucleotide-free conformation of CCT. Upon binding and hydrolysis of ATP, CCT undergoes a conformational change that simultaneously creates a lid over the folding cavity and releases the nascent protein into the cavity. The protein is then allowed to fold in this confined space isolated from other proteins in the cell. Dissociation of phosphate and ADP allows CCT to relax back into its open conformation and release the protein if it has achieved its native conformation. Recent studies suggest that CCT initially extends and unfolds at least some substrate proteins by binding them at sites on opposite sides of the folding cavity. Interestingly, subunits on one side of the CCT ring bind and hydrolyze ATP more effectively than subunits on the other side, suggesting a sequential release mechanism first from one side and then the other, effectively dictating the folding trajectory of the protein. The CCT folding process is facilitated by co-chaperones that deliver substrates for folding, stabilize folding intermediates or promote substrate release. In this manner, CCT and its co-chaperones accommodate the folding of many protein substrates with diverse folding patterns.

**Keywords:** Protein folding, Molecular chaperone, Chaperonin, Substrate recognition, Folding mechanism, Co-chaperone, Phosducin-like protein.

* **Corresponding author Barry M. Willardson:** Department of Chemistry and Biochemistry, Brigham Young University, Provo, Utah, USA; Tel: 801-422-2785; Email: bmwillardson@chem.byu.edu

# INTRODUCTION

A protein's amino acid sequence serves as a code for the structure of the protein in its native state. However, there is a vast gap of information between the linear protein chain that exits the ribosome during translation and the final form of the protein with its secondary, tertiary, and quaternary levels of structural organization. A protein traverses this gap during folding by exploring the potential energy landscape of a multitude of interactions, attempting to reach the overall most energetically favorable state [1]. Small proteins (<100 amino acids) quickly and efficiently complete this process on their own [2]. However, larger proteins are susceptible to pitfalls that result in partially folded, misfolded, and aggregated products [3]. The intense crowding of the cytosolic environment amplifies these difficulties [4]. There are upwards of 10,000 unique proteins in mammalian cells, and most of them cannot circumvent these folding pitfalls on their own [5].

## Molecular Chaperones and Proteostasis

The problem of protein folding is overcome by molecular chaperones, a large class of 200 plus proteins that assist unfolded polypeptides to successfully navigate the folding process. These chaperones are often termed heat-shock proteins (HSPs), because they were initially discovered by their up-regulation during heat shock or other cellular stress responses, and are numbered according to their molecular weight. There are three classes of chaperones that participate generally in de novo protein folding: Hsp70, Hsp90 and the Hsp60 chaperonins [6] (Fig. **1**). The Hsp70 class has a C-terminal peptide-binding domain shaped like a clamp that identifies and binds short, hydrophobic-rich regions of nascent proteins [7]. The clamping action is driven by ATP hydrolysis and enhanced by the presence of the small chaperone Hsp40 [8, 9]. Hsp70 can bind and release the same substrate repeatedly, each time protecting hydrophobic regions from aggregating and also inducing local conformational remodeling that furthers the folding process. Hsp70 targets fully translated proteins and also acts co-translationally on large protein domains as they exit the ribosome [10]. Over 20% of eukaryotic proteins depend on Hsp70 for folding assistance [11]. A number of these are signaling proteins that are handed off from Hsp70 to Hsp90 for the final steps of folding. Hsp90 is a homodimer that, like Hsp70, is driven by ATP hydrolysis to bind and fold its substrates in a clamping manner [12]. It is also highly effective at stabilizing mutated protein variants and preventing their degradation [13]. Another class of chaperones that receives substrates from Hsp70 is the Hsp60, or chaperonin, complex. Individual Hsp60 proteins combine to form a stacked double-ring toroid of 14–16 subunits that creates a secluded space for a partially folded protein to achieve its native conformation [14]. Substrate proteins enter the folding cavity, and the chaperonin subunits each bind and hydrolyze ATP in a cycle that closes the cavity [15] and in some cases also actively furthers folding [16].

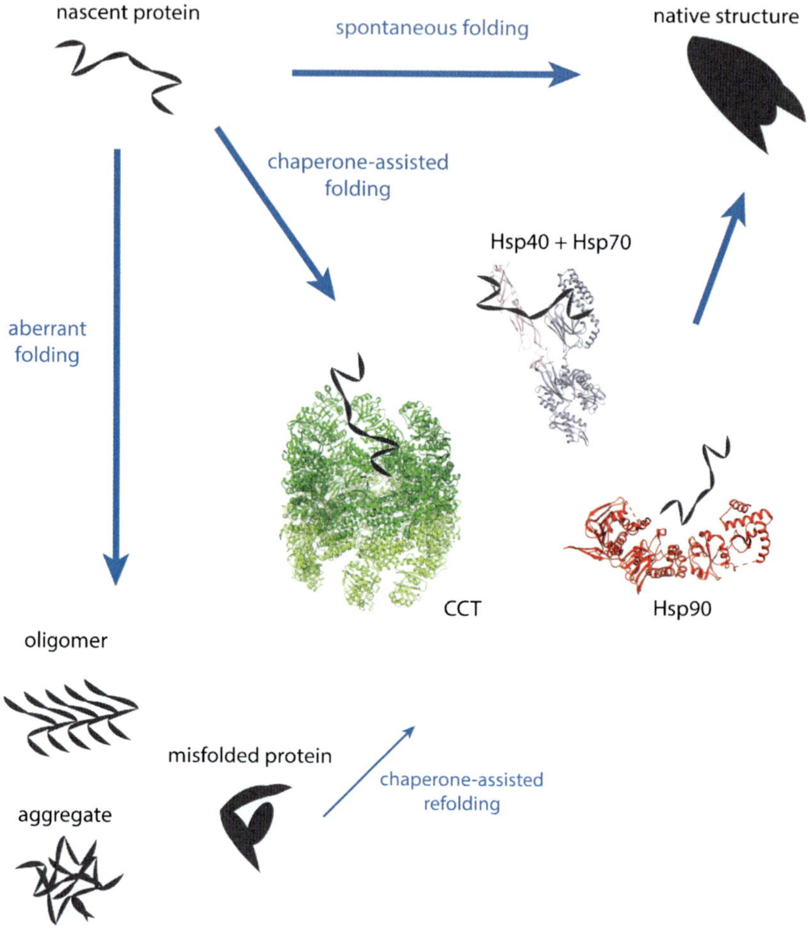


**Fig. (1).** Eukaryotic protein folding. Nascent proteins are often assisted in the folding process by chaperones to reach their native state. Chaperones help nascent proteins bypass the formation of inactive misfolded proteins and toxic aggregates. Structures of the three main chaperone classes are shown: Hsp40/70, CCT (Hsp60), and Hsp90 (adapted from PDB entries 2KHO, 1NLT, 4B2T, and 2IOQ). Chaperones can also help misfolded proteins to refold into their correct conformations.

Chaperones' contributions to the structural integrity of individual proteins are crucial in the larger proteostasis network of the cell. The ubiquitin, proteasome, and autophagy pathways combine with chaperones to regulate protein lifetime, recycle old proteins, rescue misfolded or mutated proteins, and keep aggregation and oligomerization below toxic levels [17, 18]. Aberrant proteostasis produces disease and cell death. Toxic aggregates result from general hydrophobic affinities [19] and sequence-specific oligomerization [20]. Aggregated proteins create harmful loss-of-function states, as occurs in cystic fibrosis, as well as toxic gain-of-function states, as occurs in Huntington's, Parkinson's, and Alzheimer's disease [1]. Proteostasis and longevity are regulated hand-in-hand [21] and end up cyclically contributing to each other: protein failures further aging, which decreases proteostasis, and so on.

This chapter focuses on the structure and function of the chaperonins and their ability to fold large proteins with more intricate folding patterns, paying particular attention to recent structural studies. For a more general treatment of chaperone function, we refer the reader to excellent reviews of the chaperone network [1, 22].

## THE CHAPERONINS

There are two classes of Hsp60 chaperonins: class I found in prokaryotes, mitochondria and chloroplasts, and class II found in the eukaryotic cytosol and in Archaea. The *E. coli* chaperonin GroEL is a prototypical class I chaperonin that has been extensively studied [23]. The GroEL complex consists of seven identical chaperonin subunits that form a ring structure [24, 25]. Two of these rings stack back to back, creating a barrel shape with two folding cavities. Each subunit consists of three domains: an equatorial domain, an intermediate domain and an apical domain. The equatorial domain sits at the interfaces between the rings and contains all the inter-ring contacts. This domain binds and hydrolyzes ATP, driving conformational changes within the rings that are important for substrate protein folding. The intermediate domain transfers the conformational change from the equatorial domain to the apical domain. The apical domain contains a hydrophobic patch that creates a hydrophobic surface around the interior of the folding cavity in the GroEL complex [26]. This surface is exposed in the open form of GroEL, with no nucleotide bound to the equatorial domain, allowing unfolded protein substrates to bind GroEL through hydrophobic interactions. ATP binding causes a conformational change in the apical domain that creates the binding site for GroES, a seven-membered complex of Hsp10 subunits that forms a cap over the folding cavity [27, 28]. GroES binding induces further rotation of the apical domains, dramatically changing the surface of the folding cavity from hydrophobic to hydrophilic [27, 29]. This change in the substrate-binding surface causes the substrate to release into the folding cavity where it can fold in the isolated environment created by GroES binding. Upon ATP hydrolysis, GroES is released, the subunits return to their nucleotide-free conformation and the folded substrate is released from the cavity [29, 30].

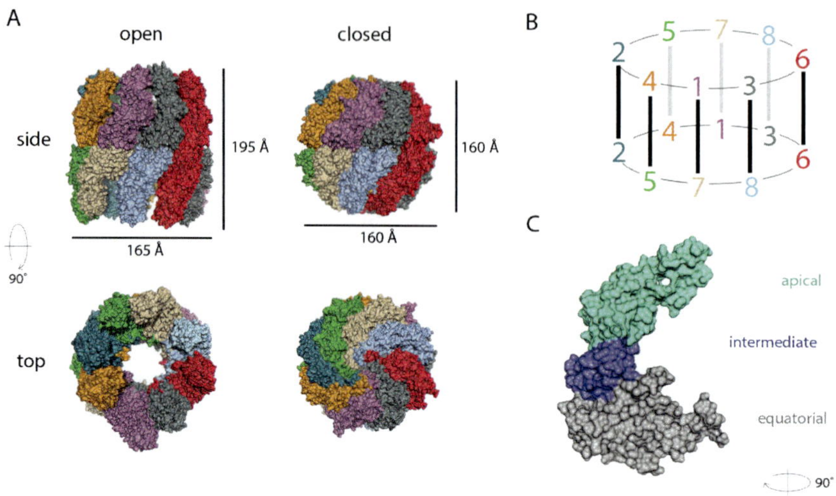


**Fig. (2). Chaperonin architecture. A)** The eukaryotic CCT complex is shown in its open and closed forms. Structures are adapted from PDB entries 4B2T and 4V8R. **B)** The ordering of the CCT subunits as determined by chemical crosslinking and statistical analysis [31, 32]. **C)** The three domains of the CCT2 subunit, shown from the side and interior of the folding cavity, respectively.

If during this process the substrate does not bury its hydrophobic regions in the protein core and reach a stable fold, then it can participate in additional rounds of folding via ATP binding and hydrolysis by GroEL until a native fold is achieved [23].

Class II chaperonins fold proteins in a manner generally similar to GroEL, with some important differences. Group II chaperonins also form double ring structures with a central folding cavity in each ring [33 - 37]. The two rings of archaeal group II chaperonins (called thermosomes) are made of one, two or three different subunits with either eight or nine subunits per ring [38]. The eukaryotic chaperonin, called chaperonin containing tailless polypeptide 1 (CCT, also referred to as tailless polypeptide 1 ring complex, TRiC), has the most complex composition with eight different subunits found in each ring (Fig. **2**). The subunits are paralogous, exhibiting 30% sequence identity among them [39]. Like GroEL, each CCT subunit has an equatorial, intermediate and apical domain arrangement, but the differences in amino acid composition among the eight subunits create a folding cavity with a chemically diverse surface made up of charged, hydrophilic and hydrophobic residues [32, 40]. The diversity of the substrate-binding surface allows CCT to interact with substrates in a much more selective and specific manner than GroEL, increasing the complexity of proteins that can be folded by CCT.

Like GroEL, CCT uses ATP binding and hydrolysis to encapsulate unfolded proteins and assist them to fold, but the mechanism of encapsulation is different. ATP hydrolysis in the equatorial domain creates a conformational change that is propagated through the intermediate domain to the apical domain. However, the CCT apical domains do not interact with a GroES-like complex, but rather the conformational change causes helical extensions in the apical domains, not found in GroEL, to form a lid over the folding cavity and the encapsulated substrate protein [35, 41, 42]. Phosphate dissociation from the nucleotide binding site reverses the conformational change, opening up the folding cavity and releasing the folded substrate [36]. We will now examine the details of CCT-dependent protein folding.

## CCT STRUCTURE

### General Features

Insights into the structure of CCT came initially from homology comparisons to the high-resolution structures of GroEL [24, 25, 27, 28] and the thermosome [33, 43] and from low-resolution cryo-EM structures [44, 45]. In the last 5 years, several structures of CCT in different nucleotide-bound states have been determined to near atomic resolution [34, 36, 42, 46]. The partially open state of CCT, bound to the non-hydrolyzable ATP analog ATPγS, was determined to 5.5 Å resolution by X-ray crystallography [36]. This structure shows the usual double-ring cylinder with a diameter of 165 Å and a height of 195 Å (Fig. **2**). As expected, the CCT domain structure is the same as archaeal group II chaperonins with equatorial, intermediate and apical domains having similar secondary and tertiary structures as the thermosome. The subunits are oriented roughly parallel to the axis running the length of the CCT barrel. The equatorial domains of CCT subunits from opposite rings interact with each other in phase, unlike class I chaperonins in which the equatorial domains interact in a staggered manner [24, 25, 27, 28]. These differences in inter-ring interactions reflect differences in allosteric communication between rings [38].

The equatorial domains are also responsible for all intra-ring contacts between CCT subunits, leaving gaps between subunits in the apical domains [36]. This lack of intra-ring interactions results in considerable conformational flexibility in the apical domains, to the point that several of the apical domains in one ring were not resolved in the CCT crystal structure [36]. This conformational flexibility likely plays a role in accommodating the many varied CCT substrates in the folding cavity. Substrates bind CCT in this open form, with the apical domains separated and an important substrate-binding region exposed between subunits [40, 47].

During the folding process, ATP hydrolysis in the equatorial domain induces the closed form, in which the helical extensions of the apical domains fold over the ends of the CCT barrel, closing the cavity and trapping the substrate protein inside. These movements were observed in a 4 Å cryo-EM structure of bovine CCT bound to the ATP hydrolysis transition state analog ADP-$AlF_x$ [34] and in a 3.8 Å resolution x-ray crystal structure of yeast CCT bound to another transition state analog ADP-$BeF_x$ [42]. In these structures, the folding cavities of both rings are closed, creating a nearly spherical shape with a diameter of ~ 160 Å (Fig. **2**). The interior of the folding cavity has a volume of ~ 145,000 Å$^3$, enough to encapsulate globular proteins up to ~ 70 kDa in size [48]. The bottom of the cavity is closed by a septum between the two rings, formed by both the N- and C-termini of the subunits, extending 10-20 residues from the equatorial domains. These termini pair with adjacent subunits to form β-sheets that separate the two folding cavities [42].

ATP hydrolysis closes the top of the cavity by inducing a 35° inward tilt of the CCT apical domains and a 30° counterclockwise rotation parallel to the central axis [38, 42, 49]. These movements bring the helical extensions in close contact with each other to cap the cavity in an iris-like motion. It is within this enclosed space that the substrate protein folds into a stable conformation. The time spent in the closed conformation has not been measured directly for CCT, but is believed to be longer than the ~ 10 sec (at 25°C) estimated for GroES [50, 51]. After phosphate release, the apical domains open up and the substrate is released from the folding cavity if it has achieved its native state [49]. If not, the substrate remains bound to the cavity and participates in another round of folding [52].

## Subunit Arrangement

The sequence differences between the eight CCT subunits create a diversity in substrate recognition and ATP hydrolysis not seen in Group I chaperonins or in the thermosome. As a result of these differences, the position of each subunit in the two rings could have important consequences in the function of CCT. Questions of whether the subunit orientation is fixed and if so, what that orientation might be remained unanswered until recently. Several methods to determine the position of the subunits, including analysis of the subunit composition of CCT sub-complexes [53], antibody tagging coupled with cryo-electron microscopy [54], and fitting of the subunits into the 4 Å electron densities from cryo-EM [34] or X-ray crystallography determinations [42] could not unambiguously determine the subunit order within the CCT rings.

To resolve the question, an emerging structural technique of chemical cross-linking coupled with mass spectrometry was applied [55]. In this method, the purified, intact CCT complex was cross-linked using disuccinimyl suberate, a homobifunctional lysine cross-linker, and the positions of the cross-links were determined by high-resolution mass spectrometry [31, 32]. This approach produced hundreds of inter-subunit cross-links that were used in a combinatorial homology-modeling scheme to find the subunit orientation that best fit the cross-link data. One study of bovine CCT and yeast CCT [32] and another study of bovine CCT [31] both arrived at the same intra-ring and inter-ring subunit orientation shown in Fig. (**2**). Furthermore, a reanalysis of the X-ray crystallography data, assessing the fit of all possible subunit orientations in the electron density, gave the same optimal subunit orientation as the cross-linking analysis [56]. These findings strongly support a single preferred subunit orientation for CCT that is conserved between yeast and mammals. This orientation reveals potential asymmetry in substrate binding and ATPase activity of the CCT complex that is discussed in the next section.

## SUBSTRATE RECOGNITION

The scope of CCT substrates is broad and encompasses highly diverse proteins, both in their functional role in the cell and by their structural and biochemical properties. CCT is believed to act principally on nascent polypeptides during their biogenesis and not on denatured proteins after cell stress [57]. This notion is consistent with the fact that expression of CCT subunits is not induced by heat shock [58]. The major cytoskeletal proteins actin and tubulin [59 - 65] were the first CCT substrates to be discovered and were initially thought to be CCT's only substrates. However, subsequent reports indicate that CCT assists in the folding of many nascent polypeptides *in vivo* [11, 66, 67]. Proteomic screens suggest that CCT substrates tend to be aggregation-prone proteins with complex folding patterns [66]. Accordingly, CCT substrates are enriched in β-sheet structures and multi-domain folds [66]. An important subclass of CCT substrates is the WD40-repeat proteins that form β-propeller tertiary structures, including G protein β subunits (Gβ) [68, 69], the cdc20 and cdh1 components of the anaphase-promoting complex [70], the Bardet-Biedl syndrome protein 7 (BBS7) subunit of the BBSome ciliary transport protein complex [71, 72], protein phosphatase 2A regulatory subunits [73] and the telomerase co-factor TCAB1 [74]. In addition to these substrates, there are a number of others directly involved in neurodegenerative diseases and cancer such as huntingtin, the transcriptional activator STAT3, and the tumor suppressor proteins p53 and VHL [75 - 79].

### Substrate Binding Sites on CCT

It has proved to be far easier to identify CCT's substrates than to determine how those substrates are recognized from among the myriad of cytosolic proteins in the process of folding. Investigation of the nature of CCT substrate recognition has been ongoing over the past 20 years and continues today. These investigations show that CCT binds its substrates more specifically than the general hydrophobic interactions observed with GroEL and the thermosome [80]. The specificity stems from the amino acid differences among the eight CCT subunits that create a chemically diverse surface for substrate binding on the inside of the CCT folding cavity (Fig. **3**). This surface displays a mosaic of hydrophilic and hydrophobic residues with conserved partitioning of charged residues between opposite sides of the ring [32]. The side of the ring consisting of CCT5, CCT2 and CCT4 has a high negative charge density while the side consisting of CCT8, CCT6 and CCT3 has a high positive charge density [32]. The functional significance of this charge partitioning is not yet clear, but the chemical diversity of the interior of the folding cavity suggests specific binding interactions between different substrates at different sites within the cavity. This diversity has led researchers to speculate that the appearance of CCT in eukaryotes may have played an important role in enabling the increased complexity of folds observed in eukaryotic proteins [40]. These observations provide a general idea of how substrates interact with the CCT folding cavity, but recent structural studies have begun to reveal the details of substrate recognition by CCT.

Early cryo-EM studies of the complexes between actin and tubulin showed that these substrates interacted with the apical domains of the CCT subunits, as is the case for GroEL substrates [44, 45, 80]. However until recently, specific substrate interaction sites within the apical domains have not been clearly identified. An NMR structural analysis of the interaction between the small 54 residue HIV protein p6 and the CCT3 apical domain identified a p6 recognition site on CCT3 [40]. This site includes helix 11 of the CCT3 apical domain, which is situated near the base of the helical protrusion, and a short loop just below helix 11, which is termed the proximal loop (Fig. **3**). This binding region encompasses a shallow groove between helix 11 and the proximal loop with an area of 1100 Å$^2$ and includes hydrophobic, polar, basic, and charged residues. Interestingly, in the open form of CCT this region sits on the side of the CCT3 apical domain facing the CCT6 subunit and not the center of the folding cavity (Fig. **3**). However, the lack of contacts between the apical domains in the open form makes this region fully accessible for substrate binding. In the transition to the closed form, contacts between the CCT3 and CCT6 apical domains make this region inaccessible (Fig. **3**), suggesting a mechanism of release of the substrate into the CCT folding cavity upon ATP hydrolysis and closure of the cavity.

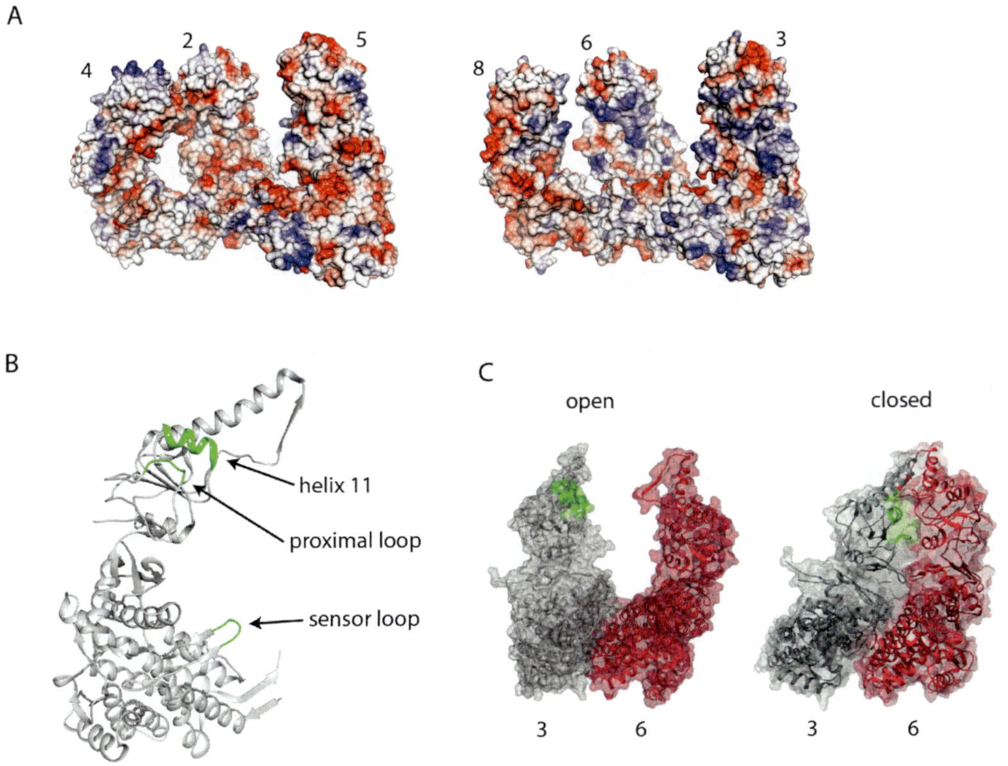


**Fig. (3). The nature of substrate binding to CCT.** A) Electrostatic charge distribution of opposite sides of the CCT ring, as viewed from the interior of the CCT cavity. Numbers identify the CCT subunits. B) Substrate binding sites are shown in green in the apical domain at the interface of helix 11 and the proximal loop and in the equatorial domain at the sensor loop between β-strands 2 and 3. Binding sites are mapped on the CCT3 subunit. C) The substrate binding sites in the helix 11 and the proximal loop region are only accessible in the open form of CCT. The exterior faces of CCT subunits 3 and 6 are shown in the open and closed forms, with the substrate binding site mapped on CCT3 in green. Structures are adapted from PDB entries 4B2T and 4V8R.

Other CCT substrates also bind in the helix 11/proximal loop region of the apical domain. Modeling and alanine mutagenesis experiments identified this region of CCT1 as a recognition site for the nine amino acid Box 1 peptide of the substrate VHL, shown previously to bind the CCT1 apical domain [40]. Furthermore, XL-MS found cross-links between the intact CCT complex and full-length substrates tubulin, actin, and the HIV Gag protein that were proximal to this same region of other CCT subunits [40]. Collectively, these findings identify helix 11 and the proximal loop as a major substrate-binding site on the apical domain of CCT. The shape of the shallow groove may be important for recognition by multiple substrates with diverse sequences because its openness allows substrate surfaces to probe the groove for binding complementarity.

In addition to this binding site in the apical domain, there is also evidence for substrate interactions with the CCT equatorial domains and the base of the folding cavity (Fig. **3**). In the 5.5 Å crystal structure of tubulin bound to the open form of CCT, tubulin is in contact with the sensor loop, a β-hairpin within the equatorial domain near the bottom of the CCT cavity that is proximal to the ATP binding site in the equatorial domain [36]. This interaction suggests a connection between substrate binding and ATP hydrolysis. Two other independent studies employing XL-MS found several cross-links between two β-propeller proteins, the G protein β subunit and the PP2A regulatory subunit 2ABG, and N- and C-terminal residues at the bottom of the folding cavity [73, 81]. Each of these studies examined purified complexes that could have potentially undergone several rounds of ATP binding and hydrolysis prior to cell lysis and purification. Thus, the complexes likely represent later intermediates in the folding process.

## CCT Binding Sites on Substrates

The chemical diversity of the substrate binding sites on CCT suggests that the CCT binding sites on substrates would also show diversity, much more than the very hydrophobic sites of substrates that bind GroEL [80]. This notion has proven true for substrates with clearly identified binding sites. For example, NMR studies showed that residues $S_{41}LFGN_{45}$ of p6 interacted directly with CCT3 [40]. This binding site has two central hydrophobic amino acids flanked by hydrophilic residues on the ends. Another well-defined binding site is the Box 1 sequence of VHL ($Y_{112}RGHLWLFR_{120}$) that binds CCT1 [75]. This site is more hydrophobic than the p6 site, matching the hydrophobicity of its binding site on CCT1 [40]. The helix 11/proximal loop region of CCT1 is the most hydrophobic of all the CCT subunits [40]. A third example can be found in the CCT3 binding site of Gβ, which was determined by site-specific, unnatural amino acid cross-linking [81]. This site resides along an N-terminal amphipathic helix of Gβ that forms a coiled-coil interaction with its obligate binding partner Gγ. The three hydrophobic residues involved in CCT3 contacts (I18, A26 and I33) flank other hydrophilic residues that make up the amphipathic helix [81]. These examples paint a picture in which substrates bind specific CCT subunits through complimentary interactions that are both hydrophobic and hydrophilic in nature.

The interaction of the Gβ coiled-coil region with CCT3 points out another common property of substrate binding sites. These sites are often involved in other interactions with physiological binding partners, and association of these interaction-prone sites with CCT may protect them from spurious interactions until the appropriate binding partner is present [40]. In the case of Gβ, binding of its coiled-coil domain to the CCT3 apical domain may sequester the site until Gγ is available to bind Gβ [81]. A similar situation is seen with β-tubulin. XL-MS and modeling suggested an interaction near K324 in a loop in the β-tubulin C-terminal domain with the CCT6 apical domain [40]. This loop interacts with the tubulin assembly co-factor A and is also part of the tubulin αβ dimer interface [82]. Like Gβ, CCT may also protect this binding surface from aggregation or other incorrect interactions.

## MECHANISM OF FOLDING

Gaining a clear understanding of the mechanism by which substrate folding is facilitated by CCT has been a major challenge in the chaperonin field for some time. CCT has been proposed to act in two different ways: i) by passively protecting the substrate from aggregation by encapsulating it and allowing it to fold in a sequestered "Anfinsen cage" environment, or ii) by actively promoting productive folding pathways through dynamic interactions with the substrate during the CCT ATPase cycle. Insight into the CCT folding mechanism can be gleaned from kinetic studies of GroEL-mediated folding. In the case of the GroEL substrate rhodanase, the rate of refolding after dilution from denaturant was the same in the presence or absence of GroEL [83]. In contrast, the rate of refolding of the TIM barrel protein DapA was accelerated 30-fold by GroEL and the folding trajectory was changed from a slow, highly concerted path in the absence of GroEL to a rapid, step-wise path in its presence [84]. This later study proposed that GroEL permits stepwise folding by confining DapA in a small negatively charged cage, thus restricting its conformational possibilities and favoring local and long-range intramolecular contacts [84]. Thus with GroEL, whether the folding mechanism is passive or active depends on the substrate.

Whether CCT employs a similar folding mechanism as GroEL is unknown, but understanding of its folding pathways is beginning to emerge from recent studies. Several substrates have been shown to bind the open form of CCT at two or more sites involving subunits on opposite sides of the folding cavity [40, 44, 45]. This arrangement stretches the substrate across the cavity and opens up partially folded intermediates (Fig. **4**). In this manner, CCT may act as an unfoldase by initially disrupting unproductive folding pathways. Evidence for CCT-dependent stretching of a substrate has been observed in fluorescence energy transfer studies with actin [85]. Folding begins upon ATP binding and hydrolysis, which transitions CCT to the closed conformation. In the closed conformation, the substrate-binding site in the helix 11/proximal loop region is masked by interactions between adjacent CCT subunits (Fig. **3C**). This masking would sterically hinder substrate binding in this region, resulting in release of the substrate into the folding cavity [40].

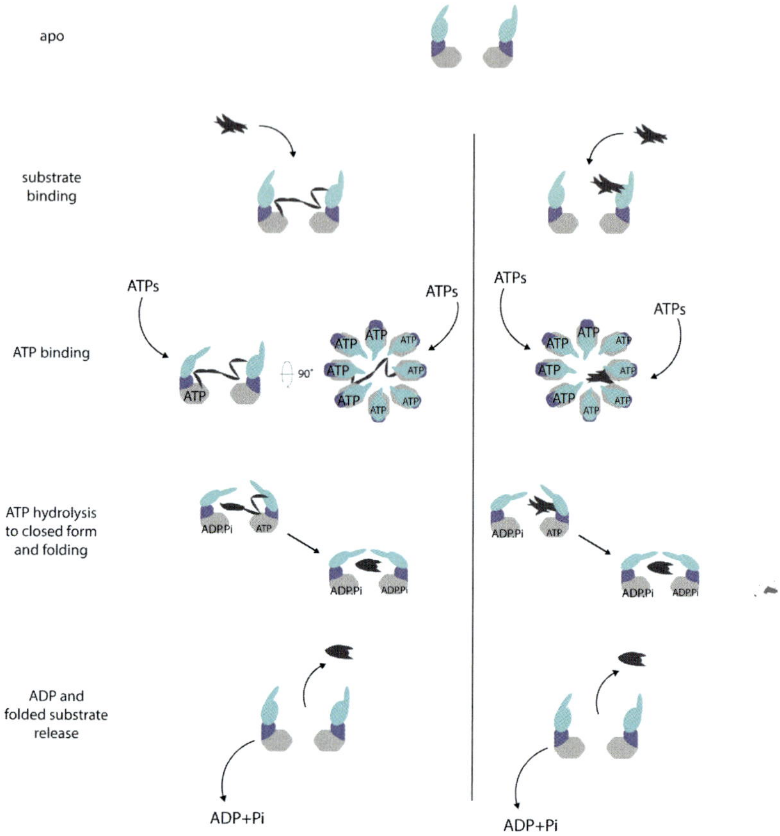


**Fig. (4). Stepwise CCT folding process.** The scheme represents a single ring of CCT as substrate folding occurs within the folding cavity. The substrate-free apo form of CCT is in an open conformation. Substrate binding can occur across the CCT folding cavity in a manner that unfolds the partially folded substrate (left panel) or without significant unfolding of the substrate (right panel). ATP binding begins on the side of the ring with the high affinity ATP subunits (ATP) and proceeds to the low affinity subunits on the other side of the ring (ATP) in a non-concerted manner. ATP hydrolysis induces closure of the cavity and release of the substrate into the cavity, first on the high ATP affinity side of the ring and then on the low ATP affinity side. This sequential release into the folding cavity permits individual domains to fold in a stepwise manner. Phosphate release opens the CCT cavity, allowing the folded substrate to exit CCT into the cytosol.

However, not all CCT subunits are equally effective in hydrolyzing ATP and transitioning to the closed conformation [86]. In fact, the CCT ring is partitioned with regard to the ATP hydrolysis capacity of its subunits. Strong ATPase subunits, CCT7, CCT5, CCT2, CCT4 and CCT1 are found on one side of the ring while weak ATPase subunits, CCT3, CCT6 and CCT8 are found on the other side of the ring [56, 87]. Phylogenetic analysis of CCT subunits suggest that the differences in ATPase activity arose from multiple gene duplication events that resulted in four subgroups represented by I) CCT6, II) CCT3, III) CCT8 and IV) CCT1/2/4/5/7, with subgroup I being the most ancient and subgroup IV the most recent [87].

Partitioning of the ATPase activity has several important implications for folding. Substrates like tubulin, which is proposed to bind CCT2 on one side of the ring and CCT6 on the other side of the ring [40, 44], may first release from CCT2 on the strong ATPase side of the ring before releasing from CCT6 on the weak ATPase side of the ring. This sequential release mechanism would alter the folding trajectory by releasing one part of the substrate into the folding cavity before the other part. Interestingly, X-ray structural analyses of tubulin-CCT [36] and actin-CCT substrate-bound complexes [42], which both have binding contacts across the ring, showed the substrate associated with only the weak ATPase side of the ring. Furthermore, each of these substrates appeared have reached a quasi-folded state within the ring. Thus it is possible that they achieved this folded state by releasing from the strong ATPase side of the ring, while maintaining their contacts with the weak ATPase side of the ring. In a sequential substrate-folding model, these structures would represent late intermediates in the folding process, and subsequent ATP binding and hydrolysis from the weak side of the ring would permit final folding and release of the substrate from the folding cavity. A model depicting the sequential release mechanism for substrate folding is shown in Fig. (**4**) (left panel).

It is not clear what fraction of CCT substrates is folded by a sequential release mechanism. The CCT substrate Gβ, which is one of many β-propeller proteins that are folded by CCT, has had binding sites identified at the bottom of the ring and in the apical domain of the CCT3 subunit on the weak ATPase side of the ring, but none have yet been found on the strong ATPase side of the ring [81]. Thus, it is unclear whether the unfoldase activity initiated by binding across the CCT ring is necessary for Gβ folding. However, it seems likely that encapsulation by CCT would greatly accelerate Gβ folding (Fig. **4** right panel). This assertion is based on similarities between the β-propeller structure of Gβ and the TIM barrel structure of DapA. Both join disparate regions from the sequence into a circular structure, and encapsulation by GroEL is believed to be responsible for accelerating DapA folding into its circular conformation [84]. Collectively, these observations provide a glimpse of what is happening inside the CCT folding cavity but do not give a clear picture of the mechanism of CCT-dependent protein folding. To directly address the mechanism, sophisticated kinetic experiments like those done with GroEL and DapA are needed to understand how CCT folds substrates from different structural classes, whether by sequential release or by simple encapsulation.

## CCT CO-CHAPERONES

### Hsc70

CCT does not work alone in folding nascent polypeptide chains, but acts in conjunction with other chaperones as part of the larger chaperone network. Key components of this network are the cytosolic Hsp70 proteins, in particular the constitutive isoform Hsc70. This chaperone binds its substrates during or shortly after their synthesis, protecting them from aggregation and promoting local folding [1]. Some of these substrates are then delivered to CCT for more extensive folding [88, 89]. The structure of Hsc70 bound to CCT suggests how this process might occur [89]. The nucleotide-binding domain of Hsc70 binds the exterior of the CCT ring in a position that would allow substrate transfer from Hsc70 to the CCT folding cavity (Fig. **5**). However, how this transfer occurs is still an open question.

### PFD

Prefoldin (PFD) is another chaperone that brings substrates to CCT for folding. The PFD complex consists of six homologous polypeptides that form a unique structure well suited for its function (Fig. **5**). The structure resembles a jellyfish. with a body made up of a double β-barrel and six tentacles consisting of long coiled-coils from each of the six subunits [91].

The coiled-coils contain hydrophobic patches near their tips that bind non-native substrates in an extended conformation, as seen in the EM structure of PFD bound to actin [90]. A cryo-EM structure of PFD bound to CCT shows PFD interacting with the CCT apical domains above the folding cavity with its coils extending into the cavity. This orientation would position substrates for direct transfer from PFD to CCT [90]. Actin and tubulin are important PFD substrates [92, 93], but recent studies suggest that PFD has other substrates. A chaperone interactome screen identified Gβ subunits and components of the cytoplasmic RNA polymerase assembly complex R2TP as PFD interacting partners [94]. Gβ is a known CCT substrate and could also be delivered to CCT by PFD.

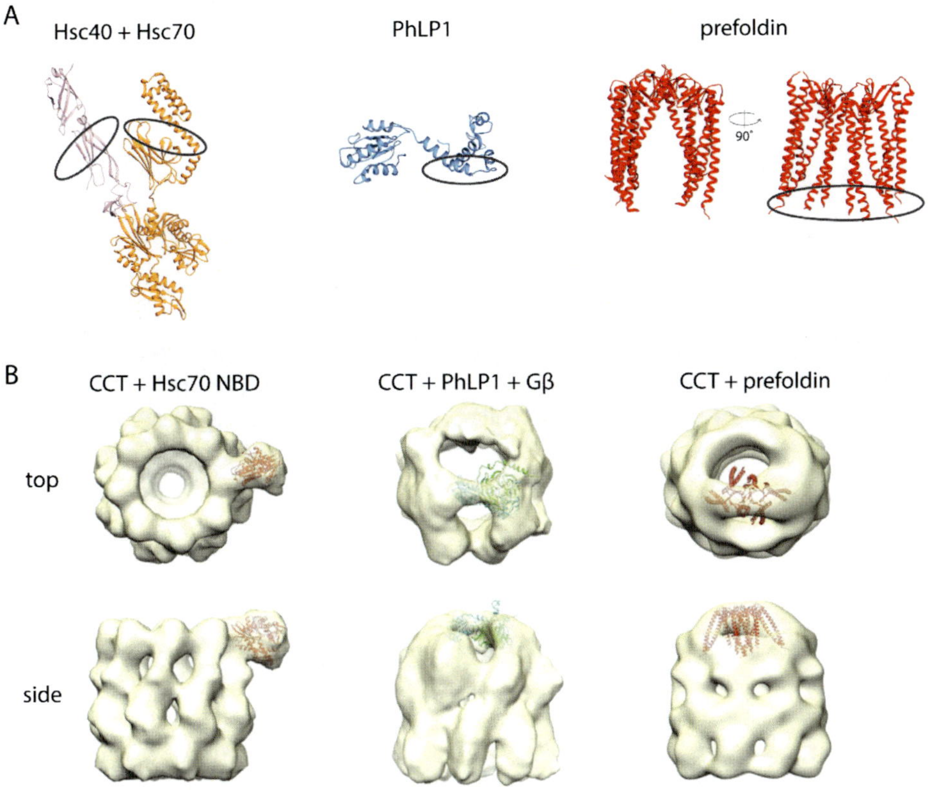


**Fig. (5). CCT co-chaperones.** Several proteins act as co-chaperones to assist in substrate folding by CCT. Shown here are the Hsc40-70 complex, PhLP1, and prefoldin. **A)** The substrate-binding region of each co-chaperone is circled on the crystal structure. Structures are adapted from PDB entries 2KHO, 1NLT, 2TRC, and 1FXK. **B)** Top and side views are shown of each co-chaperone binding to CCT, with the co-chaperone atomic structure docked into a cryo-EM reconstruction of the co-chaperone in complex with CCT [81, 89, 90]. The reconstruction with Hsc70 and CCT contains only the N-terminal, nucleotide binding domain of Hsc70. The reconstruction with PhLP1 and CCT includes PhLP1 and the folding substrate $G\beta_1$, whose atomic structure is also present in the docking.

## PhLP1

Another CCT co-chaperone, phosducin-like protein 1 (PhLP1), plays a critical role in Gβ subunit folding. Unlike co-chaperones that deliver substrates to CCT, PhLP1 participates in final folding and release of Gβ subunits from CCT [68, 69, 81, 95 - 97]. A recent structural analysis revealed important ways in which PhLP1 participates in Gβ folding and assembly with the G protein γ subunit (Gγ). The Gβ-CCT and PhLP1-Gβ-CCT structures were investigated using hybrid structural techniques of cryo-EM, XL-MS and site-specific unnatural amino acid cross-linking [81]. In the Gβ-CCT structure, Gβ binds within the CCT folding cavity in association with the CCT3 subunit on the weak ATPase side of the cavity. Gβ appears to have reached a near native state, but the β-propeller is not completely closed in a stable structure. The Gβ N-terminal region that forms a coiled-coil with Gγ is in close contact with the apical domain of CCT3. This contact creates a stable Gβ-CCT complex in which Gβ remains tightly bound to CCT in the absence of PhLP1.

Subsequent analysis of the PhLP1-Gβ-CCT structure showed how PhLP1 catalyzed the release of Gβ from CCT. PhLP1 binds Gβ from above the CCT folding cavity, making contacts with both Gβ and CCT (Fig. **5**). PhLP1 spans the cavity with its C-terminal domain interacting with CCT2 and its N-terminal, Gβ binding domain interacting with CCT3. This arrangement places the PhLP1 N-terminal domain in position to bind Gβ in the CCT folding cavity. Binding of PhLP1 to Gβ stabilizes its β-propeller structure and disrupts interactions of the N-terminal coiled-coil with CCT3, which allows the release of a PhLP1-Gβ complex from CCT. This release frees the N-terminal region to make a coiled-coil interaction with Gγ and assemble the Gβγ dimer. In this manner, PhLP1 acts as a Gβ release factor to allow the formation of Gβγ, an essential component of G protein signaling systems.

### Pdcd5

Pdcd5 is a tumor suppressor protein with a pro-apoptotic function. However, the mechanism by which Pdcd5 triggers apoptosis is not clear. Pdcd5 has been shown to bind CCT and inhibit β-tubulin folding [98]. Interestingly, Pdcd5 interacts with the apical domain of CCT2 and projects above the folding cavity without entering it. CCT2 also harbors one of the β-tubulin binding sites on CCT [40]. Thus, Pdcd5 may inhibit β-tubulin folding by sterically interfering with β-tubulin binding to CCT. The extent to which this interference contributes to the tumor suppressor function of Pdcd5 has yet to be determined, but given the importance of β-tubulin in cell division and proliferation, Pdcd5-mediated inhibition of β-tubulin folding could play a significant role.

## CONCLUDING REMARKS

CCT performs an essential role in maintaining the proteome in eukaryotes by folding nascent polypeptides with complex tertiary structures and by assisting in the assembly of polypeptides into multi-protein complexes. The advanced protein folding capabilities of CCT and its co-chaperones parallel the increased complexity of the eukaryotic proteome, suggesting a co-evolutionary relationship. Recent structural studies have increased our understanding of the subunit organization within the CCT oligomer, of substrate recognition by CCT and of the contribution of co-chaperones to the folding process. Future work will likely focus on atomic resolution structures of bound substrates to identify specific contacts and on detailed kinetic analyses to determine the mechanism by which CCT assists in the folding of different substrate classes. The knowledge gained from this work will make CCT a more accessible therapeutic target to treat diseases associated with misfolding of its substrates.

## CONSENT FOR PUBLICATION

Not applicable.

## CONFLICT OF INTEREST

The author declares no conflict of interest, financial or otherwise.

## ACKNOWLEDGEMENTS

This work was supported by US NIH grant EY012287 to BMW and grant BFU2016-44202 (AEI/FEDER, UE) to JMV.

## REFERENCES

[1] Hartl FU, Bracher A, Hayer-Hartl M. Molecular chaperones in protein folding and proteostasis. Nature 2011; 475(7356): 324-32.
[http://dx.doi.org/10.1038/nature10317] [PMID: 21776078]

[2] Kubelka J, Hofrichter J, Eaton WA. The protein folding 'speed limit'. Curr Opin Struct Biol 2004; 14(1): 76-88.
[http://dx.doi.org/10.1016/j.sbi.2004.01.013] [PMID: 15102453]

[3] Bartlett AI, Radford SE. An expanding arsenal of experimental methods yields an explosion of insights into protein folding mechanisms. Nat Struct Mol Biol 2009; 16(6): 582-8.
[http://dx.doi.org/10.1038/nsmb.1592] [PMID: 19491935]

[4] van den Berg B, Ellis RJ, Dobson CM. Effects of macromolecular crowding on protein folding and aggregation. EMBO J 1999; 18(24): 6927-33.
[http://dx.doi.org/10.1093/emboj/18.24.6927] [PMID: 10601015]

[5] Lodish HF, Berk A, Kaiser CA, *et al.* Molecular Cell Biology. 6th ed., New York: W.H. Freeman 2007.

[6] Hartl FU. Chaperone-assisted protein folding: the path to discovery from a personal perspective. Nat Med 2011; 17(10): 1206-10.
[http://dx.doi.org/10.1038/nm.2467] [PMID: 21989011]

[7] Rüdiger S, Buchberger A, Bukau B. Interaction of Hsp70 chaperones with substrates. Nat Struct Biol 1997; 4(5): 342-9.
[http://dx.doi.org/10.1038/nsb0597-342] [PMID: 9145101]

[8] Mayer MP. Gymnastics of molecular chaperones. Mol Cell 2010; 39(3): 321-31.
[http://dx.doi.org/10.1016/j.molcel.2010.07.012] [PMID: 20705236]

[9] Kampinga HH, Craig EA. The HSP70 chaperone machinery: J proteins as drivers of functional specificity. Nat Rev Mol Cell Biol 2010; 11(8): 579-92.
[http://dx.doi.org/10.1038/nrm2941] [PMID: 20651708]

[10] Frydman J, Erdjument-Bromage H, Tempst P, Hartl FU. Co-translational domain folding as the

structural basis for the rapid de novo folding of firefly luciferase. Nat Struct Biol 1999; 6(7): 697-705. [http://dx.doi.org/10.1038/10754] [PMID: 10404229]

[11] Thulasiraman V, Yang CF, Frydman J. *In vivo* newly translated polypeptides are sequestered in a protected folding environment. EMBO J 1999; 18(1): 85-95. [http://dx.doi.org/10.1093/emboj/18.1.85] [PMID: 9878053]

[12] Shiau AK, Harris SF, Southworth DR, Agard DA. Structural Analysis of E. coli hsp90 reveals dramatic nucleotide-dependent conformational rearrangements. Cell 2006; 127(2): 329-40. [http://dx.doi.org/10.1016/j.cell.2006.09.027] [PMID: 17055434]

[13] Rutherford SL, Lindquist S. Hsp90 as a capacitor for morphological evolution. Nature 1998; 396(6709): 336-42. [http://dx.doi.org/10.1038/24550] [PMID: 9845070]

[14] Horwich AL, Fenton WA. Chaperonin-mediated protein folding: using a central cavity to kinetically assist polypeptide chain folding. Q Rev Biophys 2009; 42(2): 83-116. [http://dx.doi.org/10.1017/S0033583509004764] [PMID: 19638247]

[15] Tang YC, Chang HC, Roeben A, *et al.* Structural features of the GroEL-GroES nano-cage required for rapid folding of encapsulated protein. Cell 2006; 125(5): 903-14. [http://dx.doi.org/10.1016/j.cell.2006.04.027] [PMID: 16751100]

[16] Chakraborty K, Chatila M, Sinha J, *et al.* Chaperonin-catalyzed rescue of kinetically trapped states in protein folding. Cell 2010; 142(1): 112-22. [http://dx.doi.org/10.1016/j.cell.2010.05.027] [PMID: 20603018]

[17] Arndt V, Rogon C, Höhfeld J. To be, or not to be--molecular chaperones in protein degradation. Cell Mol Life Sci 2007; 64(19-20): 2525-41. [http://dx.doi.org/10.1007/s00018-007-7188-6] [PMID: 17565442]

[18] Arndt V, Dick N, Tawo R, *et al.* Chaperone-assisted selective autophagy is essential for muscle maintenance. Curr Biol 2010; 20(2): 143-8. [http://dx.doi.org/10.1016/j.cub.2009.11.022] [PMID: 20060297]

[19] Bolognesi B, Kumita JR, Barros TP, *et al.* ANS binding reveals common features of cytotoxic amyloid species. ACS Chem Biol 2010; 5(8): 735-40. [http://dx.doi.org/10.1021/cb1001203] [PMID: 20550130]

[20] Behrends C, Langer CA, Boteva R, *et al.* Chaperonin TRiC promotes the assembly of polyQ expansion proteins into nontoxic oligomers. Mol Cell 2006; 23(6): 887-97. [http://dx.doi.org/10.1016/j.molcel.2006.08.017] [PMID: 16973440]

[21] Morley JF, Morimoto RI. Regulation of longevity in Caenorhabditis elegans by heat shock factor and molecular chaperones. Mol Biol Cell 2004; 15(2): 657-64. [http://dx.doi.org/10.1091/mbc.E03-07-0532] [PMID: 14668486]

[22] Saibil H. Chaperone machines for protein folding, unfolding and disaggregation. Nat Rev Mol Cell Biol 2013; 14(10): 630-42. [http://dx.doi.org/10.1038/nrm3658] [PMID: 24026055]

[23] Saibil HR, Fenton WA, Clare DK, Horwich AL. Structure and allostery of the chaperonin GroEL. J Mol Biol 2013; 425(9): 1476-87. [http://dx.doi.org/10.1016/j.jmb.2012.11.028] [PMID: 23183375]

[24] Braig K, Otwinowski Z, Hegde R, *et al.* The crystal structure of the bacterial chaperonin GroEL at 2.8 A. Nature 1994; 371(6498): 578-86. [http://dx.doi.org/10.1038/371578a0] [PMID: 7935790]

[25] Braig K, Adams PD, Brünger AT. Conformational variability in the refined structure of the chaperonin GroEL at 2.8 A resolution. Nat Struct Biol 1995; 2(12): 1083-94. [http://dx.doi.org/10.1038/nsb1295-1083] [PMID: 8846220]

[26] Fenton WA, Kashi Y, Furtak K, Horwich AL. Residues in chaperonin GroEL required for polypeptide binding and release. Nature 1994; 371(6498): 614-9. [http://dx.doi.org/10.1038/371614a0] [PMID: 7935796]

[27] Xu Z, Horwich AL, Sigler PB. The crystal structure of the asymmetric GroEL-GroES-(ADP)7 chaperonin complex. Nature 1997; 388(6644): 741-50. [http://dx.doi.org/10.1038/41944] [PMID: 9285585]

[28] Chaudhry C, Farr GW, Todd MJ, *et al.* Role of the gamma-phosphate of ATP in triggering protein folding by GroEL-GroES: function, structure and energetics. EMBO J 2003; 22(19): 4877-87. [http://dx.doi.org/10.1093/emboj/cdg477] [PMID: 14517228]

[29] Ranson NA, Clare DK, Farr GW, Houldershaw D, Horwich AL, Saibil HR. Allosteric signaling of ATP hydrolysis in GroEL-GroES complexes. Nat Struct Mol Biol 2006; 13(2): 147-52. [http://dx.doi.org/10.1038/nsmb1046] [PMID: 16429154]

[30] Rye HS, Roseman AM, Chen S, *et al.* GroEL-GroES cycling: ATP and nonnative polypeptide direct alternation of folding-active rings. Cell 1999; 97(3): 325-38. [http://dx.doi.org/10.1016/S0092-8674(00)80742-4] [PMID: 10319813]

[31] Kalisman N, Adams CM, Levitt M. Subunit order of eukaryotic TRiC/CCT chaperonin by cross-linking, mass spectrometry, and combinatorial homology modeling. Proc Natl Acad Sci USA 2012; 109(8): 2884-9. [http://dx.doi.org/10.1073/pnas.1119472109] [PMID: 22308438]

[32] Leitner A, Joachimiak LA, Bracher A, *et al.* The molecular architecture of the eukaryotic chaperonin TRiC/CCT. Structure 2012; 20(5): 814-25. [http://dx.doi.org/10.1016/j.str.2012.03.007] [PMID: 22503819]

[33] Ditzel L, Löwe J, Stock D, *et al.* Crystal structure of the thermosome, the archaeal chaperonin and homolog of CCT. Cell 1998; 93(1): 125-38. [http://dx.doi.org/10.1016/S0092-8674(00)81152-6] [PMID: 9546398]

[34] Cong Y, Baker ML, Jakana J, *et al.* 4.0-A resolution cryo-EM structure of the mammalian chaperonin TRiC/CCT reveals its unique subunit arrangement. Proc Natl Acad Sci USA 2010; 107(11): 4967-72. [http://dx.doi.org/10.1073/pnas.0913774107] [PMID: 20194787]

[35] Pereira JH, Ralston CY, Douglas NR, *et al.* Crystal structures of a group II chaperonin reveal the open and closed states associated with the protein folding cycle. J Biol Chem 2010; 285(36): 27958-66. [http://dx.doi.org/10.1074/jbc.M110.125344] [PMID: 20573955]

[36] Muñoz IG, Yébenes H, Zhou M, *et al.* Crystal structure of the open conformation of the mammalian chaperonin CCT in complex with tubulin. Nat Struct Mol Biol 2011; 18(1): 14-9. [http://dx.doi.org/10.1038/nsmb.1971] [PMID: 21151115]

[37] Huo Y, Hu Z, Zhang K, *et al.* Crystal structure of group II chaperonin in the open state. Structure 2010; 18(10): 1270-9. [http://dx.doi.org/10.1016/j.str.2010.07.009] [PMID: 20947016]

[38] Yébenes H, Mesa P, Muñoz IG, Montoya G, Valpuesta JM. Chaperonins: two rings for folding. Trends Biochem Sci 2011; 36(8): 424-32. [http://dx.doi.org/10.1016/j.tibs.2011.05.003] [PMID: 21723731]

[39] Kubota H, Hynes G, Willison K. The chaperonin containing t-complex polypeptide 1 (TCP-1). Multisubunit machinery assisting in protein folding and assembly in the eukaryotic cytosol. Eur J Biochem 1995; 230(1): 3-16. [http://dx.doi.org/10.1111/j.1432-1033.1995.tb20527.x] [PMID: 7601114]

[40] Joachimiak LA, Walzthoeni T, Liu CW, Aebersold R, Frydman J. The structural basis of substrate recognition by the eukaryotic chaperonin TRiC/CCT. Cell 2014; 159(5): 1042-55. [http://dx.doi.org/10.1016/j.cell.2014.10.042] [PMID: 25416944]

[41] Zhang J, Baker ML, Schröder GF, *et al.* Mechanism of folding chamber closure in a group II chaperonin. Nature 2010; 463(7279): 379-83.
[http://dx.doi.org/10.1038/nature08701] [PMID: 20090755]

[42] Dekker C, Roe SM, McCormack EA, Beuron F, Pearl LH, Willison KR. The crystal structure of yeast CCT reveals intrinsic asymmetry of eukaryotic cytosolic chaperonins. EMBO J 2011; 30(15): 3078-90.
[http://dx.doi.org/10.1038/emboj.2011.208] [PMID: 21701561]

[43] Shomura Y, Yoshida T, Iizuka R, Maruyama T, Yohda M, Miki K. Crystal structures of the group II chaperonin from Thermococcus strain KS-1: steric hindrance by the substituted amino acid, and inter-subunit rearrangement between two crystal forms. J Mol Biol 2004; 335(5): 1265-78.
[http://dx.doi.org/10.1016/j.jmb.2003.11.028] [PMID: 14729342]

[44] Llorca O, Martín-Benito J, Ritco-Vonsovici M, *et al.* Eukaryotic chaperonin CCT stabilizes actin and tubulin folding intermediates in open quasi-native conformations. EMBO J 2000; 19(22): 5971-9.
[http://dx.doi.org/10.1093/emboj/19.22.5971] [PMID: 11080144]

[45] Llorca O, McCormack EA, Hynes G, *et al.* Eukaryotic type II chaperonin CCT interacts with actin through specific subunits. Nature 1999; 402(6762): 693-6.
[http://dx.doi.org/10.1038/45294] [PMID: 10604479]

[46] Cong Y, Schröder GF, Meyer AS, *et al.* Symmetry-free cryo-EM structures of the chaperonin TRiC along its ATPase-driven conformational cycle. EMBO J 2012; 31(3): 720-30.
[http://dx.doi.org/10.1038/emboj.2011.366] [PMID: 22045336]

[47] Douglas NR, Reissmann S, Zhang J, *et al.* Dual action of ATP hydrolysis couples lid closure to substrate release into the group II chaperonin chamber. Cell 2011; 144(2): 240-52.
[http://dx.doi.org/10.1016/j.cell.2010.12.017] [PMID: 21241893]

[48] Rüßmann F, Stemp MJ, Mönkemeyer L, Etchells SA, Bracher A, Hartl FU. Folding of large multidomain proteins by partial encapsulation in the chaperonin TRiC/CCT. Proc Natl Acad Sci USA 2012; 109(52): 21208-15.
[http://dx.doi.org/10.1073/pnas.1218836109] [PMID: 23197838]

[49] Meyer AS, Gillespie JR, Walther D, Millet IS, Doniach S, Frydman J. Closing the folding chamber of the eukaryotic chaperonin requires the transition state of ATP hydrolysis. Cell 2003; 113(3): 369-81.
[http://dx.doi.org/10.1016/S0092-8674(03)00307-6] [PMID: 12732144]

[50] Reissmann S, Parnot C, Booth CR, Chiu W, Frydman J. Essential function of the built-in lid in the allosteric regulation of eukaryotic and archaeal chaperonins. Nat Struct Mol Biol 2007; 14(5): 432-40.
[http://dx.doi.org/10.1038/nsmb1236] [PMID: 17460696]

[51] Kim YE, Hipp MS, Bracher A, Hayer-Hartl M, Hartl FU. Molecular chaperone functions in protein folding and proteostasis. Annu Rev Biochem 2013; 82: 323-55.
[http://dx.doi.org/10.1146/annurev-biochem-060208-092442] [PMID: 23746257]

[52] Farr GW, Scharl EC, Schumacher RJ, Sondek S, Horwich AL. Chaperonin-mediated folding in the eukaryotic cytosol proceeds through rounds of release of native and nonnative forms. Cell 1997; 89(6): 927-37.
[http://dx.doi.org/10.1016/S0092-8674(00)80278-0] [PMID: 9200611]

[53] Liou AK, Willison KR. Elucidation of the subunit orientation in CCT (chaperonin containing TCP1) from the subunit composition of CCT micro-complexes. EMBO J 1997; 16(14): 4311-6.
[http://dx.doi.org/10.1093/emboj/16.14.4311] [PMID: 9250675]

[54] Martín-Benito J, Grantham J, Boskovic J, *et al.* The inter-ring arrangement of the cytosolic chaperonin CCT. EMBO Rep 2007; 8(3): 252-7.
[http://dx.doi.org/10.1038/sj.embor.7400894] [PMID: 17304242]

[55] Politis A, Stengel F, Hall Z, *et al.* A mass spectrometry-based hybrid method for structural modeling of protein complexes. Nat Methods 2014; 11(4): 403-6.

[http://dx.doi.org/10.1038/nmeth.2841] [PMID: 24509631]

[56] Kalisman N, Schröder GF, Levitt M. The crystal structures of the eukaryotic chaperonin CCT reveal its functional partitioning. Structure 2013; 21(4): 540-9. [http://dx.doi.org/10.1016/j.str.2013.01.017] [PMID: 23478063]

[57] Frydman J. Folding of newly translated proteins *in vivo*: the role of molecular chaperones. Annu Rev Biochem 2001; 70: 603-47. [http://dx.doi.org/10.1146/annurev.biochem.70.1.603] [PMID: 11395418]

[58] Richter K, Haslbeck M, Buchner J. The heat shock response: life on the verge of death. Mol Cell 2010; 40(2): 253-66. [http://dx.doi.org/10.1016/j.molcel.2010.10.006] [PMID: 20965420]

[59] Frydman J, Nimmesgern E, Erdjument-Bromage H, Wall JS, Tempst P, Hartl FU. Function in protein folding of TRiC, a cytosolic ring complex containing TCP-1 and structurally related subunits. EMBO J 1992; 11(13): 4767-78. [PMID: 1361170]

[60] Sternlicht H, Farr GW, Sternlicht ML, Driscoll JK, Willison K, Yaffe MB. The t-complex polypeptide 1 complex is a chaperonin for tubulin and actin *in vivo*. Proc Natl Acad Sci USA 1993; 90(20): 9422-6. [http://dx.doi.org/10.1073/pnas.90.20.9422] [PMID: 8105476]

[61] Yaffe MB, Farr GW, Miklos D, Horwich AL, Sternlicht ML, Sternlicht H. TCP1 complex is a molecular chaperone in tubulin biogenesis. Nature 1992; 358(6383): 245-8. [http://dx.doi.org/10.1038/358245a0] [PMID: 1630491]

[62] Gao Y, Thomas JO, Chow RL, Lee GH, Cowan NJ. A cytoplasmic chaperonin that catalyzes beta-actin folding. Cell 1992; 69(6): 1043-50. [http://dx.doi.org/10.1016/0092-8674(92)90622-J] [PMID: 1351421]

[63] Chen X, Sullivan DS, Huffaker TC. Two yeast genes with similarity to TCP-1 are required for microtubule and actin function *in vivo*. Proc Natl Acad Sci USA 1994; 91(19): 9111-5. [http://dx.doi.org/10.1073/pnas.91.19.9111] [PMID: 7916460]

[64] Ursic D, Culbertson MR. The yeast homolog to mouse Tcp-1 affects microtubule-mediated processes. Mol Cell Biol 1991; 11(5): 2629-40. [http://dx.doi.org/10.1128/MCB.11.5.2629] [PMID: 1901944]

[65] Vinh DB, Drubin DG. A yeast TCP-1-like protein is required for actin function *in vivo*. Proc Natl Acad Sci USA 1994; 91(19): 9116-20. [http://dx.doi.org/10.1073/pnas.91.19.9116] [PMID: 7916461]

[66] Yam AY, Xia Y, Lin HT, Burlingame A, Gerstein M, Frydman J. Defining the TRiC/CCT interactome links chaperonin function to stabilization of newly made proteins with complex topologies. Nat Struct Mol Biol 2008; 15(12): 1255-62. [http://dx.doi.org/10.1038/nsmb.1515] [PMID: 19011634]

[67] Dekker C, Stirling PC, McCormack EA, *et al.* The interaction network of the chaperonin CCT. EMBO J 2008; 27(13): 1827-39. [http://dx.doi.org/10.1038/emboj.2008.108] [PMID: 18511909]

[68] Lukov GL, Baker CM, Ludtke PJ, *et al.* Mechanism of assembly of G protein betagamma subunits by protein kinase CK2-phosphorylated phosducin-like protein and the cytosolic chaperonin complex. J Biol Chem 2006; 281(31): 22261-74. [http://dx.doi.org/10.1074/jbc.M601590200] [PMID: 16717095]

[69] Lukov GL, Hu T, McLaughlin JN, Hamm HE, Willardson BM. Phosducin-like protein acts as a molecular chaperone for G protein betagamma dimer assembly. EMBO J 2005; 24(11): 1965-75. [http://dx.doi.org/10.1038/sj.emboj.7600673] [PMID: 15889144]

[70] Camasses A, Bogdanova A, Shevchenko A, Zachariae W. The CCT chaperonin promotes activation of

the anaphase-promoting complex through the generation of functional Cdc20. Mol Cell 2003; 12(1): 87-100.
[http://dx.doi.org/10.1016/S1097-2765(03)00244-2] [PMID: 12887895]

[71] Seo S, Baye LM, Schulz NP, *et al.* BBS6, BBS10, and BBS12 form a complex with CCT/TRiC family chaperonins and mediate BBSome assembly. Proc Natl Acad Sci USA 2010; 107(4): 1488-93.
[http://dx.doi.org/10.1073/pnas.0910268107] [PMID: 20080638]

[72] Zhang Q, Seo S, Bugge K, Stone EM, Sheffield VC. BBS proteins interact genetically with the IFT pathway to influence SHH-related phenotypes. Hum Mol Genet 2012; 21(9): 1945-53.
[http://dx.doi.org/10.1093/hmg/dds004] [PMID: 22228099]

[73] Herzog F, Kahraman A, Boehringer D, *et al.* Structural probing of a protein phosphatase 2A network by chemical cross-linking and mass spectrometry. Science 2012; 337(6100): 1348-52.
[http://dx.doi.org/10.1126/science.1221483] [PMID: 22984071]

[74] Freund A, Zhong FL, Venteicher AS, *et al.* Proteostatic control of telomerase function through TRiC-mediated folding of TCAB1. Cell 2014; 159(6): 1389-403.
[http://dx.doi.org/10.1016/j.cell.2014.10.059] [PMID: 25467444]

[75] Feldman DE, Spiess C, Howard DE, Frydman J. Tumorigenic mutations in VHL disrupt folding *in vivo* by interfering with chaperonin binding. Mol Cell 2003; 12(5): 1213-24.
[http://dx.doi.org/10.1016/S1097-2765(03)00423-4] [PMID: 14636579]

[76] Feldman DE, Thulasiraman V, Ferreyra RG, Frydman J. Formation of the VHL-elongin BC tumor suppressor complex is mediated by the chaperonin TRiC. Mol Cell 1999; 4(6): 1051-61.
[http://dx.doi.org/10.1016/S1097-2765(00)80233-6] [PMID: 10635329]

[77] Trinidad AG, Muller PA, Cuellar J, *et al.* Interaction of p53 with the CCT complex promotes protein folding and wild-type p53 activity. Mol Cell 2013; 50(6): 805-17.
[http://dx.doi.org/10.1016/j.molcel.2013.05.002] [PMID: 23747015]

[78] Tam S, Spiess C, Auyeung W, *et al.* The chaperonin TRiC blocks a huntingtin sequence element that promotes the conformational switch to aggregation. Nat Struct Mol Biol 2009; 16(12): 1279-85.
[http://dx.doi.org/10.1038/nsmb.1700] [PMID: 19915590]

[79] Kasembeli M, Lau WC, Roh SH, *et al.* Modulation of STAT3 folding and function by TRiC/CCT chaperonin. PLoS Biol 2014; 12(4): e1001844.
[http://dx.doi.org/10.1371/journal.pbio.1001844] [PMID: 24756126]

[80] Gómez-Puertas P, Martín-Benito J, Carrascosa JL, Willison KR, Valpuesta JM. The substrate recognition mechanisms in chaperonins. J Mol Recognit 2004; 17(2): 85-94.
[http://dx.doi.org/10.1002/jmr.654] [PMID: 15027029]

[81] Plimpton RL, Cuéllar J, Lai CW, *et al.* Structures of the Gβ-CCT and PhLP1-Gβ-CCT complexes reveal a mechanism for G-protein β-subunit folding and Gβγ dimer assembly. Proc Natl Acad Sci USA 2015; 112(8): 2413-8.
[http://dx.doi.org/10.1073/pnas.1419595112] [PMID: 25675501]

[82] You L, Gillilan R, Huffaker TC. Model for the yeast cofactor A-β-tubulin complex based on computational docking and mutagensis. J Mol Biol 2004; 341(5): 1343-54.
[http://dx.doi.org/10.1016/j.jmb.2004.06.081] [PMID: 15321725]

[83] Hofmann H, Hillger F, Pfeil SH, *et al.* Single-molecule spectroscopy of protein folding in a chaperonin cage. Proc Natl Acad Sci USA 2010; 107(26): 11793-8.
[http://dx.doi.org/10.1073/pnas.1002356107] [PMID: 20547872]

[84] Georgescauld F, Popova K, Gupta AJ, *et al.* GroEL/ES chaperonin modulates the mechanism and accelerates the rate of TIM-barrel domain folding. Cell 2014; 157(4): 922-34.
[http://dx.doi.org/10.1016/j.cell.2014.03.038] [PMID: 24813614]

[85] Villebeck L, Moparthi SB, Lindgren M, Hammarström P, Jonsson BH. Domain-specific chaperone-induced expansion is required for beta-actin folding: a comparison of beta-actin conformations upon

interactions with GroEL and tail-less complex polypeptide 1 ring complex (TRiC). Biochemistry 2007; 46(44): 12639-47.
[http://dx.doi.org/10.1021/bi700658n] [PMID: 17939680]

[86] Reissmann S, Joachimiak LA, Chen B, Meyer AS, Nguyen A, Frydman J. A gradient of ATP affinities generates an asymmetric power stroke driving the chaperonin TRIC/CCT folding cycle. Cell Reports 2012; 2(4): 866-77.
[http://dx.doi.org/10.1016/j.celrep.2012.08.036] [PMID: 23041314]

[87] Chagoyen M, Carrascosa JL, Pazos F, Valpuesta JM. Molecular determinants of the ATP hydrolysis asymmetry of the CCT chaperonin complex. Proteins 2014; 82(5): 703-7.
[http://dx.doi.org/10.1002/prot.24510] [PMID: 24420718]

[88] Melville MW, McClellan AJ, Meyer AS, Darveau A, Frydman J. The Hsp70 and TRiC/CCT chaperone systems cooperate *in vivo* to assemble the von Hippel-Lindau tumor suppressor complex. Mol Cell Biol 2003; 23(9): 3141-51.
[http://dx.doi.org/10.1128/MCB.23.9.3141-3151.2003] [PMID: 12697815]

[89] Cuéllar J, Martín-Benito J, Scheres SH, *et al.* The structure of CCT-Hsc70 NBD suggests a mechanism for Hsp70 delivery of substrates to the chaperonin. Nat Struct Mol Biol 2008; 15(8): 858-64.
[http://dx.doi.org/10.1038/nsmb.1464] [PMID: 18660820]

[90] Martín-Benito J, Boskovic J, Gómez-Puertas P, *et al.* Structure of eukaryotic prefoldin and of its complexes with unfolded actin and the cytosolic chaperonin CCT. EMBO J 2002; 21(23): 6377-86.
[http://dx.doi.org/10.1093/emboj/cdf640] [PMID: 12456645]

[91] Siegert R, Leroux MR, Scheufler C, Hartl FU, Moarefi I. Structure of the molecular chaperone prefoldin: unique interaction of multiple coiled coil tentacles with unfolded proteins. Cell 2000; 103(4): 621-32.
[http://dx.doi.org/10.1016/S0092-8674(00)00165-3] [PMID: 11106732]

[92] Geissler S, Siegers K, Schiebel E. A novel protein complex promoting formation of functional alpha- and gamma-tubulin. EMBO J 1998; 17(4): 952-66.
[http://dx.doi.org/10.1093/emboj/17.4.952] [PMID: 9463374]

[93] Vainberg IE, Lewis SA, Rommelaere H, *et al.* Prefoldin, a chaperone that delivers unfolded proteins to cytosolic chaperonin. Cell 1998; 93(5): 863-73.
[http://dx.doi.org/10.1016/S0092-8674(00)81446-4] [PMID: 9630229]

[94] Taipale M, Tucker G, Peng J, *et al.* A quantitative chaperone interaction network reveals the architecture of cellular protein homeostasis pathways. Cell 2014; 158(2): 434-48.
[http://dx.doi.org/10.1016/j.cell.2014.05.039] [PMID: 25036637]

[95] Howlett AC, Gray AJ, Hunter JM, Willardson BM. Role of molecular chaperones in G protein β5/regulator of G protein signaling dimer assembly and G protein betagamma dimer specificity. J Biol Chem 2009; 284(24): 16386-99.
[http://dx.doi.org/10.1074/jbc.M900800200] [PMID: 19376773]

[96] Lai CW, Kolesnikov AV, Frederick JM, *et al.* Phosducin-like protein 1 is essential for G-protein assembly and signaling in retinal rod photoreceptors. J Neurosci 2013; 33(18): 7941-51.
[http://dx.doi.org/10.1523/JNEUROSCI.5001-12.2013] [PMID: 23637185]

[97] Tracy CM, Kolesnikov AV, Blake DR, *et al.* Retinal cone photoreceptors require phosducin-like protein 1 for G protein complex assembly and signaling. PLoS One 2015; 10(2): e0117129.
[http://dx.doi.org/10.1371/journal.pone.0117129] [PMID: 25659125]

[98] Tracy CM, Gray AJ, Cuéllar J, *et al.* Programmed cell death protein 5 interacts with the cytosolic chaperonin containing tailless complex polypeptide 1 (CCT) to regulate β-tubulin folding. J Biol Chem 2014; 289(7): 4490-502.
[http://dx.doi.org/10.1074/jbc.M113.542159] [PMID: 24375412]

CHAPTER 7

# Mechanisms and Functions of the Cytosolic DNAJ-Hsp70 Chaperone System

**Imad Baaklini** and **Jason C. Young***

*Department of Biochemistry, Groupe de Recherche Axé sur la Structure des Protéines, McGill University, Montreal, Canada*

**Abstract:** The Hsp70 chaperone system is central to the proteostasis network in the cytosol of human cells. Hsp70 has roles in protein folding, the prevention and clearance of aggregates, and various other biological processes. A range of co-chaperone proteins regulate Hsp70, among these the Hsp40/DNAJ proteins are important activators of Hsp70 function. Stress inducible Hsp70 and its non-inducible form Hsc70 are ubiquitously expressed, and the most abundant DNAJs are DNAJA1 (Hdj2), DNAJA2 and DNAJB1 (Hsp40, Hdj1). Here, the mechanisms of Hsc70/Hsp70 with the DNAJs is reviewed. An overview of biological functions of the major DNAJs is then presented, particularly in the context of protein misfolding diseases.

**Keywords:** DNAJ, Hsp40, Hsp70, Ion channel, Molecular chaperone, Neurodegeneration, Protein aggregation, Protein degradation, Protein folding.

## INTRODUCTION

Human cells face an ongoing challenge from protein misfolding, during biosynthesis of polypeptides, stress conditions such as heat shock, genetic mutations, chemical modifications, or intrinsic instability of mature proteins. Misfolding can result in loss of function, and in certain cases protein aggregates can have toxic properties, as found in a number of neurodegenerative diseases. The system of protein homeostasis – proteostasis – is responsible for managing protein misfolding. Molecular chaperones maintain protein solubility, assist protein folding, and also disassemble protein aggregates. Oppositely, protein degradation by the ubiquitin-proteasome system and autophagy clear misfolded proteins and aggregates. The chaperone and degradation arms of proteostasis interact and cooperate with each other. Impairment of the proteostasis system may be an important underlying factor in phenotypes associated with aging [1 - 3].

* **Corresponding author Jason C. Young:** Department of Biochemistry, McGill University, 3655 Promenade Sir William Osler, Montreal, QC, Canada, H3G 1Y6; Tel: +1-514-398-2006; Email: jason.young2@mcgill.ca

**Mario D. Galigniana (Ed.)**

Chaperones are divided into structural classes, each with their own mechanisms and biological roles. Many chaperones are heat shock proteins, being highly expressed in response to stress, and are so named HSPs (however, some chaperones are not inducible, and not all HSPs are chaperones). The Hsp70 (70 kDa HSP), Hsp90 and Hsp60/chaperonin families use ATPase cycles to affect the conformations of substrate polypeptides. The small HSP family are ATP-independent and prevent substrate aggregation. The Hsp40/DNAJ family are key regulators, or co-chaperones, of Hsp70s. Among these families, the Hsp70 proteins are thought to have a central position in the cellular chaperone network. They can bind to a large number of substrates, and cooperate with chaperones from other families. In addition, they interact with the protein degradation machinery. By extension, the DNAJ co-chaperones are important in understanding the function of Hsp70s [4 - 6]. In this review, the mechanisms of the cytosolic Hsp70s and DNAJs will be described, followed by a survey of the biological functions of the most abundant DNAJ co-chaperones with Hsp70.

## HSP70

Heat shock protein 70 kDa (Hsp70) family members are found in all kingdoms, from archeabacteria to mammals, and they are among the most conserved proteins in evolution. In humans, six genes encode typical cytosolic forms of Hsp70 [7] and two others encode mitochondrial and endoplasmic reticulum (ER) localized Hsp70 forms [8]. Hsc70 and Hsp70 together represent the ubiquitously expressed, major cytosolic and nuclear chaperones associated with protein folding, protein degradation and protein targeting to organelles [4, 9]. Hsp70 (Hsp72, Hsp70-1) has two genes, *HSPA1A* and *HSPA1B*, which encode identical proteins. They are widely express and induced during stress [10, 11]. Hsc70 is encoded by *HSPA8* and is constitutively expressed but not highly induced during stress, and thus is the main housekeeping Hsp70 protein [12]. The three other genes encoding cytosolic Hsp70 proteins are: *HSPA6* also known as Hsp70B', strictly expressed during stress [13, 14]; *HSP70A1L* (Hsp70t) predominantly expressed in testis [15] and *HSPA2* (Hsp70-2) in testis and in skeletal muscle [16]. *HSPA5* (Grp78, BiP) and *HSPA9* (mtHsp70, Grp75, mortalin) are localized in the ER and mitochondria respectively and are constitutively expressed [17, 18]. *HSPA12A* and *HSPA12B*, two forms of Hsp70 greatly divergent from the others, are expressed in brain and endothelial cells, respectively, and little is known about their biology [19, 20]. *HSPA14* (Hsp70L1) is ubiquitously expressed but does not have the conserved ATPase mechanism of typical Hsp70s, instead it is part of a ribosome associated complex that recruits Hsp70 (HSPA1) to nascent polypeptides [21 - 23]. The focus of this review will be the typical cytosolic Hsp70s.

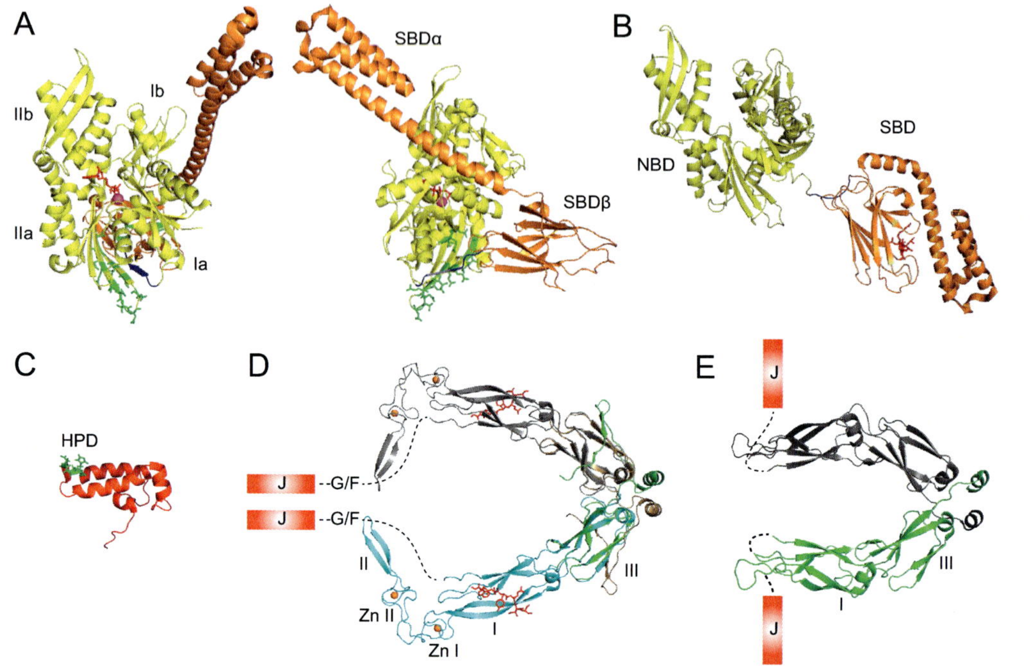


**Fig. (1).** Structures of Hsp70 and DNAJs. A, Hsp70 (DnaK) in the ATP-bound state in views rotated 90° sideways. The nucleotide-binding domain (NBD) is in yellow, the substrate binding domain (SBD) in orange, the linker in blue, ATP in red and Mg ion in magenta. Residues implicated in J domain interaction are in green. The lobes of the NBD (Ia-b and IIa-b) and SBD subdomains (α and β) are marked. From references 25, 48, 54 and 56. B, Hsp70 (DnaK) SBD with bound substrate, superimposed on DnaK in the ADP-bound state. The domains are colored as above, except peptide substrate is in red. From references 27 and 35. C, J domain of human DNAJA1. The HPD motif that interacts with Hsp70 is in green. D, Type I DNAJ (Ydj1). The central domain structure in green is superimposed on the C-terminal homodimerization structure in cyan, the opposite subunit in grey and brown. Zn ions are in orange and peptide substrate in red. The Cys(Zn) sites I and II, and subdomains I-III are marked. A proposed arrangement of the J domains and G/F linker is shown. From references 74, 75 and 88. E, Type II human DNAJB1. The central/C-terminal domain of DNAJB1 is in green and grey, subdomains I and III with homology to Type I are marked. A proposed arrangement of the J domains is shown. From references 76 and 88.

## Hsp70 Structure

A typical Hsp70 is composed of two domains, a nucleotide binding domain (NBD) and the substrate binding domain (SBD), both connected by a hydrophobic linker (Fig. **1A** and **B**). The NBD has ATPase activity, and regulates the conformation of the SBD. In the NBD structure, two lobes I and II are each one divided in two subdomains Ia-Ib, IIa-IIb [24]. The subdomains Ia and IIa join and form the nucleotide pocket and together, with other conserved amino acid from other subdomains, binds and hydrolyzes ATP [25]. The relative flexibility between subdomains allows a conformational shift from ATP to ADP state.

Hsp70 NBD structure is well conserved among Hsp70 members [26]. The SBD is also composed of two subdomains, a β-sandwich with two antiparallel four stranded β-sheets with four protruding loops, and a lid with five α-helices. Polypeptide substrate interacts with one surface of the β-sandwich subdomain including two of the loops that form a mainly hydrophobic groove within surrounding negatively charged residues [27]. Therefore, Hsp70 binds short peptide sequences with hydrophobic and moderately basic character; such sequences are found in the majority of cellular proteins [28, 29]. The α-helical subdomain is thought to act as a lid covering the binding groove, increasing affinity and lowering peptide release [25, 30]. The extreme C-terminus of Hsp70 is unstructured and in the typical cytosolic forms, contains the EEVD motif that interacts with TPR (tetratricopeptide repeat) domains found in certain co-chaperones [31].

## ATPase Cycle

In the ATP state, Hsp70 has only a poor affinity for unfolded polypeptides. However, upon ATP hydrolysis to ADP, Hsp70 has a high affinity for polypeptides, characterized by a low on- and off-rates of polypeptide binding (Fig. **2**). Cycling of Hsp70 between these two states results in polypeptide binding and release, leading to the folding of denatured polypeptides. On the one hand ATP binding in the NBD induces the opening of the α-helical lid in SBD, and on the other hand, substrate binding to the SBD leads to an increased ATP hydrolysis rate [25, 32, 33]. High resolution structures have revealed the basis of the ATP-driven Hsp70 substrate binding cycle. In the ATP-bound state, the subdomains of the SBD are separated widely and packed against the NBD, while the hydrophobic linker is packed into a groove in the NBD (Fig. **1A**) [25, 32]. In the ADP-bound state, one structure suggested different contacts between the NBD and SBD [34], however more recent work indicates that the NBD and SBD are separate from each other by the extended linker (Fig. **1B**) [35]. The structures are consistent with allosteric regulation of Hsp70 by substrate binding and nucleotide. Coordination between NBD and SBD is required in order to tune the conformational changes with productive folding. The hydrophobic linker is thought to be sensitive to conformational shifts in NBD and SBD driven by their internal subdomain flexibility upon ATP binding in the NBD [36, 37] or upon substrate binding in the SBD and to communicate these changes between the two domains [38].

In order to fulfill its functions, Hsp70 cycles between the ATP and ADP states in an essential mechanism. Although Hsp70 possesses an intrinsic ATP hydrolysis and ADP release activity, these rates are normally low and can be highly stimulated by a set of specific chaperones, the Hsp40 or DNAJ co-chaperones, and the nucleotide exchange factors (NEF), respectively [6, 39]. In human cells, and indeed most organisms, ATP is not a limiting factor because of its abundance, and Hsp70 has evolved tight regulatory mechanisms at the two states of its cycle. First, the DNAJs stimulate the Hsp70 ATP hydrolysis rate and substrate binding. Second, NEFs accelerate the slow release of ADP by Hsp70 after ATP hydrolysis, allowing ATP re-binding, release of bound substrate, and resetting of the Hsp70 cycle. Thus, both the DNAJs and the NEFs are crucial regulators of the Hsp70 cycle and function (Fig. **2**) [6].

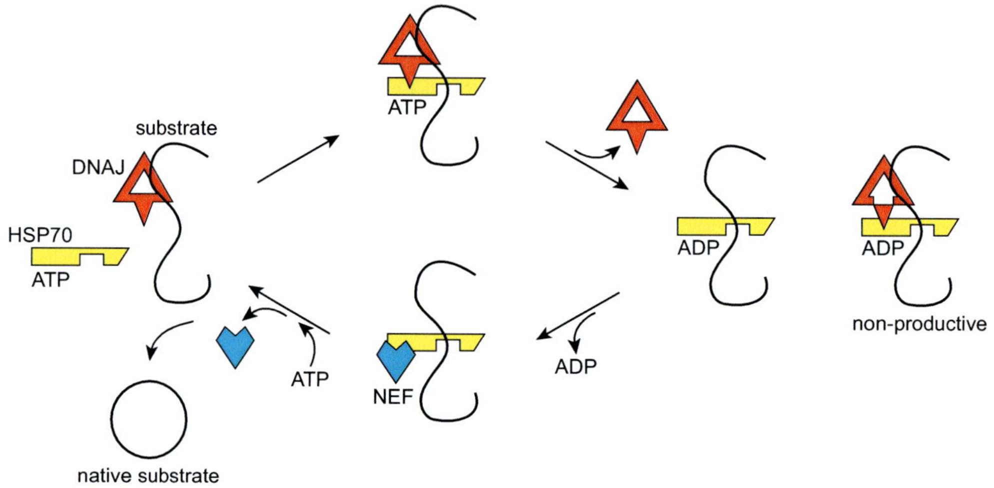


**Fig. (2).** ATPase cycle of Hsp70. From left, Type I or II DNAJ binds substrate polypeptide and Hsp70 in the ATP-bound state has low affinity for substrate. The J domain contacts Hsp70 and initiates ATP hydrolysis, in a ternary complex of DNAJ, Hsp70 and substrate. After ATP hydrolysis, the DNAJ dissociates and substrate is bound with high affinity by Hsp70 in the ADP state. Nucleotide exchange factors (NEFs) stabilize the nucleotide-free state of Hsp70 to release ADP and allow re-binding of ATP. Substrates may reach the native state, and non-native substrates return to the cycle. Right, disruption of the Cys(Zn) region of Type I DNAJs impairs transfer of substrate to Hsp70 and is non-productive for substrate folding.

# HSP40/DNAJ CO-CHAPERONES

## Classification

The DNAJ family contains the largest number of members among other families of chaperones. In humans for example, we account for at least 50 genes encoding DNAJ proteins, of which roughly half are available to cytosolic/nuclear Hsp70s [5]. DNAJs are classified in three groups based on the presence and organization of domains [40 - 42]. All members have a J domain which interacts with partner Hsp70s. Type I or A DNAJ proteins contain an N-terminal J domain followed by a glycine and phenyalanine (G/F) rich linker, and a central domain featuring a cysteine and zinc (Cys(Zn)) rich region that is continuous with a C-terminal homodimerization region. Type II or B are similar to Type I except that the Cys(Zn) region is absent, and the central/C-terminal regions are more divergent in sequence. Types I and II DNAJs are thought to bind substrates in the central/C-terminal regions. Type III or C DNAJs contain a J domain and none of the other conserved structures. The J domain can be found in various places in their sequences, and the Type III DNAJs usually contain additional domains that specify a particular function or localization; many do not bind substrates directly [5, 40].

## Domains of DNAJs

### *J Domain*

The amino acid sequences and structures of J domains are highly conserved across evolution [43]. Most of the biochemical information on DNAJs comes from studies on Hsp70 and DNAJ homologues in *Escherichia coli* (DnaK and DnaJ) and *Saccharomyces cerevisiae* (Ssa1 and Ydj1). Several structural studies of J domains by X-ray crystallography and NMR have been reported ([44 - 46], among many others). In a typical J domain, two helices facing each other are joined by a loop, which contains the absolutely conserved HPD (histidine, proline, aspartic acid) motif essential for the interaction of the J domain with Hsp70 (Fig. **1C**) [47 - 49]. No other residues are more important than the HPD to support interaction of the J domain with Hsp70, although the adjacent C-terminal residues may contribute [47]. In *E. coli* DnaJ, the mutant D35N in the HPD motif is defective, but point mutation R167H in lobe Ia of the DnaK NBD suppresses the binding defect; conversely, alanine substitution in N170 and T173 close to R167 decrease DnaJ binding to DnaK, suggesting that J domains bind on this surface (Fig. **1A**) [48]. DnaJ preferentially interacts with Hsp70 in the ATP state and the formation of DnaJ-Hsp70 complexes requires ATP hydrolysis [50]. The minimal J domain was shown to be sufficient for the interaction with DnaK and for the stimulation of ATP hydrolysis [51, 52]. An NMR study of the J domain from DnaJ shows chemical shifts on its surface around the HPD motif in the presence of DnaK-ADP and DnaK-ATP, indicating a more flexible interaction [53]. The DnaJ interaction with Hsp70 has been difficult to characterize due to its transient nature: indeed, the complexes dissociate immediately after ATP hydrolysis. Work carried out on the J domain of DNAJC6 (auxilin) and the NBD domain of Hsp70 addresses this question [54]. Auxilin is a Type III DnaJ protein with a J domain and a separate clathrin binding site, which activates mammalian Hsc70 to dissociate clathrin coated on vesicles during synaptic vesicle recycling [55]. The X-ray structure of the J domain covalently linked to the ADP-bound NBD of Hsc70 at R171, the equivalent of R167 in DnaK, shows binding to the cleft between subdomain Ia and IIa, right next to the linker [54]. In contrast, an NMR study shows a dynamic interface interaction between DnaJ and DnaK at subdomain IIa of NBD in which the J-domain scans the surface of acidic residues between 206-221 (Fig. **1A**), a distance from previously identified residue R167, and an interaction with considerable relative mobility [56]. How much of Hsp70 interacts with J domains remains in question. Earlier work showed that the J domain could not interact or stimulate the ATPase of the DnaK NBD alone, but can interact with a fragment (1-538) of DnaK missing the NBD α-helical lid [27, 57]. Experiments with the auxilin J domain and Hsc70 found that the NBD and linker was sufficient for ATPase stimulation [54]. More recently, the J domain of human DNAJB2 (Hsj1) was found to interact and activate ATP hydrolysis only with human Hsp70 containing most of the SBD including the α-helical lid. The NBD alone could not interact, based on chemical shifts observed by NMR, and J domain binding caused conformational shifts in the α-helical lid [58].

### *G/F-rich Linker*

The glycine and phenylalanine (G/F) rich linker is present in Type I and II DnaJs, which accounts for at least seventeen in the human DnaJ family. It connects the J domain to the central part of the molecule. Despite the conserved G/F residues and a DIF motif in some cases, the linker is generally unstructured and varies in sequence and in length among different proteins. Functional conservation of Type I and II G/F region among Hsp40 is observed when swapping G/F linker region between bacterial, yeast and human homologues or when introduced into a Type III Hsp40, in bacterial complementation and *in vitro* assays [59].

The role of the conserved G/F region remains unclear and was originally proposed to be a flexible link between the J domain and the central part of the molecule. Work done with bacterial DnaJ, *S. cerevisiae* Sis1 and human DNAJB6 (Mrj) and DNAJB1 (Hsp40, Hdj1) revealed a more complex function of the G/F linker. A J domain plus linker is necessary for the activation of bacterial DnaK [52], partly because the minimal J domain failed to bind to DnaK compared to a construct containing the linker [50]. When the G/F-rich region is replaced by an unrelated sequence (HMGSHM hexapeptide), DnaJ efficiently stimulates the DnaK ATPase but could not efficiently activate DnaK itself to bind the model substrate $\sigma^{32}$ [51]. This suggests that the G/F linker may be involved in a mechanism that modulates DnaK or general Hsp70 substrate binding and folding. Consistent with this idea, when the J domain of human Type I DNAJA2 is replaced by that of highly similar DNAJA1 (Hdj2), the ability of DNAJA2 to support Hsc70-mediated refolding of luciferase is maintained, but when the entire J domain and linker is similarly exchanged, this ability is lost [60, 61]. The ability of DnaJ to distinguish between states of partially folded and unfolded substrates may be linked to a functional G/F linker [62]. Mutations in the G/F region of Human DNAJB6b, in particular mutation of F89I or F93L, are responsible for Limb-Girdle Muscular Dystrophy (LGMD) [63, 64]. The anti-aggregation function of DNAJB6b is impaired in the mutant leading to a defect in clearance of TDP43 nuclear granules in mammalian cells. Similar mutations introduced in DNAJB1 impair prion *RNQ+* propagation in yeast [65]. In yeast, the G/F linker of Type II Sis1 is indispensable for the function of the chaperone [66]. Deletion of the Sis1 G/F region or point mutations N108I and D110G, results in loss of complementation of Δsis1 strains and inhibition of prion protein RNQ1 propagation [66 - 68]. A Sis1 ΔG/F protein is capable of binding substrate and stimulating Hsp70 (Ssa1) ATPase, but loses its ability to cooperate with Hsp70 to refold denatured substrate.

### *Substrate Binding and Cys(Zn) Regions*

Type I DNAJs, and some Type II and III, are thought to load substrate onto Hsp70. In this model, a typical DNAJ binds substrate itself as an ATP-independent chaperone, its J domain stimulates ATP hydrolysis by Hsp70, and as Hsp70 binds tightly to substrate the DNAJ dissociates from the complex (Fig. **2**) [6, 68, 69]. Substrate binding by DNAJs involve the central/C-terminal domains of Types I and II . However Type II DNAJs diverge in these regions. Some, like human DNAJB1 and DNAJB4 and yeast Sis1, have similarity to regions in Type I (Fig. **1D**, **E**). Others, like human DNAJB2 (Hsj1), DNAJB6 and DNAJB8 are more similar to each other or have unique features [5]. Most information about substrate binding relates to Type I and the DNAJB1/Sis1 group of Type II [51, 70 - 73]. Structures of C-terminal fragments of Type I Ydj1 and Type II Sis1 and DNAJB1 suggest how substrate is bound, although the Cys(Zn) region of Ydj1 is not found in Sis1 [74 - 76]. The Cys(Zn) consists of four CxxC motifs, the first and fourth motif binding to a Zn atom (Zn I) and the second and third binding another (Zn II, Fig. **1D**) [74, 75, 77]. A fragment of Ydj1 containing the Cys(Zn) region and surrounding sequence was crystallized with complexed GWLYEIS peptide substrate. This sequence was selected from a list of twenty peptides identified by phage display screening as the best interacting sequence with Ydj1. The 2.7 Å crystallographic structure of truncated Ydj1- GWLYEIS revealed an L-shape consisting of three β-sheet subdomains (Fig. **1D**). In the L shape, subdomains I and III form the long arm of the L and have similar β-barrel structures. Standing alone in the short L arm is subdomain II or the Cys(Zn) region, with the Zn I site forming the angle of the L. The peptide interacts with a hydrophobic pocket of subdomain I next to the Cys(Zn) region [74]. Mutation of an isoleucine residue in the hydrophobic pocket inhibited luciferase refolding by Ydj1 with yeast Hsp70 and reduced affinity for the peptide GWLYEIS [78]. A later crystallographic structure of the Ydj1 C-terminal fragment showed that domain III extends into a homodimerization interface (Fig. **1D**), but there is no evidence of substrate binding by the C-terminal fragment [79]. The structures of the Type II Sis1 and DNAJB1 peptide binding region were also determined by crystallography. A C-terminal fragment of DNAJB1 showed a dimer in a U-shaped with a large cleft formed between the two monomers. In each monomer, subdomain I is similar to subdomain I in Ydj1 and contains a hydrophobic depression involved in binding (Fig. **1E**). The homodimerization region also has similarity to that of Ydj1. The structure of the Sis1 C-terminal fragment was very similar [75, 76]. In Sis1, residues L199, F201 and F251 are essential for polypeptide binding activity and for viability of yeast strains [80]. Sequence homology suggests that the substrate binding sites in Ydj1 and Sis1 are conserved in human Type I and DNAJB1-related Type II, respectively.

In Type I DNAJs, the substrate binding site is next to the Cys(Zn) region. The function of this region is specific and conserved among the members of this group: it is essential for substrate release and transfer to Hsp70 that is linked to the ATP hydrolysis step of Hsp70. Deletion of the sequence between the Zn II motifs in human DNAJA1 or DNAJA2 significantly decreases substrate release, and inhibits Hsc70 refolding of denatured luciferase (Fig. **2**) [60]. Similar conclusions with bacterial DnaJ and yeast Ydj1 were obtained [81, 82]. In Ydj1, mutation of the cysteine residues to serines to prevent Zn binding caused a defect in substrate transfer to Hsp70 (Ssa1) resulting in loss of refolding function [81]. Deletion of the Cys(Zn) or mutation of the Zn II cysteines in DnaJ also affected the capacity of DnaK folding function [82, 83]. For all of these Type I homologues, disruption of Cys(Zn) did not affect Hsp70 ATPase stimulation nor substrate binding to the co-chaperones [60, 81, 82]. In addition to substrate transfer onto Hsp70, the Cys(Zn) motifs in DnaJ have a disulfide reductase activity that assists unfolding of improperly oxidized substrates by DnaK, and the human Type I DNAJs could function similarly [84].

The C-terminus of DNAJA1 is modified by farnesylation [85], similar to Ydj1, and both DNAJA2 and DNAJA4 are also thought to be modified. The importance of farnesylation is not certain. Hsp70 cooperates with the chaperone Hsp90 in many functions, and farnesylation of Ydj1 was found to be important for interaction with Hsp90 substrates [86]. Consistent with this, farnesylation is required for DNAJA1 to assist the folding of activation induced cytidine deaminase (AID, see below), which also requires Hsp90. However, in studies of Hsp90-independent luciferase refolding by DNAJA2 and Hsc70/Hsp70, farnesylation of DNAJA2 was not necessary for complete activity [60, 61, 70].

### *Quaternary Structure*

The first low resolution structure of a full length DNAJ was determined by small-angle X-ray scattering (SAXS) for DNAJA1 and DNAJB4 [87]. Both DNAJA1 and DNAJB4 are dimers as confirmed by analytical ultracentrifugation and SAXS analysis. DNAJB4 has an elongated shape where its J domains extend laterally from opposite sides of the central domain that appears as a dimer (Fig. **1E**). DNAJA1 is more compact and its J domains extend from one end of the central domain that appears triangular in the dimer (Fig. **1D**). Importantly, the model is consistent with the Ydj1 crystallographic structures of the central and dimerization fragments previously published [74, 79]. Together, these form an overall triangular shape of the two L shaped subunits, and potentially with an empty space between them. When NMR and crystallographic structures of the J domain and central/C-terminal fragments are modeled into the SAXS envelope in order to visualize relative positions of the domains, the Cys(Zn) arms point toward each other in the dimer and may interact. Low resolution structures of Ydj1 and Sis1 by SAXS also revealed triangular shapes for the central/C-terminal domains of both proteins, but with very different orientations of the J domains: they extend longitudinally in parallel for Ydj1, and laterally from opposite sides for Sis1 [88]. The arrangements are similar to what was observed previously for DNAJA1 and DNAJB4 [87]. In another study, a chimera was constructed with the Cys(Zn) region of Ydj1 replacing the corresponding part of Sis1, and in the SAXS envelope the J domains now have the same orientation observed for Ydj1. In the opposite chimera with a part of Sis1 replacing the Cys(Zn) region of Ydj1, the J domains are positioned similarly to Sis1. The central and Cys(Zn) region seems to determine the orientation of the J domains [88].

## NUCLEOTIDE EXCHANGE FACTORS (NEF)

The Nucleotide Exchange Factors (NEF) increase the Hsp70 ATPase rate by facilitating ADP release and favouring ATP binding to the NBD (Fig. **2**). In humans, a number of proteins were found to act as NEFs. Although they play a similar role in modulating the Hsp70 nucleotide state, they are structurally diverse, and many may have different functions in cells [89]. The NEFs are divided into four structural families: Hsp110/Grp170 proteins, HspBP1, BAG-domain proteins and the GrpE family. In the Hsp110 family are cytosolic Hsp110 (Hsp105, *HSPH1*), Apg1 (*HSP4AL*) and Apg2 (*HSPA4*), and an ER form Grp170 (*HYOU1*). For the HspBP1 family, only HspBP1 (Hsp70 binding protein) is known to be a cytosolic Hsp70 NEF, the other is ER localized and named Sil1 (BAP). The largest group is represented by the BAG-domain proteins named after the original member Bag1 (Bcl2-associated athanogene) and with four other members, Bag2 to Bag5, all found only in the cytosol. The GrpE-like proteins GRPEL1 and GRPEL2 are conserved with the *E. coli* NEF GrpE and found only in mitochondria.

Biochemical and structural studies revealed the mode of function of the Hsp70

NEFs in humans and yeast. Eukaryotic Hsp70 (Hsc70) has an ADP off-rate higher by one order of magnitude than *E. coli* DnaK, and the closest homologue to the bacterial NEF GrpE is only found in mitochondria (and chloroplasts in plants). It was therefore thought that eukaryotic Hsp70s do not require NEFs and it was a surprise when the first human NEF, Bag1, was discovered [90]. By accelerating ADP release from Hsc70, Bag1 increases the Hsc70 steady state ATPase rate (in the presence of DNAJB1) substantially [90]. A Bag1 homologue in yeast, Snl1, was discovered some time afterwards [91]. All five members of the BAG family share a conserved BAG domain at the C-terminus [92]. Bag1L and isoforms Bag1M and Bag1S (large, medium and small) generated by alternative translation initiation from the same mRNA, contain a ubiquitin-like (UBL) domain involved in interaction with the 26S proteasome [93]. Two α-helices of the three in the BAG domain contact the surface of subdomains Ib and IIb of Hsc70 NBD through electrostatic interactions with specific amino acid residues [94]. In comparison to the ATP bound state, NBD subdomain IIb has a rotation of 14° outwards upon BAG binding [94], similarly to what was observed for the complex of GrpE and NBD of DnaK [95]. The BAG domain of Bag4, as determined by NMR, is also composed of three α-helices but shorter than those of Bag1 [96]. Residues of Bag4 interacting with Hsc70 were mapped by mutagenesis and determined by yeast two-hybrid assay. Alignment of different BAG domain sequences based on Bag4 revealed that the closely related Bag3, Bag4 and Bag5 formed a subfamily of BAG domains, with Bag1 and Bag2 more divergent [97]. Bag5 contains five predicted BAG domains of which only the fifth C-terminal one binds to the Hsc70 NBD. In co-crystal structures of the fifth domain with the Hsc70 NBD, the NBD has two conformations: one with an open state similar to that observed with Bag1, and another open state where the nucleotide pocket is distorted [98]. Bag2 binding to the Hsc70 NBD also induces a different conformation than for Bag1 and Bag5, where the entire lobe II rotates [99].

Shortly after Bag1, mammalian HspBP1 and the yeast homologue Fes1 were identified as NEFs for Hsp70 [100, 101]. Mammalian Hsp110 and its yeast homologue Sse1 [102] were also identified to have NEF function. The crystallographic structure of Fes1 in complex with the NBD of Ssa1 (Hsp70), revealed a structure of four Armadillo repeats (tandem repeats of two α-helices forming a hairpin) that embraces lobe II and a displaces lobe I of Hsp70 NBD [103]. However, another structure showed that Sil1, the yeast ER homologue of HspBP1, acts as a clamp causing BiP (Hsp70) subdomain IIb to rotate by 13.5° and subdomain Ib to swing in the opposite direction by 3.5° [104]. The Hsp110/Grp170 NEFs are homologous to Hsp70 and share a similar domain composition with an NBD and an SBD, however they are very divergent from the typical Hsp70s [105, 106]. Nevertheless, there is broad structural similarity between Hsp110 and Hsp70. In a structure of yeast Sse1 its SBD and NBD could be superimposed well onto the homologous domains in DnaK, with the exception that the lid of the Sse1 SBD is completely opened [106]. Again, crystallographic structures of the Sse1-Hsp70 complex disclose the Hsp110 nucleotide exchange mechanism [107, 108]. In the complex, the SBD of Sse1 is open with the β-sandwich separated from the α-helical lid as in the structure of Sse1 alone, but both the α-helical lid and the Sse1 NBD contact the NBD of Hsp70 to envelope it. The interaction of Sse1 with Hsp70 at these different interfaces with conserved residues, opens lobe IIb of the Hsp70 NBD outwards substantially. This most likely weakens the ADP interactions in the Hsp70 NBD and causes its release. The ATP-bound Hsp110 is the only conformation that provides the contacts with Hsp70 [108, 109].

Not all NEFs function equally with Hsp70. Human Hsp110, HspBP1 and the BAG domain of Bag1 (C-Bag) were compared to each other for stimulation of the Hsc70 ATPase and refolding of denatured luciferase *in vitro*. Hsp110 was the most efficient in stimulating the ATPase of Hsc70 at low concentrations of NEF, in the presence of DNAJA1, whereas higher concentrations of C-Bag and HspBP1 were needed. Interestingly, at increasing NEF concentration, Hsp110 becomes inhibitory of the Hsc70 ATPase [61]. At optimal concentrations, both Hsp110 and C-Bag enhanced luciferase refolding similarly, but HspBP1 only inhibited refolding [61]. Similar experiments were done with Bag1, 2 and 3 comparing their affinities with and stimulation of Hsc70 function [110]. Bag3 outcompeted Bag 1 and 2, and Bag1 was better than Bag2 in stimulating Hsc70 functions. As observed previously for Hsp110, increasing concentrations of BAG proteins inhibited the Hsc70 ATPase and refolding functions, at an 8-fold molar ratio over Hsc70 for Bag3 and Bag1, or 128-fold for Bag2. In comparison, an equimolar ratio of Hsp110 to Hsc70 was enough to inhibit Hsc70 function [61]. Overall, Hsp110 appears to be much more efficient than the BAG proteins as an NEF for cytosolic Hsc70, perhaps because of its larger interface with the Hsc70 NBD which might better facilitate ADP release. In general, increasing NEF concentration acts as an inhibitor of Hsc70 function, probably by trapping the Hsc70 NBD in an open nucleotide-free conformation, or even possibly by accelerating the Hsc70 NBD conformational changes from the ADP to ATP bound states faster than what is required for productive coordination between the NBD and SBD. This hypothesis remains to be tested.

## CHAPERONE FUNCTION

From newly synthesized to native states, proteins can face obstacles causing them to misfold or aggregate [111]. In the native folded state proteins are usually stable but upon stress such as heat shock, they are likely to misfold and aggregate. Compared to native states, aggregates are thought to be in a low free energy state and unfolded polypeptides in higher states of free energy; misfolded states may be in local free energy minima where they are kinetically trapped [4]. Chaperones, in particular the Hsp70 system, are thought to work through a process where they bring the misfolded or aggregated substrate to a more unfolded state, allowing them to fold upon release from the chaperone [112]. Therefore, the folding function of Hsc70 chaperones system is to change the free energy state of substrate, through conformational changes. Hsc70 itself goes through conformational changes in the folding process. Hydrolysis of ATP to ADP in the Hsp70 NBD triggers conformational change in the Hsp70 SBD, thus increasing the affinity of Hsp70 for the substrate [36, 106]. Also, DNAJ co-chaperones bind and release substrate in coordination with Hsp70. DNAJ substrate release is crucial for folding and requires a J domain contact with Hsc70-ATP, ATP hydrolysis and a contribution of the Cys(Zn) region in the case of Type I DNAJs [60, 113]. Finally, polypeptides are released from Hsp70 when ADP is exchanged for ATP by NEFs. Once released, the polypeptide can fold to the native conformation, or become misfolded again and reenter the cycle for subsequent rounds of refolding (Fig. **2**).

Changes in substrate conformation in the presence of single or combined bacterial DnaJ, DnaK and GrpE, was assayed by single molecule spectroscopy [114]. Binding of DnaJ to denatured rhodanese substrate resulted in a compact and denatured ensemble of substrate conformations. Following DnaK-ATP binding, however, expansion of denatured rhodanese was observed, consistent with unfolding by multiple DnaK molecules binding to the substrate at once. In another study, the action of the DnaK chaperone system on tandem immunoglobulin-like domains of titin was studied by Atomic Force Microscopy (AFM) with the substrate attached to a solid support [115]. DnaJ appears to work by binding substrate to limit misfolding but does not unfold substrate; it is proposed that DnaJ restricts the range of unfolded substrate conformations. When DnaK is added, the population of unfolded states increases as well as that of the native state, suggesting that unfolding by DnaK probably leads to folding. Interestingly, the addition of NEF GrpE did not change the amounts of folded states but instead shifts the unfolded to a misfolded state, in agreement with a role of the NEF in releasing substrate from DnaK.

The function of human (mammalian) cytosolic Hsc70/Hsp70 in folding, aggregation prevention and disaggregation of polypeptides involves the most abundant DNAJ co-chaperones DNAJA1, DNAJA2, DNAJA4 and DNAJB1. DNAJB1 knockout mice are viable, but it is required for thermotolerance in the early phase of recovery following heat shock [116] and it suppresses misfolding associated with aggregation [117]. All four co-chaperones, DNAJA1, DNAJA2, DNAJA4 and DNAJB1 were comparable in suppressing aggregation of denatured luciferase substrate *in vitro* and in cells [118, 119]. Stimulation of Hsc70 ATPase by the J domains is essential in this process and is also comparable between these different DNAJs. Since the J domain is well conserved and is solely responsible for ATPase stimulation, it is then not surprising that Hsc70 ATPase rates are equivalent when stimulated by DNAJA1, DNAJA2 and DNAJA4 in single turnover or steady state assays [60, 70]. Variation in steady state Hsc70 ATPase rates is rather observed in the presence of different types of NEF [61, 110]. But, not all four DNAJs are equivalent in the folding process. With purified Hsc70,

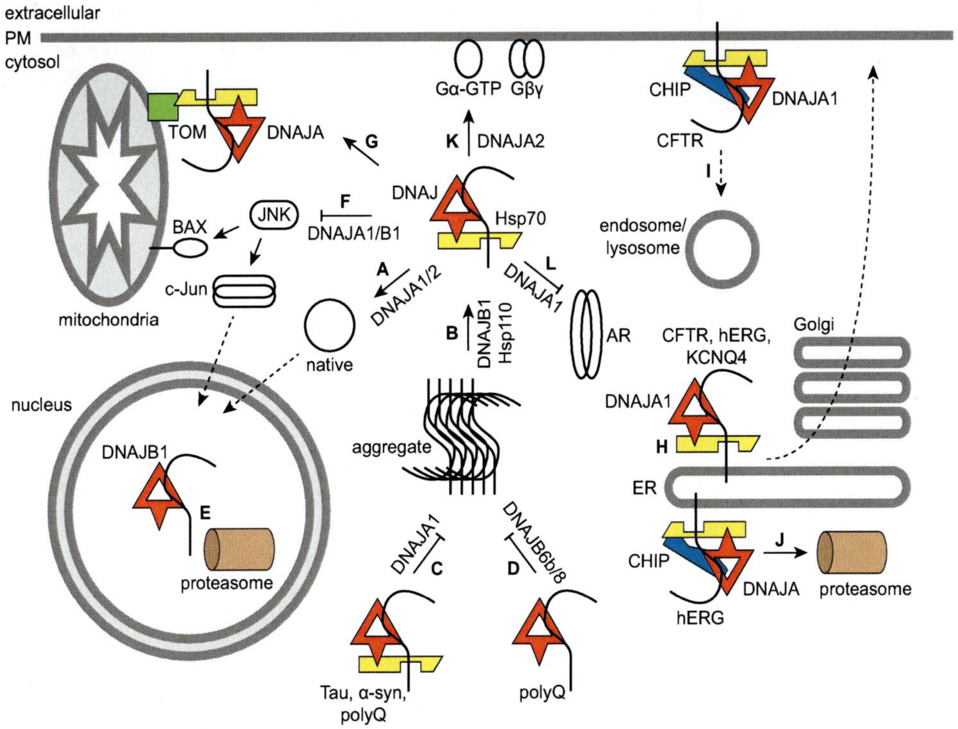


**Fig. (3).** Functions of the major DNAJs. A, DNAJA1 and DNAJA2 support Hsp70-assisted protein folding, including of AID which then acts in the nucleus. B, DNAJB1 works with Hsp70 and Hsp110/Apg2 to dissociate protein aggregates. C, DNAJA1 assists Hsp70 in preventing aggregation of Tau, α-synuclein and possibly polyQ proteins. D, DNAJB6b and DNAJB8 prevent polyQ aggregation independently of Hsp70. E, DNAJB1 brings unfolded proteins to the nucleus for degradation by proteasomes. F, DNAJA1 suppresses c-Jun, and prevents BAX translocation to mitochondria, most likely by upstream inhibition of JNK. G, DNAJA1, DNAJA2 and DNAJA4 assist Hsp70 in targeting some proteins to the mitochondria for import. H, DNAJA1 assists the folding of the CFTR, hERG and KCNQ4 ion channels at the ER, allowing them to traffic to the plasma membrane. I, at the plasma membrane, DNAJA1 assists CHIP-mediated internalization and lysosomal degradation of misfolded CFTR. J, Excess DNAJA1 and DNAJA2 promote hERG degradation by the CHIP E3 ubiquitin ligase. K, DNAJA2 and CSPα promote trimeric G protein signaling by activation of GTP-bound $G_\alpha$. L, DNAJA1 inhibits AR and DNAJA1 knockout upregulates AR. See text for references.

DNAJA1 does not support refolding of denatured luciferase above the background level with Hsc70 alone, and DNAJA4 moderately promotes refolding. In contrast, DNAJA2 is very efficient in promoting luciferase refolding in the presence of Hsc70, almost to the level observed with the model cytosol, reticulocyte lysate (Fig. **3A**) [61, 70]. DNAJB1 also stimulates Hsc70 refolding function but is markedly less efficient than DNAJA2 [110, 120]. It is worth mentioning that luciferase refolding *in vitro* by Hsc70 is increased in the presence of additional NEF but they are not absolutely necessary for refolding. Refolding of heat denatured luciferase in cells was also dependent on DNAJA2 and DNAJA4 but not on DNAJA1 [60, 119]. After heat shock, around half of luciferase was lost from soluble material into aggregates and not degraded since treatment with proteasome inhibitor MG132 does not restore the luciferase level; some resolubilization of luciferase is observed together with refolding. Although DNAJA1 is ineffective for luciferase refolding, it appears that different substrates require either DNAJA1 or DNAJA2 for their folding, as described below.

DNAJB1, distinct from the human Type I co-chaperones, works in synergy with Hsp70/HSPA1 and Hsp110 to resolubilize protein aggregates *in vitro* (Fig. **3B**). Using aggregates of luciferase in the presence of yeast Hsp26 from the small HSP family, DNAJB1 in combination with Hsp70 and Hsp110 or Apg2 was much more efficient at disaggregation then DNAJA2 with Hsc70 and Hsp110/Apg2, and Bag1 was also less effective than Hsp110/Apg2 [121]. These amorphous aggregates are distinct from the amyloid-like aggregates in neurodegenerative diseases. Recently, DNAJB1 and DNAJA2 were found to work cooperatively with Hsp70 and Hsp110 to disaggregate proteins, more effectively than DNAJB1 by itself [122]. DNAJB1 was more efficient at promoting solubilization of large aggregates, while DNAJA2 had more activity on small aggregates. However, both co-chaperones together had synergistic activity over the range of aggregate sizes. Cross-linking and fluorescence studies showed that DNAJB1 and DNAJA2 form transient, specific complexes with each other for this function, with the J domain of one binding next to the substrate binding site of the other, and vice versa, but in a way that leaves the HPD motifs exposed. The binding sites are conserved in metazoa, and mutation abolished the cooperative function. These findings suggest that heterocomplexes of other DNAJs could greatly expand the function of the Hsp70 system.

How the DNAJs differ mechanistically is under investigation. One difference between DNAJA1 and DNAJA2 is in the timing of substrate release triggered by Hsc70 [60]. Release from DNAJA1 and DNAJA2 is inhibited in presence of Hsc70-AMPPNP, a non-hydrolyzable ATP analogue. Whereas DNAJA2 only releases in the presence of Hsc70-ATP, DNAJA1 was capable of releasing in both the presence of Hsc70-ATP and ADP [60]. Because the ADP state is formed after

ATP hydrolysis, this suggests that DNAJA1 binds longer or releases more slowly than DNAJA2 [60]. This may be consistent with the idea that DNAJA1 has a better affinity for some polypeptides than DNAJA2, although there is little information about substrate affinity and specificity between the different DNAJ co-chaperones. Studies comparing relative binding of various polypeptides to DNAJA1, DNAJA2 and DNAJA4 show that DNAJA1 has stronger binding for many of the polypeptides tested, although for others, binding was similar to DNAJA2 and DNAJA4 [61, 70]. While the transfer of substrate from DNAJ to Hsp70 may be a determinant of refolding function, how it happens mechanistically is also unclear. One hypothesis to explain substrate release, is a transfer of substrate from DNAJ to Hsp70 upon ATP hydrolysis in a two-step mechanism, in which the polypeptide sequence bound first by the DNAJ is passed to Hsc70 during ATP hydrolysis [123]. The mechanism underlying substrate transfer remains to be defined, but the model suggests that the DNAJ must release from the substrate sequence before Hsp70 binds to the same sequence. Hsp70 in the ADP state may have stronger affinity for a binding site and can outcompete DNAJ binding to that site. This model suggests that allosteric changes in the DNAJ are not necessary for substrate transfer. The extensive variation present in the Hsp40 family argues against a unified allosteric mechanism: substrate transfer has been observed for Type II and III DNAJs [113, 124]. Other evidence suggests a model where Hsp70 and the DNAJ bind at different or adjacent sites [125]. Indeed, simultaneous binding of DNAJA2 or DNAJA1 and Hsc70 to the same substrate has been observed, including when substrate transfer is prevented [60, 125]. The unique role of the Type I Cys(Zn) region in substrate release may be consistent with an allosteric shift in the DNAJ that reduces substrate affinity in the polypeptide-binding domain. While an allosteric mechanism would apply to Type I DNAJs, it would not be a necessity for others. However, an allosteric mechanism could be behind the greater folding function of the Type I co-chaperones.

## FUNCTIONS IN CELLS AND TISSUES

Hsc70 and Hsp70/HSPA1 mRNA is found in all human tissues but mRNA of the four major DNAJ co-chaperones show more specific tissue distribution [126]: In a genome-wide survey of mRNA expression in different human tissues, Hsc70 is very abundant in many brain tissues, in testis, and in several other tissues including certain white blood cell types [127]. Similarly to Hsc70, DNAJA1 and DNAJA2 mRNA are also abundant in numerous brain tissues [127]. DNAJB1 mRNA is abundant in testis, as well as brain, lung and T cells; DNAJA4 is in lower abundance except in certain tissues including skeletal muscle [127]. The protein levels of these chaperones have also been examined. In HeLa cells, chaperones represents up to 9% of the mass of the cells, based on quantitative mass spectrometry [128]. Together, Hsc70 and Hsp70 copy number in the HeLa cytosol is close to 2 x $10^8$ which is much more abundant than DNAJA1, DNAJB1 and DNAJA2 with respectively 9.2, 5.8 and 4.5 x $10^6$ copies [128]. The copy numbers of the Hsp70 system in ER and mitochondria were also quantified and also show an excess of Hsp70 compared to its co-chaperones [128]. In studies of protein expression by Western blot, DNAJB1 protein is most abundant in lung, DNAJA1 is almost ubiquitous with higher amounts in brain and kidney, DNAJA2 is expressed most highly in heart and skeletal muscle, and DNAJA4 is most abundant in heart and testis and moderately in brain and ovary [126, 129]. The expression of some Hsp70 system proteins is induced by the transcription factor Heat Shock Factor 1 (HSF1) upon stress. Dependent on HSF1, the level of protein increase for Hsp70 is between 2- and 3.3-fold, for DNAJA1 by around 1.7-fold and DNAJB1 between 2 and 8.7-fold, as determined by mass spectrometry. DNAJA4 is inducible to around 1.7-fold, but DNAJA2 is not induced at all [130, 131]. Differences in the sub-cellular localization between DNAJs is also observed but in general they are expressed in cytosol or nucleus [119]. After heat stress, DNAJB1 levels increase dramatically and it is mostly translocated in the nuclear compartment [116], whereas DNAJA1, DNAJA2 and DNAJA4 levels remains relatively constant and are evenly spread throughout the cell with some increase of co-localization of DNAJA4 with Hsc70 in the nucleus [119].

## Neurological Diseases

Abnormal accumulation and aggregation of proteins in neurological disease, for instance Tau, α-synuclein and polyglutamine (polyQ), were found to interact with Hsc70 and DNAJA1. In neurons, Tau accumulation leads to its hyperphosphorylation and proteolytic cleavage, a mark of several diseases including Alzheimer (AD) and Parkinson disease (PD) . In mouse brains with a Tau transgenic model of AD, levels of DNAJA1 are reduced and inversely correlate with the level of Tau. In cultured cells, Tau level is reduced by almost 50% when DNAJA1 is overexpressed, but Tau levels seem to increase when DNAJA1 is knocked down (Fig. **3C**) [132]. Fibrillar α-synuclein (α-syn) is a principal component of Lewy bodies present in the central nervous system of PD and synucleinopathy, accounting for loss of up to 70% of dopaminergic neurons in severe cases. DNAJA1 associates with the α-syn aggregates and in the presence of Hsc70, is capable of reducing and delaying the aggregation of α-syn (Fig. **3C**). Interestingly, Hsc70 in a nucleotide-free state with a passive interaction with substrate has similar effect, suggesting that the DNAJA1 effect mediated through Hsc70 might be inhibitory [133]. However, in the presence of ATP, another co-chaperone HIP (Hsp70 interacting protein, *ST13*) stabilizes the ADP-bound state of Hsc70 and enhances prevention of α-syn aggregation [134]. Hsp104, a yeast AAA+ family chaperone with no orthologue in humans or metazoa, was also found to reduce α-syn assembly in a rat brain model and *in vitro* with a stronger effect in the presence of DNAJA1 and Hsc70 [135].

PolyQ expansion in various proteins is also responsible for inherited neurological pathologies such as Huntington disease (HD), dentatorubral-pallidoluysian atrophy (DRPLA), spinal and bulbar muscular atrophy (SBMA) and several types of spinocerebellar ataxias (SCA1, 2, 3, 6, 7 and 17). In Ataxin-1 (SCA1) patients and in Purkinje neurons of mice, DNAJA1 concentrates with Ataxin-1 nuclear inclusions instead of its normal diffuse localization throughout the cell [136]. When DNAJA1 is overexpressed, SCA1 nuclear inclusions are significantly reduced and the deletion of the DNAJA1 J domain abolishes this effect (Fig. **3C**). The 20S proteasome, ubiquitin and Hsp70 colocalize with SCA1 nuclear inclusions [136]. Nuclear polyQ inclusions induced by expression of either Htt-74Q (huntingtin exon-1 fragment with 74 Glu residues) in human neuroblastoma (SH-SY5Y), rat phaeochromocytoma (PC12) and african green monkey (COS-7) cells [137], or by tNhtt (truncated huntingtin, 150Q) in neuro2a cells [138], were shown to colocalize with 20S proteasome or ubiquitin, DNAJA1 and Hsc70 [137, 138]. As seen previously for SCA1 aggregates, DNAJA1 overexpression reduces these polyQ Htt-74Q inclusions and its effect is abolished in the presence of proteasome inhibitor lactacystin [137]. DNAJA1 and Hsc70 were shown to preferentially interact with longer polyQ (tNhtt 150Q) rather than with shorter length polyQ (tNhtt 16Q or 60Q) [138]. In HeLa cells, cytoplasmic polyQ aggregates induced by the expression of ARQ48 (Androgen Receptor with 48 Glu residues, causal for SBMA) were also found to be sensitive to DNAJA1 overexpression [139].

The aggregates related to neurodegeneration are thought to be amyloid or prion like, accounting for their toxic properties. However, there are likely roles for the DNAJs in other aggregation of misfolded proteins in the brain. When cerebral ischemia is provoked in rat brain, neurons stained for ubiquitin show a change in distribution from evenly diffuse throughout the cells to discrete inclusions at the periphery. Overexpression of Hsp70 in these neurons reverses the effect [140]. In astrocytes deprived of glucose, ubiquitin staining shows similar inclusions or aggregates throughout the cytoplasm [141] and this phenomenon in astrocytes is prevented by DNAJA1 overexpression, which maintain diffuse ubiquitin in cells and reduces cell death.

Other DNAJs may be specialized to prevent or clear amyloid-like aggregation. DNAJB6b (the cytosolic splice isoform) is mostly found in brain and skeletal muscle, and closely related DNAJB8 is neuron specific. Both are substantially more effective in preventing toxic polyQ aggregation than the Type I DNAJs (Fig. **3D**) [142 - 144]. In addition to stimulating Hsc70, they act independently of

Hsc70 as chaperones by themselves; as noted above, their C-terminal regions diverge from that of DNAJB1, and may be specialized to prevent polyQ or amyloid-like aggregation. Both DNAJB6b and DNAJB8 are targets of hisitone deacetylases (HDAC), and removal of acetyl modifications from C-terminal lysines is necessary for their function [143]. There is evidence that they interact directly with the polyQ peptides that cannot be digested by proteasomes, as opposed to flanking regions [144]. DNAJB6b also binds to aggregates of amyloid-β (Aβ42), which is implicated in AD pathology upstream of Tau, and prevents growth into longer fibrils [145]. Interestingly, DNAJB6b has been observed along with α-syn in Lewy bodies [146], suggesting a role in managing the aggregates. DNAJB2a, which is also expressed in brain, has ubiquitin-interacting motifs (UIMs) at its C-terminus and targets substrates for proteasomal degradation [147]. DNAJB2a has been found to increase clearance of polyQ aggregates, and also of SOD1 mutant G93A, causal for familial amyotrophic lateral sclerosis (ALS) [148 - 150]. After heat shock, DNAJB1 localizes to the nucleus, where it is thought to deliver misfolded proteins for proteasomal degradation (Fig. **3E**); sequestration of DNAJB1 in polyQ aggregates interferes with this function [151].

## Cancer

A limited amount of information is available on the involvement of DNAJA1, DNAJA2 and other DNAJs in cancers. However, Hsp90 has been established as an important factor in many cancers, and it cooperates with Hsp70 in many of its functions. As abundant co-chaperones of Hsp70, the involvement of Type I DNAJs in this multifaceted disease is implied [152, 153]. DNAJA1 has been shown to be down-regulated at the level of mRNA in a range of prostate carcinoma cell lines [154] and at the level of protein in tissue samples from a patient with chronic pancreatitis [155]. In MiaPaCa2 pancreatic cancer cells, DNAJA1 expression suppresses c-Jun activation in response to induced stress (UV treatment or anisomycin) and diminishes survival of cancer cells (Fig. **3F**). In several cancer cell types, c-Jun has been shown to inhibit apoptosis, leading to the uncontrolled growth typical of cancer [156]. It should be noted that Hsp70 inhibits JNK (c-Jun N-terminal kinase) upstream of c-Jun in certain conditions [157, 158] and DNAJA1 may act at this step. DNAJA1 overexpression increases resistance to irradiation in radiosensitized glioblastoma pretreated with a farnesyltransferase inhibitor (FTI) targeting K-Ras as well as DNAJA1, and the converse is observed when DNAJA1 expression is reduced by knockdown [159]. Apparently, irradiation causes nuclear translocation of DNAJA1 and this is inhibited by pre-treatment with FTI; DNAJA1 may be an important secondary target of possibly therapeutic FTI. DNAJA1 has also been a biomarker target for pancreatic cancer to evaluate the effects of FTI [160].

Hsp70 has roles in folding and degradation of tumour suppressor p53, independent of its usual regulatory E3 ubiquitin ligase Mdm2 [161, 162]. Hsc70/Hsp70-mediated degradation of p53 is through the E3 ligase CHIP (C-terminus of Hsp70 interacting protein, *STUB1*), which has a TPR domain that binds Hsp70 or Hsp90, and ubiquitinates misfolded proteins [163]. However, DNAJB1 independently stabilizes Mdm2 but inhibits the Mdm2 E3 ligase activity towards p53, supporting tumour suppressor function in MCF7 breast cancer cells [164]. The tumour suppressor PDCD5 acts in the DNA damage pathway to activate p53, and DNAJB1 was recently found to induce PDCD5 degradation in HCT116 colon cancer cells, preventing p53-mediated apoptosis [165]. Finally, in the Warburg effect cells shift from respiration to glycolysis, determined in part by upregulation of pyruvate kinase M2 isoform (PKM2) and downregulation of other isoforms. In several cancer cell lines, DNAJB1 promotes degradation of PKM2 and thus regulates glycolysis, leading to effects on cell proliferation [166]. It seems likely that these degradation activities of DNAJB1 are also through CHIP.

## Mitochondrial Import

Most mitochondrial proteins are synthesized in the cytosol and imported into the organelle through the translocase of the outer membrane (TOM) complex together with the translocase of the inner membrane (TIM) complexes [167]. The dependence on cytosolic chaperones for import differs between mitochondrial preproteins, with metabolite carriers of the inner membrane the most dependent on Hsp70 and Hsp90. On the outer membrane, the Tom70 receptor interacts with Hsp90 and Hsc70/Hsp70 through a TPR domain to facilitate targeting and translocation of these carrier proteins into mitochondria [55, 168 - 170]. In addition, DNAJA1, DNAJA2 and DNAJA4 bind to the proteins in a complex with Hsc70, and cooperate in targeting them to Tom70, whereas DNAJB1 appears not to be involved (Fig. **3G**) [70]. Mitochondrial proteins with presequences for matrix targeting do not require chaperone interaction with Tom70, and depend instead on the Tom20 receptor. Chaperones may be still involved in maintaining solubility before import. Early reports indicated that depletion of DNAJA1 from rabbit reticulocyte lysate prevents import of ornithine transcarbamylase precursor (pOTC) into rat liver mitochondria [126, 171]. While farnesylation of DNAJA1 was suggested to be required for this function in mitochondrial import [85], another study found no requirement [171].

The mitochondrial pathway of apoptosis requires the insertion of apoptotic factors into the mitochondrial outer membrane, a process independent of the TOM machinery. BAX is cytosolic, but upon induction of apoptosis, it undergoes a conformational shift and localizes into the mitochondria outer membrane resulting in cytochrome C release and activation of caspases. It was found that Hsc70 and

DNAJA1 or DNAJB1 can prevent nitric oxide-mediated apoptosis upstream of caspase 3 by inhibiting the translocation of pro-apoptotic BAX into the mitochondria membrane [172, 173]. While the chaperones are suggested to bind BAX directly, another study finds that Hsp70 acts instead by inhibiting JNK signaling upstream of BAX activation (Fig. **3F**) [174].

## Ion Channels

As a result of genetic mutations, trafficking to the plasma membrane of several ion channels is impaired, resulting in diseases. Typically, mutant misfolded channels in the immature forms are retained and degraded at the endoplasmic reticulum (ER), decreasing the population of mature channels at the plasma membrane. Hsc70 and co-chaperones DNAJA1, DNAJA2 and DNAJA4 were identified as important modulators of the maturation of these ion channels [175]. Hsc70 is also involved in degradation of the channels by interacting with the CHIP E3 ubiquitin ligase [163], distinct from the ER-associated E3 ligases responsible for most ER-associated degradation [176].

Mutations in the CFTR (cystic fibrosis transmembrane conductance regulator, *ABCC7*) chloride channel are responsible for cystic fibrosis, and the CFTR ΔF508 mutation is the most common [177]. Hsp90 is a major determinant of CFTR folding and ER-associated degradation [178, 179]. In addition, Hsc70, DNAJB1 and DNAJA1 interact with the immature form of wild-type CFTR and in a more prolonged manner with immature ΔF508 [180, 181]. DNAJA1, or Hsc70 together with DNAJA1 or DNAJB1 are capable of preventing aggregation of a denatured fragment (nucleotide binding domain 1) of CFTR *in vitro* [181]. Overexpression of DNAJA1 increases CFTR biogenesis in HEK293 cells [182], and inversely, DNAJA1 knockdown decreases both biogenesis and stability of both CFTR and immature ΔF508 (Fig. **3H**) [183]. In contrast, DNAJC5 (CSPα) and DNAJB12 promote ER-associated degradation of CFTR ΔF508 by CHIP-dependent and -independent pathways [182 - 185]. Mature CFTR ΔF508 can accumulate at low temperature (26°C) but is drastically lost when cells are returned to the non-permissive temperature (37°). CHIP-mediated ubiquitination and internalization from the plasma membrane are responsible for targeting the rescued ΔF508 mutant to lysosomal degradation [186]. Knockdown of Hsc70, Hsp90α and DNAJA1 all reduce ubiquitination, and stabilize mature CFTR ΔF508. Notably, silencing of DNAJA1 was the most efficient in inhibiting ubiquitination and internalization of rΔF508, whereas silencing of DNAJB1 was ineffective (Fig. **3I**) [186].

Mutations in the *KCNH2* gene encoding the hERG/Kv11.1 (human ether-a-go-go related gene) potassium channel α-subunits have been linked to cardiac long QT syndrome type 2 (LQT2), characterized by a prolongation of ventricular action potential with a tendency to cause arrhythmia and sudden cardiac arrest. Mutants of hERG with trafficking deficiency accounts for most of the loss of function for this potassium channel [187]. hERG interacts with Hsp90 and Hsc70/Hsp70 chaperones, and Hsp90 is essential for hERG maturation; the Hsp90 co-chaperone FKBP38 also assists maturation [188, 189]. DNAJA1, DNAJA2 and DNAJA4 were found to be key modulators for the hERG degradation pathway. Overexpression of these co-chaperones negatively affect hERG trafficking by increasing its proteasomal degradation through CHIP-mediated ubiquitination (Fig. **3J**). However, knockdown of DNAJA1 but not DNAJA2 reduces hERG trafficking, indicating that DNAJA1 is important for Hsc70 to assist hERG folding (Fig. **3H**) [190]. DNAJA2 or CHIP overexpression also reduces maturation of mutant G601S hERG at 26°C, a temperature at which trafficking deficiency of this mutant is restored [190].

Mutations in the KCNQ4/Kv7.4 potassium channel cause loss of function that leads to autosomal dominant deafness type 2A (DFNA2A), a deafness characterized by a sensorineural loss of hearing. The majority of KCNQ4 mutants identified decrease the presence and functional currents of the channel at the cell surface [191]. Hsp90α/β, Hsc70 and DNAJA1 were shown to interact with KCNQ4 and to affect its stability (Fig. **3H**). Overexpression of Hsc70 or DNAJA1 both increase KCNQ4 stability, and knockdown of either one of these chaperones has the opposite result [192]. This effect of Hsc70 and DNAJA1 on KCNQ4 seems to involve degradation mediated through the CHIP pathway as observed for hERG.

The major functions of CSPα are in the presynaptic terminals of neurons, where it associates with synaptic vesicles. Together with Hsc70/Hsp70 and the TPR domain co-chaperone SGT, it has various functions including assisting the SNARE vesicle fusion machinery, limiting voltage- and calcium-activated BK channel density, and regulating N-type calcium channels through trimeric G proteins [193, 194]. These functions more likely involve conformational switches than folding or unfolding events. CSPα promotes $G_{\alpha}$ activation by GTP binding, leading to disassociation from the $G_{\beta\gamma}$ subunits and inhibition of N-type channels [194]. DNAJA2 has a parallel role to CSPα in the activation of $G_{\alpha}$ signaling (Fig. **3K**), and acts in complex with Hsc70, Hsp110, Hsp90, HIP and the co-chaperone HOP (Hsp organizing protein), which links Hsc70 and Hsp90 through separate TPR domains [195].

## Androgen Receptor

In DNAJA1 -/- mice lifespan is not affected and protein levels of DNAJA2 and

DNAJB1 were moderately increased and decreased, respectively. The average weight and fertility are reduced in male DNAJA1 -/- mice. In males, testis weight was also reduced by almost 50%, spermatozoa number reduced by 7% and motility was lost for 95% of population. In female DNAJA1 -/- mice fertility was not affected. The default in spermatogenesis and in particularly in spermatid differentiation with lack of elongation was observed in Sertoli cells. In these cells, the levels of androgen receptor (AR) and glucocorticoid receptor (GR) proteins were both increased inducing the expression of androgen responsive genes such as Pem, Testin, ABP (androgen binding protein) and transferrin; overall, DNAJA1 appears to inhibit AR (Fig. **3L**) [196].

## Activation-Induced Cytidine Deaminase

Hsc70 and cochaperones DNAJA1, DNAJA2 and DNAJA3 (Tid1, the cytosolic fraction of the mitochondrial matrix co-chaperone), as well as Bag2, Hsp90α/β and the Hsp90 co-chaperone Aha1 were co purified with AID (activation-induced cytidine deaminase) and identified by mass spectrometry [197]. AID induces mutations in DNA by removing amine groups on cytosine to create uracil, which in turn triggers somatic hypermutations, genetic conversion and class-switch recombination (CSR) of immunoglobulin genes. DNAJA1 and DNAJA2 are major interactors with AID. Changes in the amounts of DNAJA1 but not DNAJA2 affect AID levels and function in cells (Fig. **3A**). Upon DNAJA1 overexpression AID amounts increase and the protein is stabilized resulting in increased CSR. When DNAJA1 is silenced, AID amounts decrease by almost 50% and as a consequence CSR is also reduced [198]. In mouse DNAJA1-/- splenic B cells [199], AID levels are reduced as well as CSR upon mouse immunization at primary and secondary response. Transfection of DNAJA1 in mouse DNAJA1-/- splenic B cells complements this default. DNAJA1 farnesylation is crucial for interaction with AID, since removal of the farnesylation site on DNAJA1 results in loss of interaction with AID and abolishes complementation of CSR in DNAJA1-/- mice [198].

## Viruses

DNAJB1 and Hsp70 were found to interact with Nef, a protein crucial for HIV pathogenesis [200, 201]. Nef seems to be involved in regulation of DNAJB1 expression which increases after virus infection and decreases in the absence of Nef [200]. DNAJB1 enhances HIV-LTR mediated expression and its down-regulation by siRNA inhibits viral replication. As observed during stress conditions, the amount of DNAJB1 increases in the nucleus upon HIV infection and in addition interacts with Cdk9 in a Cdk9-Nef-DNAJB1 complex [200]. With Hsp70, opposite effects are observed: HIV replication decreases upon Hsp70 down-regulation and HIV replication increases when Hsp70 is overexpressed [201]. Hsp70 interaction with Nef seems to cause a decrease of Cdk9-phosphorylation and RNA pol III recruitment at the viral LTR promoter [201]. Among several factors that affect reovirus replication, DNAJA1 silencing in HeLa Cells reduces HIV titer by 50% [202].

Dengue virus was recently shown to require Hsp70 and multiple DNAJ proteins including DNAJA2 at several steps of its life cycle [203]. Hsp70 inhibitors were first used to identify chaperone-dependent steps, and shRNA knockdown screening of DNAJs determined what specific roles the co-chaperones have, if any. Type III DNAJs (DNAJC9, C16 and C18) are involved in virus entry, DNAJA2 is needed after entry, Type II DNAJB6 and B11 are associated with RNA replication and capsid formation [203]. These findings are a good illustration of how DNAJs direct the Hsp70 system to specific substrates and functions.

## OUTLOOK

There has been much progress in understanding the DNAJ-Hsp70 system, but much remains to be explored. High resolution structures of full-length Type I and Type II DNAJs are lacking and could provide insight into contacts between the J domain, G/F rich linker and other parts of the proteins. The exact role of the DNAJs in Hsp70-mediated folding needs further investigation. Growing evidence indicates that the major DNAJ co-chaperones are not entirely interchangeable, and studies comparing the DNAJs are more common recently. Also, there are many functions of Hsc70/Hsp70 identified, for example in cancer [204], for which the relevant DNAJs have not been determined. It is likely that more substrates and processes that depend on the major DNAJs will be identified, by inhibition of Hsp70 [205], interaction studies, and knockdown approaches. Manipulation of the DNAJs to address aggregation-related neurodegeneration, or genetic diseases of misfolding, is an interesting prospect.

## CONSENT FOR PUBLICATION

Not applicable.

## CONFLICT OF INTEREST

The author declares no conflict of interest, financial or otherwise.

## ACKNOWLEDGEMENTS

JCY is supported by the Canadian Institutes of Health Research and Cystic Fibrosis Canada, and holds a Canada Research Chair in Molecular Chaperones,

Tier II. Patrick Kim Chiaw is thanked for careful reading of the manuscript.

## REFERENCES

[1] Chiti F, Dobson CM. Protein misfolding, functional amyloid, and human disease. Annu Rev Biochem 2006; 75: 333-66.
[http://dx.doi.org/10.1146/annurev.biochem.75.101304.123901] [PMID: 16756495]

[2] Powers ET, Morimoto RI, Dillin A, Kelly JW, Balch WE. Biological and chemical approaches to diseases of proteostasis deficiency. Annu Rev Biochem 2009; 78: 959-91.
[http://dx.doi.org/10.1146/annurev.biochem.052308.114844] [PMID: 19298183]

[3] Hipp MS, Park SH, Hartl FU. Proteostasis impairment in protein-misfolding and -aggregation diseases. Trends Cell Biol 2014; 24(9): 506-14.
[http://dx.doi.org/10.1016/j.tcb.2014.05.003] [PMID: 24946960]

[4] Hartl FU, Bracher A, Hayer-Hartl M. Molecular chaperones in protein folding and proteostasis. Nature 2011; 475(7356): 324-32.
[http://dx.doi.org/10.1038/nature10317] [PMID: 21776078]

[5] Kampinga HH, Craig EA. The HSP70 chaperone machinery: J proteins as drivers of functional specificity. Nat Rev Mol Cell Biol 2010; 11(8): 579-92.
[http://dx.doi.org/10.1038/nrm2941] [PMID: 20651708]

[6] Mayer MP, Bukau B. Hsp70 chaperones: cellular functions and molecular mechanism. Cell Mol Life Sci 2005; 62(6): 670-84.
[http://dx.doi.org/10.1007/s00018-004-4464-6] [PMID: 15770419]

[7] Daugaard M, Rohde M, Jäättelä M. The heat shock protein 70 family: Highly homologous proteins with overlapping and distinct functions. FEBS Lett 2007; 581(19): 3702-10.
[http://dx.doi.org/10.1016/j.febslet.2007.05.039] [PMID: 17544402]

[8] Stricher F, Macri C, Ruff M, Muller S. HSPA8/HSC70 chaperone protein: structure, function, and chemical targeting. Autophagy 2013; 9(12): 1937-54.
[http://dx.doi.org/10.4161/auto.26448] [PMID: 24121476]

[9] Kettern N, Dreiseidler M, Tawo R, Höhfeld J. Chaperone-assisted degradation: multiple paths to destruction. Biol Chem 2010; 391(5): 481-9.
[http://dx.doi.org/10.1515/bc.2010.058] [PMID: 20302520]

[10] Milner CM, Campbell RD. Structure and expression of the three MHC-linked HSP70 genes. Immunogenetics 1990; 32(4): 242-51.
[http://dx.doi.org/10.1007/BF00187095] [PMID: 1700760]

[11] Wu B, Hunt C, Morimoto R. Structure and expression of the human gene encoding major heat shock protein HSP70. Mol Cell Biol 1985; 5(2): 330-41.
[http://dx.doi.org/10.1128/MCB.5.2.330] [PMID: 2858050]

[12] Dworniczak B, Mirault ME. Structure and expression of a human gene coding for a 71 kd heat shock 'cognate' protein. Nucleic Acids Res 1987; 15(13): 5181-97.
[http://dx.doi.org/10.1093/nar/15.13.5181] [PMID: 3037489]

[13] Leung TK, Rajendran MY, Monfries C, Hall C, Lim L. The human heat-shock protein family. Expression of a novel heat-inducible HSP70 (HSP70B′) and isolation of its cDNA and genomic DNA. Biochem J 1990; 267(1): 125-32.
[http://dx.doi.org/10.1042/bj2670125] [PMID: 2327978]

[14] Noonan EJ, Place RF, Giardina C, Hightower LE. Hsp70B′ regulation and function. Cell Stress Chaperones 2007; 12(3): 219-29.
[http://dx.doi.org/10.1379/CSC-278.1] [PMID: 17915554]

[15] Fourie AM, Peterson PA, Yang Y. Characterization and regulation of the major histocompatibility

complex-encoded proteins Hsp70-Hom and Hsp70-1/2. Cell Stress Chaperones 2001; 6(3): 282-95. [http://dx.doi.org/10.1379/1466-1268(2001)006<0282:CAROTM>2.0.CO;2] [PMID: 11599570]

[16] Bonnycastle LL, Yu CE, Hunt CR, *et al.* Cloning, sequencing, and mapping of the human chromosome 14 heat shock protein gene (HSPA2). Genomics 1994; 23(1): 85-93. [http://dx.doi.org/10.1006/geno.1994.1462] [PMID: 7829106]

[17] Bhattacharyya T, Karnezis AN, Murphy SP, *et al.* Cloning and subcellular localization of human mitochondrial hsp70. J Biol Chem 1995; 270(4): 1705-10. [http://dx.doi.org/10.1074/jbc.270.4.1705] [PMID: 7829505]

[18] Hendershot LM, Valentine VA, Lee AS, Morris SW, Shapiro DN. Localization of the gene encoding human BiP/GRP78, the endoplasmic reticulum cognate of the HSP70 family, to chromosome 9q34. Genomics 1994; 20(2): 281-4. [http://dx.doi.org/10.1006/geno.1994.1166] [PMID: 8020977]

[19] Han Z, Truong QA, Park S, Breslow JL. Two Hsp70 family members expressed in atherosclerotic lesions. Proc Natl Acad Sci USA 2003; 100(3): 1256-61. [http://dx.doi.org/10.1073/pnas.252764399] [PMID: 12552099]

[20] Hu G, Tang J, Zhang B, *et al.* A novel endothelial-specific heat shock protein HspA12B is required in both zebrafish development and endothelial functions in vitro. J Cell Sci 2006; 119(Pt 19): 4117-26. [http://dx.doi.org/10.1242/jcs.03179] [PMID: 16968741]

[21] Hundley HA, Walter W, Bairstow S, Craig EA. Human Mpp11 J protein: ribosome-tethered molecular chaperones are ubiquitous. Science 2005; 308(5724): 1032-4. [http://dx.doi.org/10.1126/science.1109247] [PMID: 15802566]

[22] Jaiswal H, Conz C, Otto H, *et al.* The chaperone network connected to human ribosome-associated complex. Mol Cell Biol 2011; 31(6): 1160-73. [http://dx.doi.org/10.1128/MCB.00986-10] [PMID: 21245388]

[23] Otto H, Conz C, Maier P, *et al.* The chaperones MPP11 and Hsp70L1 form the mammalian ribosome-associated complex. Proc Natl Acad Sci USA 2005; 102(29): 10064-9. [http://dx.doi.org/10.1073/pnas.0504400102] [PMID: 16002468]

[24] Flaherty KM, DeLuca-Flaherty C, McKay DB. Three-dimensional structure of the ATPase fragment of a 70K heat-shock cognate protein. Nature 1990; 346(6285): 623-8. [http://dx.doi.org/10.1038/346623a0] [PMID: 2143562]

[25] Kityk R, Kopp J, Sinning I, Mayer MP. Structure and dynamics of the ATP-bound open conformation of Hsp70 chaperones. Mol Cell 2012; 48(6): 863-74. [http://dx.doi.org/10.1016/j.molcel.2012.09.023] [PMID: 23123194]

[26] Wisniewska M, Karlberg T, Lehtiö L, *et al.* Crystal structures of the ATPase domains of four human Hsp70 isoforms: HSPA1L/Hsp70-hom, HSPA2/Hsp70-2, HSPA6/Hsp70B′, and HSPA5/BiP/GRP78. PLoS One 2010; 5(1): e8625. [http://dx.doi.org/10.1371/journal.pone.0008625] [PMID: 20072699]

[27] Zhu X, Zhao X, Burkholder WF, *et al.* Structural analysis of substrate binding by the molecular chaperone DnaK. Science 1996; 272(5268): 1606-14. [http://dx.doi.org/10.1126/science.272.5268.1606] [PMID: 8658133]

[28] Rüdiger S, Buchberger A, Bukau B. Interaction of Hsp70 chaperones with substrates. Nat Struct Biol 1997; 4(5): 342-9. [http://dx.doi.org/10.1038/nsb0597-342] [PMID: 9145101]

[29] Van Durme J, Maurer-Stroh S, Gallardo R, Wilkinson H, Rousseau F, Schymkowitz J. Accurate prediction of DnaK-peptide binding via homology modelling and experimental data. PLOS Comput Biol 2009; 5(8): e1000475. [http://dx.doi.org/10.1371/journal.pcbi.1000475] [PMID: 19696878]

[30] Moro F, Fernández-Sáiz V, Muga A. The lid subdomain of DnaK is required for the stabilization of

the substrate-binding site. J Biol Chem 2004; 279(19): 19600-6.
[http://dx.doi.org/10.1074/jbc.M400921200] [PMID: 14985342]

[31] Scheufler C, Brinker A, Bourenkov G, *et al.* Structure of TPR domain-peptide complexes: critical elements in the assembly of the Hsp70-Hsp90 multichaperone machine. Cell 2000; 101(2): 199-210.
[http://dx.doi.org/10.1016/S0092-8674(00)80830-2] [PMID: 10786835]

[32] Qi R, Sarbeng EB, Liu Q, *et al.* Allosteric opening of the polypeptide-binding site when an Hsp70 binds ATP. Nat Struct Mol Biol 2013; 20(7): 900-7.
[http://dx.doi.org/10.1038/nsmb.2583] [PMID: 23708608]

[33] Schlecht R, Erbse AH, Bukau B, Mayer MP. Mechanics of Hsp70 chaperones enables differential interaction with client proteins. Nat Struct Mol Biol 2011; 18(3): 345-51.
[http://dx.doi.org/10.1038/nsmb.2006] [PMID: 21278757]

[34] Jiang J, Prasad K, Lafer EM, Sousa R. Structural basis of interdomain communication in the Hsc70 chaperone. Mol Cell 2005; 20(4): 513-24.
[http://dx.doi.org/10.1016/j.molcel.2005.09.028] [PMID: 16307916]

[35] Bertelsen EB, Chang L, Gestwicki JE, Zuiderweg ER. Solution conformation of wild-type E. coli Hsp70 (DnaK) chaperone complexed with ADP and substrate. Proc Natl Acad Sci USA 2009; 106(21): 8471-6.
[http://dx.doi.org/10.1073/pnas.0903503106] [PMID: 19439666]

[36] Vogel M, Bukau B, Mayer MP. Allosteric regulation of Hsp70 chaperones by a proline switch. Mol Cell 2006; 21(3): 359-67.
[http://dx.doi.org/10.1016/j.molcel.2005.12.017] [PMID: 16455491]

[37] Vogel M, Mayer MP, Bukau B. Allosteric regulation of Hsp70 chaperones involves a conserved interdomain linker. J Biol Chem 2006; 281(50): 38705-11.
[http://dx.doi.org/10.1074/jbc.M609020200] [PMID: 17052976]

[38] Zhuravleva A, Clerico EM, Gierasch LM. An interdomain energetic tug-of-war creates the allosterically active state in Hsp70 molecular chaperones. Cell 2012; 151(6): 1296-307.
[http://dx.doi.org/10.1016/j.cell.2012.11.002] [PMID: 23217711]

[39] Young JC. Mechanisms of the Hsp70 chaperone system. Biochem Cell Biol 2010; 88(2): 291-300.
[http://dx.doi.org/10.1139/O09-175] [PMID: 20453930]

[40] Ohtsuka K, Hata M. Mammalian HSP40/DNAJ homologs: cloning of novel cDNAs and a proposal for their classification and nomenclature. Cell Stress Chaperones 2000; 5(2): 98-112.
[http://dx.doi.org/10.1379/1466-1268(2000)005<0098:MHDHCO>2.0.CO;2] [PMID: 11147971]

[41] Qiu XB, Shao YM, Miao S, Wang L. The diversity of the DnaJ/Hsp40 family, the crucial partners for Hsp70 chaperones. Cell Mol Life Sci 2006; 63(22): 2560-70.
[http://dx.doi.org/10.1007/s00018-006-6192-6] [PMID: 16952052]

[42] Kampinga HH, Hageman J, Vos MJ, *et al.* Guidelines for the nomenclature of the human heat shock proteins. Cell Stress Chaperones 2009; 14(1): 105-11.
[http://dx.doi.org/10.1007/s12192-008-0068-7] [PMID: 18663603]

[43] Hennessy F, Cheetham ME, Dirr HW, Blatch GL. Analysis of the levels of conservation of the J domain among the various types of DnaJ-like proteins. Cell Stress Chaperones 2000; 5(4): 347-58.
[http://dx.doi.org/10.1379/1466-1268(2000)005<0347:AOTLOC>2.0.CO;2] [PMID: 11048657]

[44] Cheetham ME, Caplan AJ. Structure, function and evolution of DnaJ: conservation and adaptation of chaperone function. Cell Stress Chaperones 1998; 3(1): 28-36.
[http://dx.doi.org/10.1379/1466-1268(1998)003<0028:SFAEOD>2.3.CO;2] [PMID: 9585179]

[45] Hennessy F, Nicoll WS, Zimmermann R, Cheetham ME, Blatch GL. Not all J domains are created equal: implications for the specificity of Hsp40-Hsp70 interactions. Protein Sci 2005; 14(7): 1697-709.
[http://dx.doi.org/10.1110/ps.051406805] [PMID: 15987899]

[46] Qian YQ, Patel D, Hartl FU, McColl DJ. Nuclear magnetic resonance solution structure of the human Hsp40 (HDJ-1) J-domain. J Mol Biol 1996; 260(2): 224-35.
[http://dx.doi.org/10.1006/jmbi.1996.0394] [PMID: 8764402]

[47] Genevaux P, Schwager F, Georgopoulos C, Kelley WL. Scanning mutagenesis identifies amino acid residues essential for the in vivo activity of the Escherichia coli DnaJ (Hsp40) J-domain. Genetics 2002; 162(3): 1045-53.
[PMID: 12454054]

[48] Suh WC, Burkholder WF, Lu CZ, Zhao X, Gottesman ME, Gross CA. Interaction of the Hsp70 molecular chaperone, DnaK, with its cochaperone DnaJ. Proc Natl Acad Sci USA 1998; 95(26): 15223-8.
[http://dx.doi.org/10.1073/pnas.95.26.15223] [PMID: 9860950]

[49] Tsai J, Douglas MG. A conserved HPD sequence of the J-domain is necessary for YDJ1 stimulation of Hsp70 ATPase activity at a site distinct from substrate binding. J Biol Chem 1996; 271(16): 9347-54.
[http://dx.doi.org/10.1074/jbc.271.16.9347] [PMID: 8621599]

[50] Suh WC, Lu CZ, Gross CA. Structural features required for the interaction of the Hsp70 molecular chaperone DnaK with its cochaperone DnaJ. J Biol Chem 1999; 274(43): 30534-9.
[http://dx.doi.org/10.1074/jbc.274.43.30534] [PMID: 10521435]

[51] Wall D, Zylicz M, Georgopoulos C. The NH2-terminal 108 amino acids of the Escherichia coli DnaJ protein stimulate the ATPase activity of DnaK and are sufficient for lambda replication. J Biol Chem 1994; 269(7): 5446-51.
[PMID: 8106526]

[52] Wall D, Zylicz M, Georgopoulos C. The conserved G/F motif of the DnaJ chaperone is necessary for the activation of the substrate binding properties of the DnaK chaperone. J Biol Chem 1995; 270(5): 2139-44.
[http://dx.doi.org/10.1074/jbc.270.5.2139] [PMID: 7836443]

[53] Greene MK, Maskos K, Landry SJ. Role of the J-domain in the cooperation of Hsp40 with Hsp70. Proc Natl Acad Sci USA 1998; 95(11): 6108-13.
[http://dx.doi.org/10.1073/pnas.95.11.6108] [PMID: 9600925]

[54] Jiang J, Maes EG, Taylor AB, *et al.* Structural basis of J cochaperone binding and regulation of Hsp70. Mol Cell 2007; 28(3): 422-33.
[http://dx.doi.org/10.1016/j.molcel.2007.08.022] [PMID: 17996706]

[55] Young JC, Barral JM, Ulrich Hartl F. More than folding: localized functions of cytosolic chaperones. Trends Biochem Sci 2003; 28(10): 541-7.
[http://dx.doi.org/10.1016/j.tibs.2003.08.009] [PMID: 14559183]

[56] Ahmad A, Bhattacharya A, McDonald RA, *et al.* Heat shock protein 70 kDa chaperone/DnaJ cochaperone complex employs an unusual dynamic interface. Proc Natl Acad Sci USA 2011; 108(47): 18966-71.
[http://dx.doi.org/10.1073/pnas.1111220108] [PMID: 22065753]

[57] Gässler CS, Buchberger A, Laufen T, *et al.* Mutations in the DnaK chaperone affecting interaction with the DnaJ cochaperone. Proc Natl Acad Sci USA 1998; 95(26): 15229-34.
[http://dx.doi.org/10.1073/pnas.95.26.15229] [PMID: 9860951]

[58] Gao XC, Zhou CJ, Zhou ZR, Wu M, Cao CY, Hu HY. The C-terminal helices of heat shock protein 70 are essential for J-domain binding and ATPase activation. J Biol Chem 2012; 287(8): 6044-52.
[http://dx.doi.org/10.1074/jbc.M111.294728] [PMID: 22219199]

[59] Cajo GC, Horne BE, Kelley WL, Schwager F, Georgopoulos C, Genevaux P. The role of the DIF motif of the DnaJ (Hsp40) co-chaperone in the regulation of the DnaK (Hsp70) chaperone cycle. J Biol Chem 2006; 281(18): 12436-44.
[http://dx.doi.org/10.1074/jbc.M511192200] [PMID: 16533811]

[60] Baaklini I, Wong MJ, Hantouche C, Patel Y, Shrier A, Young JC. The DNAJA2 substrate release mechanism is essential for chaperone-mediated folding. J Biol Chem 2012; 287(50): 41939-54. [http://dx.doi.org/10.1074/jbc.M112.413278] [PMID: 23091061]

[61] Tzankov S, Wong MJ, Shi K, Nassif C, Young JC. Functional divergence between co-chaperones of Hsc70. J Biol Chem 2008; 283(40): 27100-9. [http://dx.doi.org/10.1074/jbc.M803923200] [PMID: 18684711]

[62] Perales-Calvo J, Muga A, Moro F. Role of DnaJ G/F-rich domain in conformational recognition and binding of protein substrates. J Biol Chem 2010; 285(44): 34231-9. [http://dx.doi.org/10.1074/jbc.M110.144642] [PMID: 20729526]

[63] Couthouis J, Raphael AR, Siskind C, *et al.* Exome sequencing identifies a DNAJB6 mutation in a family with dominantly-inherited limb-girdle muscular dystrophy. Neuromuscul Disord 2014; 24(5): 431-5. [http://dx.doi.org/10.1016/j.nmd.2014.01.014] [PMID: 24594375]

[64] Sarparanta J, Jonson PH, Golzio C, *et al.* Mutations affecting the cytoplasmic functions of the co-chaperone DNAJB6 cause limb-girdle muscular dystrophy. Nat Genet 2012; 44(4): 450-5, S1-2.

[65] Stein KC, Bengoechea R, Harms MB, Weihl CC, True HL. Myopathy-causing mutations in an HSP40 chaperone disrupt processing of specific client conformers. J Biol Chem 2014; 289(30): 21120-30. [http://dx.doi.org/10.1074/jbc.M114.572461] [PMID: 24920671]

[66] Yan W, Craig EA. The glycine-phenylalanine-rich region determines the specificity of the yeast Hsp40 Sis1. Mol Cell Biol 1999; 19(11): 7751-8. [http://dx.doi.org/10.1128/MCB.19.11.7751] [PMID: 10523664]

[67] Sondheimer N, Lopez N, Craig EA, Lindquist S. The role of Sis1 in the maintenance of the [RNQ+] prion. EMBO J 2001; 20(10): 2435-42. [http://dx.doi.org/10.1093/emboj/20.10.2435] [PMID: 11350932]

[68] Liberek K, Wall D, Georgopoulos C. The DnaJ chaperone catalytically activates the DnaK chaperone to preferentially bind the sigma 32 heat shock transcriptional regulator. Proc Natl Acad Sci USA 1995; 92(14): 6224-8. [http://dx.doi.org/10.1073/pnas.92.14.6224] [PMID: 7603976]

[69] Laufen T, Mayer MP, Beisel C, *et al.* Mechanism of regulation of hsp70 chaperones by DnaJ cochaperones. Proc Natl Acad Sci USA 1999; 96(10): 5452-7. [http://dx.doi.org/10.1073/pnas.96.10.5452] [PMID: 10318904]

[70] Bhangoo MK, Tzankov S, Fan AC, Dejgaard K, Thomas DY, Young JC. Multiple 40-kDa heat-shock protein chaperones function in Tom70-dependent mitochondrial import. Mol Biol Cell 2007; 18(9): 3414-28. [http://dx.doi.org/10.1091/mbc.E07-01-0088] [PMID: 17596514]

[71] Goffin L, Georgopoulos C. Genetic and biochemical characterization of mutations affecting the carboxy-terminal domain of the Escherichia coli molecular chaperone DnaJ. Mol Microbiol 1998; 30(2): 329-40. [http://dx.doi.org/10.1046/j.1365-2958.1998.01067.x] [PMID: 9791178]

[72] Lu Z, Cyr DM. Protein folding activity of Hsp70 is modified differentially by the hsp40 co-chaperones Sis1 and Ydj1. J Biol Chem 1998; 273(43): 27824-30. [http://dx.doi.org/10.1074/jbc.273.43.27824] [PMID: 9774392]

[73] Lu Z, Cyr DM. The conserved carboxyl terminus and zinc finger-like domain of the co-chaperone Ydj1 assist Hsp70 in protein folding. J Biol Chem 1998; 273(10): 5970-8. [http://dx.doi.org/10.1074/jbc.273.10.5970] [PMID: 9488737]

[74] Li J, Qian X, Sha B. The crystal structure of the yeast Hsp40 Ydj1 complexed with its peptide substrate. Structure 2003; 11(12): 1475-83. [http://dx.doi.org/10.1016/j.str.2003.10.012] [PMID: 14656432]

[75] Sha B, Lee S, Cyr DM. The crystal structure of the peptide-binding fragment from the yeast Hsp40 protein Sis1. Structure 2000; 8(8): 799-807. [http://dx.doi.org/10.1016/S0969-2126(00)00170-2] [PMID: 10997899]

[76] Hu J, Wu Y, Li J, Qian X, Fu Z, Sha B. The crystal structure of the putative peptide-binding fragment from the human Hsp40 protein Hdj1. BMC Struct Biol 2008; 8: 3. [http://dx.doi.org/10.1186/1472-6807-8-3] [PMID: 18211704]

[77] Martinez-Yamout M, Legge GB, Zhang O, Wright PE, Dyson HJ. Solution structure of the cysteine-rich domain of the Escherichia coli chaperone protein DnaJ. J Mol Biol 2000; 300(4): 805-18. [http://dx.doi.org/10.1006/jmbi.2000.3923] [PMID: 10891270]

[78] Li J, Sha B. Structure-based mutagenesis studies of the peptide substrate binding fragment of type I heat-shock protein 40. Biochem J 2005; 386(Pt 3): 453-60. [http://dx.doi.org/10.1042/BJ20041050] [PMID: 15500443]

[79] Wu Y, Li J, Jin Z, Fu Z, Sha B. The crystal structure of the C-terminal fragment of yeast Hsp40 Ydj1 reveals novel dimerization motif for Hsp40. J Mol Biol 2005; 346(4): 1005-11. [http://dx.doi.org/10.1016/j.jmb.2004.12.040] [PMID: 15701512]

[80] Lee S, Fan CY, Younger JM, Ren H, Cyr DM. Identification of essential residues in the type II Hsp40 Sis1 that function in polypeptide binding. J Biol Chem 2002; 277(24): 21675-82. [http://dx.doi.org/10.1074/jbc.M111075200] [PMID: 11919183]

[81] Fan CY, Ren HY, Lee P, Caplan AJ, Cyr DM. The type I Hsp40 zinc finger-like region is required for Hsp70 to capture non-native polypeptides from Ydj1. J Biol Chem 2005; 280(1): 695-702. [http://dx.doi.org/10.1074/jbc.M410645200] [PMID: 15496404]

[82] Linke K, Wolfram T, Bussemer J, Jakob U. The roles of the two zinc binding sites in DnaJ. J Biol Chem 2003; 278(45): 44457-66. [http://dx.doi.org/10.1074/jbc.M307491200] [PMID: 12941935]

[83] Banecki B, Liberek K, Wall D, *et al.* Structure-function analysis of the zinc finger region of the DnaJ molecular chaperone. J Biol Chem 1996; 271(25): 14840-8. [http://dx.doi.org/10.1074/jbc.271.25.14840] [PMID: 8662861]

[84] Mattoo RU, Farina Henriquez Cuendet A, Subanna S, *et al.* Synergism between a foldase and an unfoldase: reciprocal dependence between the thioredoxin-like activity of DnaJ and the polypeptide-unfolding activity of DnaK. Front Mol Biosci 2014; 1: 7. [http://dx.doi.org/10.3389/fmolb.2014.00007] [PMID: 25988148]

[85] Kanazawa M, Terada K, Kato S, Mori M. HSDJ, a human homolog of DnaJ, is farnesylated and is involved in protein import into mitochondria. J Biochem 1997; 121(5): 890-5. [http://dx.doi.org/10.1093/oxfordjournals.jbchem.a021670] [PMID: 9192730]

[86] Flom GA, Lemieszek M, Fortunato EA, Johnson JL. Farnesylation of Ydj1 is required for in vivo interaction with Hsp90 client proteins. Mol Biol Cell 2008; 19(12): 5249-58. [http://dx.doi.org/10.1091/mbc.E08-04-0435] [PMID: 18829866]

[87] Borges JC, Fischer H, Craievich AF, Ramos CH. Low resolution structural study of two human HSP40 chaperones in solution. DJA1 from subfamily A and DJB4 from subfamily B have different quaternary structures. J Biol Chem 2005; 280(14): 13671-81. [http://dx.doi.org/10.1074/jbc.M408349200] [PMID: 15661747]

[88] Ramos CH, Oliveira CL, Fan CY, Torriani IL, Cyr DM. Conserved central domains control the quaternary structure of type I and type II Hsp40 molecular chaperones. J Mol Biol 2008; 383(1): 155-66. [http://dx.doi.org/10.1016/j.jmb.2008.08.019] [PMID: 18723025]

[89] Bracher A, Verghese J, Grp E. GrpE, Hsp110/Grp170, HspBP1/Sil1 and BAG domain proteins: nucleotide exchange factors for Hsp70 molecular chaperones. Subcell Biochem 2015; 78: 1-33. [http://dx.doi.org/10.1007/978-3-319-11731-7_1] [PMID: 25487014]

[90] Höhfeld J, Jentsch S. GrpE-like regulation of the hsc70 chaperone by the anti-apoptotic protein BAG-1. EMBO J 1997; 16(20): 6209-16.
[http://dx.doi.org/10.1093/emboj/16.20.6209] [PMID: 9321400]

[91] Sondermann H, Ho AK, Listenberger LL, *et al.* Prediction of novel Bag-1 homologs based on structure/function analysis identifies Snl1p as an Hsp70 co-chaperone in Saccharomyces cerevisiae. J Biol Chem 2002; 277(36): 33220-7.
[http://dx.doi.org/10.1074/jbc.M204624200] [PMID: 12105220]

[92] Takayama S, Krajewski S, Krajewska M, *et al.* Expression and location of Hsp70/Hsc-binding anti-apoptotic protein BAG-1 and its variants in normal tissues and tumor cell lines. Cancer Res 1998; 58(14): 3116-31.
[PMID: 9679980]

[93] Alberti S, Esser C, Höhfeld J. BAG-1--a nucleotide exchange factor of Hsc70 with multiple cellular functions. Cell Stress Chaperones 2003; 8(3): 225-31.
[http://dx.doi.org/10.1379/1466-1268(2003)008<0225:BNEFOH>2.0.CO;2] [PMID: 14984055]

[94] Sondermann H, Scheufler C, Schneider C, Hohfeld J, Hartl FU, Moarefi I. Structure of a Bag/Hsc70 complex: convergent functional evolution of Hsp70 nucleotide exchange factors. Science 2001; 291(5508): 1553-7.
[http://dx.doi.org/10.1126/science.1057268] [PMID: 11222862]

[95] Harrison CJ, Hayer-Hartl M, Di Liberto M, Hartl F, Kuriyan J. Crystal structure of the nucleotide exchange factor GrpE bound to the ATPase domain of the molecular chaperone DnaK. Science 1997; 276(5311): 431-5.
[http://dx.doi.org/10.1126/science.276.5311.431] [PMID: 9103205]

[96] Briknarová K, Takayama S, Brive L, *et al.* Structural analysis of BAG1 cochaperone and its interactions with Hsc70 heat shock protein. Nat Struct Biol 2001; 8(4): 349-52.
[http://dx.doi.org/10.1038/86236] [PMID: 11276257]

[97] Briknarová K, Takayama S, Homma S, *et al.* BAG4/SODD protein contains a short BAG domain. J Biol Chem 2002; 277(34): 31172-8.
[http://dx.doi.org/10.1074/jbc.M202792200] [PMID: 12058034]

[98] Arakawa A, Handa N, Ohsawa N, *et al.* The C-terminal BAG domain of BAG5 induces conformational changes of the Hsp70 nucleotide-binding domain for ADP-ATP exchange. Structure 2010; 18(3): 309-19.
[http://dx.doi.org/10.1016/j.str.2010.01.004] [PMID: 20223214]

[99] Xu Z, Page RC, Gomes MM, *et al.* Structural basis of nucleotide exchange and client binding by the Hsp70 cochaperone Bag2. Nat Struct Mol Biol 2008; 15(12): 1309-17.
[http://dx.doi.org/10.1038/nsmb.1518] [PMID: 19029896]

[100] Kabani M, Beckerich JM, Brodsky JL. Nucleotide exchange factor for the yeast Hsp70 molecular chaperone Ssa1p. Mol Cell Biol 2002; 22(13): 4677-89.
[http://dx.doi.org/10.1128/MCB.22.13.4677-4689.2002] [PMID: 12052876]

[101] Kabani M, McLellan C, Raynes DA, Guerriero V, Brodsky JL. HspBP1, a homologue of the yeast Fes1 and Sls1 proteins, is an Hsc70 nucleotide exchange factor. FEBS Lett 2002; 531(2): 339-42.
[http://dx.doi.org/10.1016/S0014-5793(02)03570-6] [PMID: 12417338]

[102] Dragovic Z, Broadley SA, Shomura Y, Bracher A, Hartl FU. Molecular chaperones of the Hsp110 family act as nucleotide exchange factors of Hsp70s. EMBO J 2006; 25(11): 2519-28.
[http://dx.doi.org/10.1038/sj.emboj.7601138] [PMID: 16688212]

[103] Shomura Y, Dragovic Z, Chang HC, *et al.* Regulation of Hsp70 function by HspBP1: structural analysis reveals an alternate mechanism for Hsp70 nucleotide exchange. Mol Cell 2005; 17(3): 367-79.
[PMID: 15694338]

[104] Yan M, Li J, Sha B. Structural analysis of the Sil1-Bip complex reveals the mechanism for Sil1 to function as a nucleotide-exchange factor. Biochem J 2011; 438(3): 447-55. [http://dx.doi.org/10.1042/BJ20110500] [PMID: 21675960]

[105] Easton DP, Kaneko Y, Subjeck JR. The hsp110 and Grp1 70 stress proteins: newly recognized relatives of the Hsp70s. Cell Stress Chaperones 2000; 5(4): 276-90. [http://dx.doi.org/10.1379/1466-1268(2000)005<0276:THAGSP>2.0.CO;2] [PMID: 11048651]

[106] Liu Q, Hendrickson WA. Insights into Hsp70 chaperone activity from a crystal structure of the yeast Hsp110 Sse1. Cell 2007; 131(1): 106-20. [http://dx.doi.org/10.1016/j.cell.2007.08.039] [PMID: 17923091]

[107] Polier S, Dragovic Z, Hartl FU, Bracher A. Structural basis for the cooperation of Hsp70 and Hsp110 chaperones in protein folding. Cell 2008; 133(6): 1068-79. [http://dx.doi.org/10.1016/j.cell.2008.05.022] [PMID: 18555782]

[108] Schuermann JP, Jiang J, Cuellar J, *et al.* Structure of the Hsp110:Hsc70 nucleotide exchange machine. Mol Cell 2008; 31(2): 232-43. [http://dx.doi.org/10.1016/j.molcel.2008.05.006] [PMID: 18550409]

[109] Andréasson C, Fiaux J, Rampelt H, Mayer MP, Bukau B. Hsp110 is a nucleotide-activated exchange factor for Hsp70. J Biol Chem 2008; 283(14): 8877-84. [http://dx.doi.org/10.1074/jbc.M710063200] [PMID: 18218635]

[110] Rauch JN, Gestwicki JE. Binding of human nucleotide exchange factors to heat shock protein 70 (Hsp70) generates functionally distinct complexes in vitro. J Biol Chem 2014; 289(3): 1402-14. [http://dx.doi.org/10.1074/jbc.M113.521997] [PMID: 24318877]

[111] Hartl FU, Hayer-Hartl M. Molecular chaperones in the cytosol: from nascent chain to folded protein. Science 2002; 295(5561): 1852-8. [http://dx.doi.org/10.1126/science.1068408] [PMID: 11884745]

[112] Priya S, Sharma SK, Goloubinoff P. Molecular chaperones as enzymes that catalytically unfold misfolded polypeptides. FEBS Lett 2013; 587(13): 1981-7. [http://dx.doi.org/10.1016/j.febslet.2013.05.014] [PMID: 23684649]

[113] Jin Y, Awad W, Petrova K, Hendershot LM. Regulated release of ERdj3 from unfolded proteins by BiP. EMBO J 2008; 27(21): 2873-82. [http://dx.doi.org/10.1038/emboj.2008.207] [PMID: 18923428]

[114] Kellner R, Hofmann H, Barducci A, Wunderlich B, Nettels D, Schuler B. Single-molecule spectroscopy reveals chaperone-mediated expansion of substrate protein. Proc Natl Acad Sci USA 2014; 111(37): 13355-60. [http://dx.doi.org/10.1073/pnas.1407086111] [PMID: 25165400]

[115] Nunes JM, Mayer-Hartl M, Hartl FU, Müller DJ. Action of the Hsp70 chaperone system observed with single proteins. Nat Commun 2015; 6: 6307. [http://dx.doi.org/10.1038/ncomms7307] [PMID: 25686738]

[116] Uchiyama Y, Takeda N, Mori M, Terada K. Heat shock protein 40/DjB1 is required for thermotolerance in early phase. J Biochem 2006; 140(6): 805-12. [http://dx.doi.org/10.1093/jb/mvj212] [PMID: 17050614]

[117] Chien V, Aitken JF, Zhang S, *et al.* The chaperone proteins HSP70, HSP40/DnaJ and GRP78/BiP suppress misfolding and formation of β-sheet-containing aggregates by human amylin: a potential role for defective chaperone biology in Type 2 diabetes. Biochem J 2010; 432(1): 113-21. [http://dx.doi.org/10.1042/BJ20100434] [PMID: 20735358]

[118] Hafizur RM, Yano M, Gotoh T, Mori M, Terada K. Modulation of chaperone activities of Hsp70 and Hsp70-2 by a mammalian DnaJ/Hsp40 homolog, DjA4. J Biochem 2004; 135(2): 193-200. [http://dx.doi.org/10.1093/jb/mvh023] [PMID: 15047721]

[119] Hageman J, van Waarde MA, Zylicz A, Walerych D, Kampinga HH. The diverse members of the mammalian HSP70 machine show distinct chaperone-like activities. Biochem J 2011; 435(1): 127-42. [http://dx.doi.org/10.1042/BJ20101247] [PMID: 21231916]

[120] Minami Y, Höhfeld J, Ohtsuka K, Hartl FU. Regulation of the heat-shock protein 70 reaction cycle by the mammalian DnaJ homolog, Hsp40. J Biol Chem 1996; 271(32): 19617-24. [http://dx.doi.org/10.1074/jbc.271.32.19617] [PMID: 8702658]

[121] Shorter J. The mammalian disaggregase machinery: Hsp110 synergizes with Hsp70 and Hsp40 to catalyze protein disaggregation and reactivation in a cell-free system. PLoS One 2011; 6(10): e26319. [http://dx.doi.org/10.1371/journal.pone.0026319] [PMID: 22022600]

[122] Nillegoda NB, Kirstein J, Szlachcic A, *et al.* Crucial HSP70 co-chaperone complex unlocks metazoan protein disaggregation. Nature 2015; 524(7564): 247-51. [http://dx.doi.org/10.1038/nature14884] [PMID: 26245380]

[123] Summers DW, Douglas PM, Ramos CH, Cyr DM. Polypeptide transfer from Hsp40 to Hsp70 molecular chaperones. Trends Biochem Sci 2009; 34(5): 230-3. [http://dx.doi.org/10.1016/j.tibs.2008.12.009] [PMID: 19359181]

[124] Petrova K, Oyadomari S, Hendershot LM, Ron D. Regulated association of misfolded endoplasmic reticulum lumenal proteins with P58/DNAJc3. EMBO J 2008; 27(21): 2862-72. [http://dx.doi.org/10.1038/emboj.2008.199] [PMID: 18923430]

[125] Terada K, Oike Y. Multiple molecules of Hsc70 and a dimer of DjA1 independently bind to an unfolded protein. J Biol Chem 2010; 285(22): 16789-97. [http://dx.doi.org/10.1074/jbc.M110.101501] [PMID: 20363747]

[126] Terada K, Mori M. Human DnaJ homologs dj2 and dj3, and bag-1 are positive cochaperones of hsc70. J Biol Chem 2000; 275(32): 24728-34. [http://dx.doi.org/10.1074/jbc.M002021200] [PMID: 10816573]

[127] Su AI, Wiltshire T, Batalov S, *et al.* A gene atlas of the mouse and human protein-encoding transcriptomes. Proc Natl Acad Sci USA 2004; 101(16): 6062-7. [http://dx.doi.org/10.1073/pnas.0400782101] [PMID: 15075390]

[128] Finka A, Goloubinoff P. Proteomic data from human cell cultures refine mechanisms of chaperone-mediated protein homeostasis. Cell Stress Chaperones 2013; 18(5): 591-605. [http://dx.doi.org/10.1007/s12192-013-0413-3] [PMID: 23430704]

[129] Abdul KM, Terada K, Gotoh T, Hafizur RM, Mori M. Characterization and functional analysis of a heart-enriched DnaJ/ Hsp40 homolog dj4/DjA4. Cell Stress Chaperones 2002; 7(2): 156-66. [http://dx.doi.org/10.1379/1466-1268(2002)007<0156:CAFAOA>2.0.CO;2] [PMID: 12380683]

[130] Ryno LM, Genereux JC, Naito T, *et al.* Characterizing the altered cellular proteome induced by the stress-independent activation of heat shock factor 1. ACS Chem Biol 2014; 9(6): 1273-83. [http://dx.doi.org/10.1021/cb500062n] [PMID: 24689980]

[131] Sharma K, Vabulas RM, Macek B, *et al.* Quantitative proteomics reveals that Hsp90 inhibition preferentially targets kinases and the DNA damage response. Molecular & cellular proteomics : MCP 2012; 11(3): M111 014654. [http://dx.doi.org/10.1074/mcp.M111.014654]

[132] Abisambra JF, Jinwal UK, Suntharalingam A, *et al.* DnaJA1 antagonizes constitutive Hsp70-mediated stabilization of tau. J Mol Biol 2012; 421(4-5): 653-61. [http://dx.doi.org/10.1016/j.jmb.2012.02.003] [PMID: 22343013]

[133] Pemberton S, Madiona K, Pieri L, Kabani M, Bousset L, Melki R. Hsc70 protein interaction with soluble and fibrillar alpha-synuclein. J Biol Chem 2011; 286(40): 34690-9. [http://dx.doi.org/10.1074/jbc.M111.261321] [PMID: 21832061]

[134] Roodveldt C, Bertoncini CW, Andersson A, *et al.* Chaperone proteostasis in Parkinson's disease:

stabilization of the Hsp70/alpha-synuclein complex by Hip. EMBO J 2009; 28(23): 3758-70. [http://dx.doi.org/10.1038/emboj.2009.298] [PMID: 19875982]

[135] Lo Bianco C, Shorter J, Régulier E, *et al.* Hsp104 antagonizes alpha-synuclein aggregation and reduces dopaminergic degeneration in a rat model of Parkinson disease. J Clin Invest 2008; 118(9): 3087-97. [http://dx.doi.org/10.1172/JCI35781] [PMID: 18704197]

[136] Cummings CJ, Mancini MA, Antalffy B, DeFranco DB, Orr HT, Zoghbi HY. Chaperone suppression of aggregation and altered subcellular proteasome localization imply protein misfolding in SCA1. Nat Genet 1998; 19(2): 148-54. [http://dx.doi.org/10.1038/502] [PMID: 9620770]

[137] Wyttenbach A, Carmichael J, Swartz J, *et al.* Effects of heat shock, heat shock protein 40 (HDJ-2), and proteasome inhibition on protein aggregation in cellular models of Huntington's disease. Proc Natl Acad Sci USA 2000; 97(6): 2898-903. [http://dx.doi.org/10.1073/pnas.97.6.2898] [PMID: 10717003]

[138] Jana NR, Tanaka M, Wang Gh, Nukina N. Polyglutamine length-dependent interaction of Hsp40 and Hsp70 family chaperones with truncated N-terminal huntingtin: their role in suppression of aggregation and cellular toxicity. Hum Mol Genet 2000; 9(13): 2009-18. [http://dx.doi.org/10.1093/hmg/9.13.2009] [PMID: 10942430]

[139] Stenoien DL, Cummings CJ, Adams HP, *et al.* Polyglutamine-expanded androgen receptors form aggregates that sequester heat shock proteins, proteasome components and SRC-1, and are suppressed by the HDJ-2 chaperone. Hum Mol Genet 1999; 8(5): 731-41. [http://dx.doi.org/10.1093/hmg/8.5.731] [PMID: 10196362]

[140] Giffard RG, Xu L, Zhao H, *et al.* Chaperones, protein aggregation, and brain protection from hypoxic/ischemic injury. J Exp Biol 2004; 207(Pt 18): 3213-20. [http://dx.doi.org/10.1242/jeb.01034] [PMID: 15299042]

[141] Qiao Y, Ouyang YB, Giffard RG. Overexpression of HDJ-2 protects astrocytes from ischemia-like injury and reduces redistribution of ubiquitin staining in vitro. J Cereb Blood Flow Metab 2003; 23(10): 1113-6. [http://dx.doi.org/10.1097/01.WCB.0000088765.02615.FE] [PMID: 14526221]

[142] Chuang JZ, Zhou H, Zhu M, Li SH, Li XJ, Sung CH. Characterization of a brain-enriched chaperone, MRJ, that inhibits Huntingtin aggregation and toxicity independently. J Biol Chem 2002; 277(22): 19831-8. [http://dx.doi.org/10.1074/jbc.M109613200] [PMID: 11896048]

[143] Hageman J, Rujano MA, van Waarde MA, *et al.* A DNAJB chaperone subfamily with HDAC-dependent activities suppresses toxic protein aggregation. Mol Cell 2010; 37(3): 355-69. [http://dx.doi.org/10.1016/j.molcel.2010.01.001] [PMID: 20159555]

[144] Gillis J, Schipper-Krom S, Juenemann K, *et al.* The DNAJB6 and DNAJB8 protein chaperones prevent intracellular aggregation of polyglutamine peptides. J Biol Chem 2013; 288(24): 17225-37. [http://dx.doi.org/10.1074/jbc.M112.421685] [PMID: 23612975]

[145] Månsson C, Arosio P, Hussein R, *et al.* Interaction of the molecular chaperone DNAJB6 with growing amyloid-beta 42 (Aβ42) aggregates leads to sub-stoichiometric inhibition of amyloid formation. J Biol Chem 2014; 289(45): 31066-76. [http://dx.doi.org/10.1074/jbc.M114.595124] [PMID: 25217638]

[146] Durrenberger PF, Filiou MD, Moran LB, *et al.* DnaJB6 is present in the core of Lewy bodies and is highly up-regulated in parkinsonian astrocytes. J Neurosci Res 2009; 87(1): 238-45. [http://dx.doi.org/10.1002/jnr.21819] [PMID: 18711724]

[147] Westhoff B, Chapple JP, van der Spuy J, Höhfeld J, Cheetham ME. HSJ1 is a neuronal shuttling factor for the sorting of chaperone clients to the proteasome. Curr Biol 2005; 15(11): 1058-64. [http://dx.doi.org/10.1016/j.cub.2005.04.058] [PMID: 15936278]

[148] Howarth JL, Kelly S, Keasey MP, *et al.* Hsp40 molecules that target to the ubiquitin-proteasome system decrease inclusion formation in models of polyglutamine disease. Molecular therapy : the journal of the American Society of Gene Therapy 2007; 15(6): 1100-5. [http://dx.doi.org/10.1038/sj.mt.6300163]

[149] Gao XC, Zhou CJ, Zhou ZR, *et al.* Co-chaperone HSJ1a dually regulates the proteasomal degradation of ataxin-3. PLoS One 2011; 6(5): e19763. [http://dx.doi.org/10.1371/journal.pone.0019763] [PMID: 21625540]

[150] Novoselov SS, Mustill WJ, Gray AL, *et al.* Molecular chaperone mediated late-stage neuroprotection in the SOD1(G93A) mouse model of amyotrophic lateral sclerosis. PLoS One 2013; 8(8): e73944. [http://dx.doi.org/10.1371/journal.pone.0073944] [PMID: 24023695]

[151] Park SH, Kukushkin Y, Gupta R, *et al.* PolyQ proteins interfere with nuclear degradation of cytosolic proteins by sequestering the Sis1p chaperone. Cell 2013; 154(1): 134-45. [http://dx.doi.org/10.1016/j.cell.2013.06.003] [PMID: 23791384]

[152] Mitra A, Shevde LA, Samant RS. Multi-faceted role of HSP40 in cancer. Clin Exp Metastasis 2009; 26(6): 559-67. [http://dx.doi.org/10.1007/s10585-009-9255-x] [PMID: 19340594]

[153] Sterrenberg JN, Blatch GL, Edkins AL. Human DNAJ in cancer and stem cells. Cancer Lett 2011; 312(2): 129-42. [http://dx.doi.org/10.1016/j.canlet.2011.08.019] [PMID: 21925790]

[154] Tang D, Khaleque MA, Jones EL, *et al.* Expression of heat shock proteins and heat shock protein messenger ribonucleic acid in human prostate carcinoma in vitro and in tumors in vivo. Cell Stress Chaperones 2005; 10(1): 46-58. [http://dx.doi.org/10.1379/CSC-44R.1] [PMID: 15832947]

[155] Crnogorac-Jurcevic T, Gangeswaran R, Bhakta V, *et al.* Proteomic analysis of chronic pancreatitis and pancreatic adenocarcinoma. Gastroenterology 2005; 129(5): 1454-63. [http://dx.doi.org/10.1053/j.gastro.2005.08.012] [PMID: 16285947]

[156] Stark JL, Mehla K, Chaika N, *et al.* Structure and function of human DnaJ homologue subfamily a member 1 (DNAJA1) and its relationship to pancreatic cancer. Biochemistry 2014; 53(8): 1360-72. [http://dx.doi.org/10.1021/bi401329a] [PMID: 24512202]

[157] Gabai VL, Mabuchi K, Mosser DD, Sherman MY. Hsp72 and stress kinase c-jun N-terminal kinase regulate the bid-dependent pathway in tumor necrosis factor-induced apoptosis. Mol Cell Biol 2002; 22(10): 3415-24. [http://dx.doi.org/10.1128/MCB.22.10.3415-3424.2002] [PMID: 11971973]

[158] Gabai VL, Yaglom JA, Volloch V, *et al.* Hsp72-mediated suppression of c-Jun N-terminal kinase is implicated in development of tolerance to caspase-independent cell death. Mol Cell Biol 2000; 20(18): 6826-36. [http://dx.doi.org/10.1128/MCB.20.18.6826-6836.2000] [PMID: 10958679]

[159] Wang CC, Liao YP, Mischel PS, Iwamoto KS, Cacalano NA, McBride WH. HDJ-2 as a target for radiosensitization of glioblastoma multiforme cells by the farnesyltransferase inhibitor R115777 and the role of the p53/p21 pathway. Cancer Res 2006; 66(13): 6756-62. [http://dx.doi.org/10.1158/0008-5472.CAN-06-0185] [PMID: 16818651]

[160] Patnaik A, Eckhardt SG, Izbicka E, *et al.* A phase I, pharmacokinetic, and biological study of the farnesyltransferase inhibitor tipifarnib in combination with gemcitabine in patients with advanced malignancies. Clin Cancer Res 2003; 9(13): 4761-71. [PMID: 14581347]

[161] Esser C, Scheffner M, Höhfeld J. The chaperone-associated ubiquitin ligase CHIP is able to target p53 for proteasomal degradation. J Biol Chem 2005; 280(29): 27443-8. [http://dx.doi.org/10.1074/jbc.M501574200] [PMID: 15911628]

[162] Muller P, Hrstka R, Coomber D, Lane DP, Vojtesek B. Chaperone-dependent stabilization and degradation of p53 mutants. Oncogene 2008; 27(24): 3371-83. [http://dx.doi.org/10.1038/sj.onc.1211010] [PMID: 18223694]

[163] Esser C, Alberti S, Höhfeld J. Cooperation of molecular chaperones with the ubiquitin/proteasome system. Biochim Biophys Acta 2004; 1695(1-3): 171-88. [http://dx.doi.org/10.1016/j.bbamcr.2004.09.020] [PMID: 15571814]

[164] Qi M, Zhang J, Zeng W, Chen X. DNAJB1 stabilizes MDM2 and contributes to cancer cell proliferation in a p53-dependent manner. Biochim Biophys Acta 2014; 1839(1): 62-9. [http://dx.doi.org/10.1016/j.bbagrm.2013.12.003] [PMID: 24361594]

[165] Cui X, Choi HK, Choi YS, *et al.* DNAJB1 destabilizes PDCD5 to suppress p53-mediated apoptosis. Cancer Lett 2015; 357(1): 307-15. [http://dx.doi.org/10.1016/j.canlet.2014.11.041] [PMID: 25444898]

[166] Huang L, Yu Z, Zhang T, Zhao X, Huang G. HSP40 interacts with pyruvate kinase M2 and regulates glycolysis and cell proliferation in tumor cells. PLoS One 2014; 9(3): e92949. [http://dx.doi.org/10.1371/journal.pone.0092949] [PMID: 24658033]

[167] Schmidt O, Pfanner N, Meisinger C. Mitochondrial protein import: from proteomics to functional mechanisms. Nat Rev Mol Cell Biol 2010; 11(9): 655-67. [http://dx.doi.org/10.1038/nrm2959] [PMID: 20729931]

[168] Young JC, Hoogenraad NJ, Hartl FU. Molecular chaperones Hsp90 and Hsp70 deliver preproteins to the mitochondrial import receptor Tom70. Cell 2003; 112(1): 41-50. [http://dx.doi.org/10.1016/S0092-8674(02)01250-3] [PMID: 12526792]

[169] Fan AC, Bhangoo MK, Young JC. Hsp90 functions in the targeting and outer membrane translocation steps of Tom70-mediated mitochondrial import. J Biol Chem 2006; 281(44): 33313-24. [http://dx.doi.org/10.1074/jbc.M605250200] [PMID: 16968702]

[170] Fan AC, Kozlov G, Hoegl A, *et al.* Interaction between the human mitochondrial import receptors Tom20 and Tom70 in vitro suggests a chaperone displacement mechanism. J Biol Chem 2011; 286(37): 32208-19. [http://dx.doi.org/10.1074/jbc.M111.280446] [PMID: 21771790]

[171] Terada K, Kanazawa M, Bukau B, Mori M. The human DnaJ homologue dj2 facilitates mitochondrial protein import and luciferase refolding. J Cell Biol 1997; 139(5): 1089-95. [http://dx.doi.org/10.1083/jcb.139.5.1089] [PMID: 9382858]

[172] Gotoh T, Terada K, Mori M. hsp70-DnaJ chaperone pairs prevent nitric oxide-mediated apoptosis in RAW 264.7 macrophages. Cell Death Differ 2001; 8(4): 357-66. [http://dx.doi.org/10.1038/sj.cdd.4400829] [PMID: 11550087]

[173] Gotoh T, Terada K, Oyadomari S, Mori M. hsp70-DnaJ chaperone pair prevents nitric oxide- and CHOP-induced apoptosis by inhibiting translocation of Bax to mitochondria. Cell Death Differ 2004; 11(4): 390-402. [http://dx.doi.org/10.1038/sj.cdd.4401369] [PMID: 14752510]

[174] Stankiewicz AR, Lachapelle G, Foo CP, Radicioni SM, Mosser DD. Hsp70 inhibits heat-induced apoptosis upstream of mitochondria by preventing Bax translocation. J Biol Chem 2005; 280(46): 38729-39. [http://dx.doi.org/10.1074/jbc.M509497200] [PMID: 16172114]

[175] Young JC. The role of the cytosolic HSP70 chaperone system in diseases caused by misfolding and aberrant trafficking of ion channels. Dis Model Mech 2014; 7(3): 319-29. [http://dx.doi.org/10.1242/dmm.014001] [PMID: 24609033]

[176] Smith MH, Ploegh HL, Weissman JS. Road to ruin: targeting proteins for degradation in the endoplasmic reticulum. Science 2011; 334(6059): 1086-90. [http://dx.doi.org/10.1126/science.1209235] [PMID: 22116878]

[177] Riordan JR. CFTR function and prospects for therapy. Annu Rev Biochem 2008; 77: 701-26. [http://dx.doi.org/10.1146/annurev.biochem.75.103004.142532] [PMID: 18304008]

[178] Loo MA, Jensen TJ, Cui L, Hou Y, Chang XB, Riordan JR. Perturbation of Hsp90 interaction with nascent CFTR prevents its maturation and accelerates its degradation by the proteasome. EMBO J 1998; 17(23): 6879-87. [http://dx.doi.org/10.1093/emboj/17.23.6879] [PMID: 9843494]

[179] Wang X, Venable J, LaPointe P, *et al.* Hsp90 cochaperone Aha1 downregulation rescues misfolding of CFTR in cystic fibrosis. Cell 2006; 127(4): 803-15. [http://dx.doi.org/10.1016/j.cell.2006.09.043] [PMID: 17110338]

[180] Yang Y, Janich S, Cohn JA, Wilson JM. The common variant of cystic fibrosis transmembrane conductance regulator is recognized by hsp70 and degraded in a pre-Golgi nonlysosomal compartment. Proc Natl Acad Sci USA 1993; 90(20): 9480-4. [http://dx.doi.org/10.1073/pnas.90.20.9480] [PMID: 7692448]

[181] Meacham GC, Lu Z, King S, Sorscher E, Tousson A, Cyr DM. The Hdj-2/Hsc70 chaperone pair facilitates early steps in CFTR biogenesis. EMBO J 1999; 18(6): 1492-505. [http://dx.doi.org/10.1093/emboj/18.6.1492] [PMID: 10075921]

[182] Zhang H, Schmidt BZ, Sun F, *et al.* Cysteine string protein monitors late steps in cystic fibrosis transmembrane conductance regulator biogenesis. J Biol Chem 2006; 281(16): 11312-21. [http://dx.doi.org/10.1074/jbc.M512013200] [PMID: 16469739]

[183] Grove DE, Fan CY, Ren HY, Cyr DM. The endoplasmic reticulum-associated Hsp40 DNAJB12 and Hsc70 cooperate to facilitate RMA1 E3-dependent degradation of nascent CFTRDeltaF508. Mol Biol Cell 2011; 22(3): 301-14. [http://dx.doi.org/10.1091/mbc.E10-09-0760] [PMID: 21148293]

[184] Schmidt BZ, Watts RJ, Aridor M, Frizzell RA. Cysteine string protein promotes proteasomal degradation of the cystic fibrosis transmembrane conductance regulator (CFTR) by increasing its interaction with the C terminus of Hsp70-interacting protein and promoting CFTR ubiquitylation. J Biol Chem 2009; 284(7): 4168-78. [http://dx.doi.org/10.1074/jbc.M806485200] [PMID: 19098309]

[185] Yamamoto YH, Kimura T, Momohara S, *et al.* A novel ER J-protein DNAJB12 accelerates ER-associated degradation of membrane proteins including CFTR. Cell Struct Funct 2010; 35(2): 107-16. [http://dx.doi.org/10.1247/csf.10023] [PMID: 21150129]

[186] Okiyoneda T, Barrière H, Bagdány M, *et al.* Peripheral protein quality control removes unfolded CFTR from the plasma membrane. Science 2010; 329(5993): 805-10. [http://dx.doi.org/10.1126/science.1191542] [PMID: 20595578]

[187] Sanguinetti MC, Tristani-Firouzi M. hERG potassium channels and cardiac arrhythmia. Nature 2006; 440(7083): 463-9. [http://dx.doi.org/10.1038/nature04710] [PMID: 16554806]

[188] Ficker E, Dennis AT, Wang L, Brown AM. Role of the cytosolic chaperones Hsp70 and Hsp90 in maturation of the cardiac potassium channel HERG. Circ Res 2003; 92(12): e87-e100. [http://dx.doi.org/10.1161/01.RES.0000079028.31393.15] [PMID: 12775586]

[189] Walker VE, Atanasiu R, Lam H, Shrier A. Co-chaperone FKBP38 promotes HERG trafficking. J Biol Chem 2007; 282(32): 23509-16. [http://dx.doi.org/10.1074/jbc.M701006200] [PMID: 17569659]

[190] Walker VE, Wong MJ, Atanasiu R, Hantouche C, Young JC, Shrier A. Hsp40 chaperones promote degradation of the HERG potassium channel. J Biol Chem 2010; 285(5): 3319-29. [http://dx.doi.org/10.1074/jbc.M109.024000] [PMID: 19940115]

[191] Gao Y, Yechikov S, Vázquez AE, Chen D, Nie L. Impaired surface expression and conductance of the KCNQ4 channel lead to sensorineural hearing loss. J Cell Mol Med 2013; 17(7): 889-900.

[http://dx.doi.org/10.1111/jcmm.12080] [PMID: 23750663]

[192] Gao Y, Yechikov S, Vazquez AE, Chen D, Nie L. Distinct roles of molecular chaperones HSP90α and HSP90β in the biogenesis of KCNQ4 channels. PLoS One 2013; 8(2): e57282. [http://dx.doi.org/10.1371/journal.pone.0057282] [PMID: 23431407]

[193] Donnelier J, Braun JE. CSPα-chaperoning presynaptic proteins. Front Cell Neurosci 2014; 8: 116. [http://dx.doi.org/10.3389/fncel.2014.00116] [PMID: 24808827]

[194] Magga JM, Jarvis SE, Arnot MI, Zamponi GW, Braun JE. Cysteine string protein regulates G protein modulation of N-type calcium channels. Neuron 2000; 28(1): 195-204. [http://dx.doi.org/10.1016/S0896-6273(00)00096-9] [PMID: 11086994]

[195] Rosales-Hernandez A, Beck KE, Zhao X, Braun AP, Braun JE. RDJ2 (DNAJA2) chaperones neural G protein signaling pathways. Cell Stress Chaperones 2009; 14(1): 71-82. [http://dx.doi.org/10.1007/s12192-008-0056-y] [PMID: 18595009]

[196] Terada K, Yomogida K, Imai T, *et al.* A type I DnaJ homolog, DjA1, regulates androgen receptor signaling and spermatogenesis. EMBO J 2005; 24(3): 611-22. [http://dx.doi.org/10.1038/sj.emboj.7600549] [PMID: 15660130]

[197] Orthwein A, Patenaude AM, Affar B, Lamarre A, Young JC, Di Noia JM. Regulation of activation-induced deaminase stability and antibody gene diversification by Hsp90. J Exp Med 2010; 207(12): 2751-65.
[http://dx.doi.org/10.1084/jem.20101321] [PMID: 21041454]

[198] Orthwein A, Zahn A, Methot SP, *et al.* Optimal functional levels of activation-induced deaminase specifically require the Hsp40 DnaJa1. EMBO J 2012; 31(3): 679-91. [http://dx.doi.org/10.1038/emboj.2011.417] [PMID: 22085931]

[199] Terada K, Yomogida K, Imai T, *et al.* A type I DnaJ homolog, DjA1, regulates androgen receptor signaling and spermatogenesis. EMBO J 2005; 24(3): 611-22. [http://dx.doi.org/10.1038/sj.emboj.7600549] [PMID: 15660130]

[200] Kumar M, Mitra D. Heat shock protein 40 is necessary for human immunodeficiency virus-1 Nef-mediated enhancement of viral gene expression and replication. J Biol Chem 2005; 280(48): 40041-50.
[http://dx.doi.org/10.1074/jbc.M508904200] [PMID: 16179353]

[201] Kumar M, Rawat P, Khan SZ, *et al.* Reciprocal regulation of human immunodeficiency virus-1 gene expression and replication by heat shock proteins 40 and 70. J Mol Biol 2011; 410(5): 944-58. [http://dx.doi.org/10.1016/j.jmb.2011.04.005] [PMID: 21763498]

[202] Dziuba N, Ferguson MR, O'Brien WA, *et al.* Identification of cellular proteins required for replication of human immunodeficiency virus type 1. AIDS Res Hum Retroviruses 2012; 28(10): 1329-39. [http://dx.doi.org/10.1089/aid.2011.0358] [PMID: 22404213]

[203] Taguwa S, Maringer K, Li X, *et al.* Defining Hsp70 Subnetworks in Dengue Virus Replication Reveals Key Vulnerability in Flavivirus Infection. Cell 2015; 163(5): 1108-23. [http://dx.doi.org/10.1016/j.cell.2015.10.046] [PMID: 26582131]

[204] Murphy ME. The HSP70 family and cancer. Carcinogenesis 2013; 34(6): 1181-8. [http://dx.doi.org/10.1093/carcin/bgt111] [PMID: 23563090]

[205] Taldone T, Ochiana SO, Patel PD, Chiosis G. Selective targeting of the stress chaperome as a therapeutic strategy. Trends Pharmacol Sci 2014; 35(11): 592-603. [http://dx.doi.org/10.1016/j.tips.2014.09.001] [PMID: 25262919]

# SUBJECT INDEX

## A

Accessory proteins 14
Activation-induced cytidine deaminase 236
Acute myeloid leukemia (AML) 8, 153
Adenine nucleotide 91, 174, 178, 179
ADP-bound state 30, 96, 216, 217
Advanced protein folding capabilities 206
Affinity chromatography 181, 182
Alanine residues 37
Allosteric mechanism 134, 229
Alpha-helix 175
Alterations 8, 144
  genetic 144
  mutational 8
Amino acid residues 94, 176, 224
Amino acid sequences 76, 80, 152, 191, 219
  protein's 191
Analytical ultracentrifugation 47, 49, 51, 54, 56, 64, 84, 223
Androgen receptor (AR) 98, 99, 121, 142, 150, 227, 231, 235, 236
Anfinsen cage 179, 180
Angle, curvature 78, 79
Angle X-ray scattering 49, 50, 64, 94
Antennapedia homeodomain protein 152
Anti-aggregation functions 221
Anti-cancer strategies 14
Anti-parallel α-helices 79, 128, 135, 136
Antp-TPR and Hsp70-targeted peptide 154
Antp-TPR peptide 153, 154
Apical domain binding site 177
Apical domains 174, 175, 176, 177, 178, 179, 180, 182, 193, 195, 196, 198, 199, 200, 201, 203, 204, 206
Apical domains fold 196
Apical domains result 177
Apical domain tops 177
Apo-protein 93
Apoptosis 10, 13, 127, 144, 145, 152, 206, 232, 233
Apoptosis-inducing factor (AIF) 13
Arachidonic acid 127, 128, 130
Astrocytes 231
Asymmetric unit 122, 128
A-syn aggregation 230
Atomic force microscopy (AFM) 226
ATPase activity 12, 13, 14, 33, 49, 53, 57, 63, 73, 74, 86, 92, 96, 97, 98, 100, 104, 190, 197, 202, 203, 216
ATPase cycle 75, 87, 215, 217, 218
ATPase domain 14, 53, 91
ATPase function 91
ATP binding 14, 25, 31, 49, 55, 95, 96, 97, 100, 144, 193, 202, 217
ATP binding and hydrolysis 145, 180, 190, 195, 200, 201, 203
ATP binding domain 92
ATP-binding proteins 145
ATP-bound state 23, 30, 96, 216, 217, 218
ATP-dependent chaperones 87
ATP-dependent molecular chaperone 86
ATP hydrolysis 53, 54, 92, 96, 97, 100, 178, 179, 180, 191, 193, 195, 196, 200, 202, 217, 218, 219, 220, 226, 229
ATP hydrolysis rate and substrate binding 218
ATP-independent chaperone 221
ATP-induced state 97
ATP molecules 179
ATP re-binding 218
AU-rich element binding proteins 7
Autoinhibited conformation of PP5 129

## B

BAG domain contact 224
BAG domain of Bag1 225
BAG-domain proteins 223
BAG domains 224
  conserved 224
  five predicted 224
BAG domain sequences 224

**Mario D. Galigniana (Ed.)**

BAG proteins 225
Bardet-Biedl syndrome protein 197
BBSome ciliary transport protein 197
Beta-sheet structures 58
Binding, nucleotide 29, 54, 55, 96, 97
Binding affinities 133, 141, 145
Binding groove 112, 114, 139, 217
Binding interface 37, 39
Binding partners 81, 103, 104, 127, 138, 200
Binding pocket 14, 81, 100, 112, 113, 114, 116, 118, 122
Binding proteins 7, 108, 223, 236
  androgen 236
Binding site 35, 91, 99, 101, 102, 115, 116, 119, 124, 131, 132, 139, 152, 153, 174, 193, 199, 200, 203, 228, 229
  functional TPR 124
  putative 115
Binding substrate proteins 49
Binding substrates 62, 221, 226
Biogenesis, iron-sulfur 23
Bioinformatics tools 80
Box E3 ligase domains 140
Brain tissues 229
Buried surface area 103

## C

Cancer chaperones 15
The cancer genome atlas (TCGA) 7, 8
Cancer models 12, 13
Cancer stem cell (CSCs) 6
Catalytic phosphatase domain 128
Catalytic site 128, 129, 130, 134
CCT 196, 197, 198, 199, 200, 201
  bovine 196, 197
  open form of 198, 199, 200, 201
  substrate binding sites on 197, 200
CCT binding sites on substrates 200
CCT-dependent protein folding 195, 203
CCT domain structure 195
CCT equatorial domains 200
CCT folding cavity 198, 202, 203, 204, 205, 206
CCT folds substrates 203
CDK inhibitory protein 10
Cell lung cancer 147, 149, 150
Cellular homeostasis 73, 74, 75, 98
Cellular processes 23, 42, 74, 76, 87
Cellular stress 143
Chaperone activity 58, 124, 150, 151
  autonomous 58
Chaperone activity specification 62
Chaperone classes 192
Chaperone cycle 23, 24, 25, 31, 41, 42, 75, 144
Chaperone cycle in SHRs 99
Chaperone-dependent steps 237
Chaperone function 29, 75, 86, 135, 144, 193, 225
  dual 86
Chaperone interaction 233
Chaperone interactome screen 204
Chaperone machine 87
Chaperone machinery 74, 98, 152
  90-based 74, 98
Chaperone molecules 25
Chaperone network 129, 193, 204, 215
  cellular 215
Chaperones 2, 5, 32, 38, 47, 50, 99, 140, 144, 215, 230, 233
  90-associated 99
  addicted to 2
  cytosolic 233
  dimeric 140
  downstream 5
  eukaryotic 47
  foldase 50
  high affinity 144
  holder 50
  homologous 38
  mitochondrial 32
  nuclear 215
  primary 5
  yeast AAA+ family 230
Chaperones conduct 140
Chaperones' contributions 192
Chaperones help nascent proteins bypass 192
Chaperones interact 131
Chaperone site 125

Chaperone systems 181, 214
molecular 53
Chaperonin architecture 194
Chaperonin field 201
Chaperoning cycle 119
Chaperonins 49, 174, 176, 178, 190, 191, 193, 194, 195, 196
best-studied Escherichia coli 174
eukaryotic 194
molecular 178
Chaperonin subunits 192
identical 193
Characterized single-domain cyclophilins 107
Chen, domain prediction 89
CHIP and C-terminal peptides of Hsp 140
CHIP dimer 136, 137, 139, 140, 141
CHIP homodimer 138
CHIP-mediated ubiquitination 234, 235
CHIP monomer 140, 141
CHIP protomers 136, 137
CHIP TPR Domain 139
Chronic lymphocytic leukemia 148, 149
Circular dichroism (CD) 84, 126, 133, 134
Cisplatin 150, 151
Class II chaperonins fold proteins 194
Class-switch recombination (CSR) 236
Client protein activation 105
Client protein folding 141
Client protein-Hsp 90-CHIP 141
Client protein maturation 143
Client proteins interactions 96
Cluster biogenesis 32
Co-chaperone activity 61
Co-chaperone atomic structure 205
Co-chaperone binding 205
Cochaperone binding, inhibiting 152
Cochaperone complexes 98
Cochaperone regulation 92
Cochaperone requirement 87
Co-chaperones, abundant 232
Cochaperones, first 100
Co-chaperones 53, 236
holder 53
mitochondrial matrix 236
Cochaperones DNAJA1 236
Co-chaperones DNAJA1 226, 234
abundant DNAJ 226
Cochaperones function 98
Cochaperones Hsp 151
Cochaperones Hsp90-interaction 75
Cochaperones modulate progression 98
Co-chaperones parallel 206
Cochaperones regulation 98
Cochaperone Sti1 99
Cochaperones work 98
Cochaperone TPR-Hsp 142
Complexed GWLYEIS peptide substrate 221
Computational methods 24, 25, 142
Conformational flexibility 195
Connecting loop (CL) 78, 90, 91
Consensus superhelical structure 131, 132
Conserved residues 84, 122, 131, 225
Conserved TPR-clamp domain 100
Core TPR domain 105, 123
Co-translational folding 4, 6
Coumarin antibiotics 150
Coupled folding-binding mechanism 134, 135
Cryo-electron microscopy 177, 196
Crystallographic structures 91, 117, 122, 129, 221, 223, 224, 225
Crystal packing 123
Crystal structures 35, 42, 79, 85, 90, 95, 102, 105, 107, 110, 111, 116, 117, 123, 127, 128, 135, 136, 177, 200, 205
C-terminal α-helix 101, 136, 137
C-terminal decapeptide 136, 138
C-terminal domain (CTD) 58, 90, 91, 92, 93, 95, 97, 102, 105, 116, 118, 125, 126, 127, 133, 150, 152, 201, 206, 216, 221, 223
C-terminal domains of Types 221
C-terminal fragments 127, 221, 223
C-terminal inhibitors 145, 154
C-terminal MEEVD sequence 108, 110
C-terminal monomers 60, 61
C-terminal mutant 59
C-terminal pentapeptide 132, 139
C-terminal peptide 104, 134, 139, 140, 153
C-terminal peptide-binding domain 191
C-terminal peptide of Hsp 103, 116
C-terminal phosphatase catalytic domain 127
C-terminal phosphatase domain 126, 127, 128

C-terminal regions 57, 58, 113, 123, 151, 152, 153, 155, 219, 232
C-terminal regions of partner proteins 106
C-terminal residues 106, 141, 200
  adjacent 219
C-terminal residues by TPR domains 134
C-terminal residues EKAAK fitting 110
C-terminal subdomain 129
C-terminal TPR2 domain 102
C-terminal TPR domain 107
C-terminal U-box 136, 137
C-terminal YbgF TPR domain 85
CTPR3 protein 83
CTPRa2 and CTPRa3 proteins 82
CTPRa3 proteins 82
CTPR family 82, 83
CTPR proteins 82
Cyclophilin domain conformations 107
Cyclophilins 106, 107, 110
CyP40 protein 132
Cyp40 TPR domain surface 111
Cytochrome 181, 233
Cytoplasm 2, 135, 231
Cytoskeletal proteins actin 197
Cytosolic Hsp 73, 204, 223
Cytotoxic activity 153, 154
Cytotoxic activity of Antp-TPR 153, 154

## D

Defective chaperone function 143
Denaturation 126, 181
  irreversible 181
  thermal 126
Denatured proteins 87, 176, 178, 180, 181, 197
  charged 176, 181
De novo protein folding 191
Depletion-mediated senescence 10
Differential scanning calorimetry (DSC) 82
Diffusion coefficient 52, 54, 62
Dimensional Ising models 82
Dimerization capability 93, 94
Dimerization domain (DD) 91, 92, 103, 150
Disassemble protein aggregates 214
Disordered Gly 57, 58
Disordered linker region 127, 128
Disrupting client protein maturation 145
Disrupting TPR domain contacts 130
DNA- and RNA-binding proteins 87
DNAJ substrate release 226
DnaK chaperone system on tandem 226
DnaK folding function 222
Docking 24, 30, 35, 36, 40, 41, 42, 97, 205
  global 35, 36, 40
  local 35, 36
Docking space 35
Domain architecture 90
Domain arrangement 58, 194
  apical 194
Domain ATPases 90
Domain binding 98, 220
Domain boundaries 94
Domain composition 224
Domain contact 226
Domain-containing protein structures 81
Domain dissociates 31
Domain evolution 118
Domain interaction 216
Domain organization 50, 91
Domains 60, 62, 63, 64, 74, 77, 80, 90, 91, 92, 103, 104, 107, 108, 109, 110, 127, 128, 129, 130, 131, 132, 150, 175, 178, 193, 195, 200, 218, 223, 225, 229
  α-helical 109
  catalytic 128, 129, 132
  central 60, 62, 63, 218, 223
  coiled-coil 200
  cyclophilin 108, 109, 110
  dimerization 91, 92, 103, 150
  distinct 127
  helical 80
  homologous 225
  intermediate 175, 178, 193, 195
  kDa 107
  multiple 64
  phosphatase 128, 129, 130, 131, 132
  polypeptide-binding 229
  respective 104
  structural 74, 90

three-TPR 77, 80
Domains arrangement of Types 58
Domains interact 94
Domains of DNAJs 219
Domains playing 174
Domain structure 112, 216
  central 216
  typical FKBP 112
Domain swapping 109
Down-regulation 5, 236, 237
Downstream 116
Dynamic light scattering (DLS) 51, 52, 54, 56, 62

## E

EEVD domain of HSP70 14
Electrostatic interactions 102, 176, 182, 224
Encapsulation 195, 203
Endoplasmic reticulum 88, 215, 234
Entropic contributions 124, 125
Epilgallocatechin-3-gallate 150, 151
ER-associated degradation 234
Eukaryotic cells 73, 74, 87
Eukaryotic Hsp 90, 91, 92, 93, 98, 224
Eukaryotic protein folding 192
Extracellular isoform functions 89

## F

Family, heat shock protein 175
Family members, functional HSP70 2
Family of chaperones in cancer 1, 15
Farnesylation 222
F-box proteins 79
FKBP52, structure of 116, 119, 120, 121
FKBP52 cochaperone, free 122
FKBP52 function 114, 123
FKBP52 TPR domain 119
FKBP domains 112, 116, 119
  first 112
  studied 112
FKBP domains of FKBP52 116
FKBP proteins 118
FK domain duplication event 114
Fluorescence resonance energy transfer (FRET) 99
Folding and stability of TPR proteins 81
Folding substrate Gβ1 205
Force field, coarse-grained UNRES 36, 37, 41
Functional antagonism 117, 120
Functional differences 93, 94
Functional domains 1, 152
Functional interactions 4, 80
Functional similarity 116, 122

## G

Gastrointestinal stromal tumor 147, 148, 149
Gβ binding domain 206
Gβ folding 203, 205
Genes encoding DnaJ proteins 218
Genetic instability 144
Glioblastoma cells 153
Globular proteins 81, 196
  encapsulate 196
Glucocorticoid receptor (GR) 99, 100, 107, 120, 121, 236
Glu residues 231
Glycine 58, 218, 220
Glycolysis 233
GroEL 87, 174, 175, 176, 177, 178, 179, 180, 181, 182, 193, 194, 195, 198, 200, 201, 203
  absence of 176, 201
  apical domains of 177, 178
  asymmetrical complex 179
  native 182
GroEL ability 175, 181
GroEL affinity 176
GroEL affinity for substrate polypeptides 176, 180, 182
GroEL affinity for substrate proteins 182
GroEL ATPase activity 178
GroEL binding 175, 180
GroEL Functioning 178
GroEL functioning, models of 179, 180
GroEL-GroES complexes 180
GroELGroES machinery 180
GroEL-interacting surface of subtilisin 176

GroEL interaction 175, 176, 177, 179, 181
GroEL ligands 174, 176, 179, 180, 182
GroEL-mediated folding 201
GroEL particle 175, 177, 178
GroEL-polypeptide 176, 177
GroEL polypeptide binding site 177
GroEL residues 177
GroEL-substrate protein 179
GroEL substrate rhodanase 201
GroEL substrates 198
GroEL subunits 175, 176, 182
GroEL surface 174
GroES 175, 176, 178, 179, 180, 181, 182, 193, 196
  co-chaperonin 175
  dissociation of 179
GroES absence 179, 180
GroES binding 178, 193
GroESGroEL 180
Groove, shallow 198, 199
Grp170 proteins 223
GrpE C-terminal 57

## H

Hairpin 136, 137, 138, 224
  anti-parallel 137
Heat capacity 176
Heat shock 5, 6, 191, 197, 214, 225, 226, 228, 232
Heat shock factors (HSFs) 143, 150
Heat-shock proteins (HSPs) 2, 7, 23, 135, 143, 191, 215
Heat shock protein seventy 1
Heat shock response 143, 152
HeLa Cells 229, 231, 237
Helical extensions 195, 196
Helical hairpins 137, 138
Helical pair 136, 137
Helices 32, 36, 91, 108, 114, 137, 219
Helix 31, 78, 79, 85, 101, 108, 122, 123, 136, 137, 198, 199, 200, 201
  capping 79
Helix-breaking residues 77
HERG trafficking 235
Heterocomplexes 75, 99, 119, 228
Heterodimer 37, 94
High-resolution structures 50, 55, 64, 195
HIV Gag protein 199
HIV replication 236, 237
Homobifunctional lysine cross-linker 197
Homodimer asymmetry 139
Homologous TPR2A domain 131
Homology models of yeast Ssq1 chaperone 40
Hop 74, 79, 98, 99, 100, 101, 102, 104, 105, 106, 108, 113, 116, 119, 122, 131, 132, 134, 139, 140, 141, 152, 153
  cochaperone 140
  co-chaperone 235
Hop co-chaperone 2
Hop domains 106
Hop protein 153
HOP TPR1 domain 103
HOP TPR2A domain representation 101
Hop-TPR domain 130
Hop TPR-domain complexes 139
Hsc70 chaperones system 226
Hsc70-interacting protein 135
Hsp 5, 6, 10, 11, 13, 14, 48, 49, 181, 215
Hsp
  bacterial 25, 30, 31, 87
  domain contacts 218
  inducible protein 1
  inducible proteins 2
  molecular chaperone 129
HSP70 1, 2, 12, 13, 14, 215, 218
  interaction of 14
  pseudogenes of 3
HSP70-accessory protein interactions 15
HSP70-accessory protein relationships 14
HSP70 and accessory proteins 14
HSP70 ATPase domain 13
Hsp70B 2, 3, 4, 5, 9
Hsp70B proteins 4
Hsp 70 chaperones 23, 32, 39, 74, 155, 181, 235
Hsp70-CHIP 141
Hsp70-dependent client proteins 141
HSP70 family 1, 2, 3, 7, 8
HSP70 family members 2, 9, 11, 12

HSP70 family of molecular chaperones 1, 7
HSP70 family proteins 2, 3
HSP70 genes 4
HSP70 homologs 8
HSP70 inhibitors 12, 13, 14
HSP70 inhibitory molecules 12
Hsp70-interacting protein 74, 98
HSP70 Nucleotide-Binding Domain 13
HSP70 paralogs 12
Hsp70 protein 13, 154
  mitochondrial 13
HSP70 protein interaction 5
HSP70-Protein Interactions 14
HSP70 PROTEINS 2, 5, 14
HSP70 Substrate-Binding Domain 12
Hsp70-targeted peptide 153, 154
Hsp90-binding TPR domains 113
Hsp90 chaperoning cycle 107
Hsp 90-client protein 145
Hsp90 client proteins 142
Hsp 90-cochaperonas interaction 75
Hsp90 co-chaperone Aha1 236
Hsp 90 cochaperones 73
Hsp90-dependent client proteins 141
Hsp90 inhibitors 73, 142, 150
HSP90 protein(s) 75, 86
Hsp90 provokes client protein degradation 142
Hsp 90-TPR domain protein interaction 75
HSP90-TPR protein interaction(s) 73, 98
Hsp 90-TPR proteins interaction 98
Hsp-organizing protein 74, 98
Human FKBP51 domains 112
Human FKBP52 C-terminal domain 118
Human FKBP52 domains 118
Huntington disease (HD) 231
Hu protein family 7
HuR, mRNA stabilizing protein 7
Hydrodynamic parameters 51, 52, 54
Hydrogen bonds 117, 122, 131
Hydrolysis 97, 145, 179, 180, 190, 194, 195, 200, 201, 203
Hydrolyze ATP 96, 190, 192
Hydrophobic clusters 174, 175, 181
Hydrophobic groups, exposed 175, 176
Hydrophobic interactions 57, 91, 102, 117, 132, 139, 175, 176, 193, 198
Hydrophobic linker 216, 217
  short 24, 25
Hydrophobic pockets 80, 221
Hydrophobic residues 176, 194, 198, 200
HydroPro software 51, 60, 62
Hypoxia 9, 11, 88, 89, 143

# I

Identification of TPR motif of TPRPRED software 112, 113
Identified key residues 84, 116
Immature ΔF508 234
Immature forms 234
Immunophilins 104, 106, 107, 117, 123, 152
Inclusion bodies 181
Inclusions, nuclear 231
Inhibition of hsp 143, 144, 154
Inhibition of Hsp90 provokes client protein degradation 142
Inhibition of TPR protein interactions 14
Interdomain communication 31
Inter-domain junctions 96
Intermediate domain transfers 193
Internal subdomain flexibility 217
Iron Sulfur Cluster (ISC) 33
Isoforms of hsp 73, 88, 93, 96
Isothermal titration calorimetry (ITC) 102, 103, 124, 133, 139, 145
Isu1 homologue proteins 34
Isu1 protein 35

# J

And Jac1 proteins, Isu1 33
J-domain and C-terminal region of DnaJ proteins 57, 58

# K

KDa heat shock protein 48, 53
KO mammary tumors 5

## L


Large immunophilins 106, 107, 108, 109
Large-scale conformational changes 175, 178, 180
Ligand binding 75, 78, 80, 87, 139
Ligand interactions 80
Ligands in GroEL Functioning 178
Limb-Girdle muscular dystrophy (LGMD) 221
Linker protein 100
Linker segment 32
Location of substrate polypeptides 177
Loops 78, 91, 92, 93, 96, 97, 108, 110, 112, 114, 118, 119, 121, 122, 128, 137, 198, 199, 200, 201, 217, 219
  catalytic 91, 96, 97
  crossing 92, 93
  disordered 137
  divergent 108, 110
  proximal 198, 199
  sensor 199, 200
  short 82, 198
Luciferase 220, 222, 225, 228
  denatured 222, 225, 228
Luciferase refolding 228


## M


Malignant cell 1, 2, 15
Malignant phenotypes 143, 144
Mass spectrometry 177, 196, 230, 236
MEEVD peptide 110, 122
MEEVD region 152, 154
Melanoma, cutaneous 7, 8
Members of PPP family 127, 128
Membrane protein λ-holin 178
Messenger ribonucleoproteins 6
Metastasis 1, 9, 10, 11, 143, 144, 146
Metastatic properties 10, 11
Mitochondrial import 233
Mitochondrial outer membrane protein Fis1 84
Mitochondrial preproteins 233
Molecular chaperone families 48
Molecular chaperones 1, 7, 23, 42, 47, 48, 49, 51, 73, 86, 87, 89, 99, 136, 142, 174, 179, 180, 190, 191, 214
Molecular chaperones Hsp 140
Molecular dynamics 23, 24, 26, 41, 42
  coarse-grained 23, 24, 42
Molecular dynamics simulations (MDS) 24, 25, 30, 34, 83
Molecular-modeling tools 41, 42
Monomeric consensus protein TPR 85
Monomers of DnaJA 59
MRNA over-expression 8, 9
Multi-chaperon system 154
Multiple DnaJ proteins 237
Multiple sequence alignment (MSA) 106, 115
Multiprotein complexes 74, 83, 86, 112, 151
Multi-protein complexes 84, 206
Multivalent binding of substrate proteins 177
MutL mismatch repair proteins 91
Mutual-type sub-domain vectors 30


## N


Nascent proteins 24, 48, 50, 53, 190, 191, 192
  folding of 24, 48, 53
  unfolded 50
Natural GroEL substrate targets 176
NBD domains 26, 30, 219
NBD-II domains return 32
NBD structure 216, 217
NBD subdomain 224
NDB domain 40
NEF in releasing substrate 226
Neurodegeneration 214, 231
Neurological diseases 230
Neurons 230, 231, 235
NMR structures of Hsp40 domains 61
NMR techniques 131, 132, 155
Non-native proteins delivery 98
Novel TPR protein design 85
NTD dimerization 96, 97
NTD lid region 95
NTD-MD orientation 95
N-terminal α-helix 120, 122

N-terminal ATPase domain 2
N-terminal coiled-coil domain 85
N-terminal domain 88, 90, 91, 94, 145
N-terminal domain (NTD) 88, 90, 91, 92, 93, 94, 95, 96, 97, 145, 147, 150, 206
N-terminal domain in position 206
N-terminal inhibitors 145, 152, 153
N-terminal monomers 60
N-terminal PPIase domain 116
N-terminal region 122, 135, 138, 145, 149, 206
N-terminal TPR domain 101, 127, 133, 134
N-terminal TPR domain of Hop 102
N-terminal TPR domain of PP5 blocks 130
Nuclear exchange factors (NEFs) 14, 24, 218, 223, 224, 225, 226, 227, 228
Nuclear magnetic resonance (NMR) 27, 47, 49, 100, 119, 131, 134, 198, 200, 219, 220, 224
Nucleotide-binding domain (NBD) 2, 23, 24, 25, 26, 27, 28, 29, 30, 31, 32, 42, 53, 55, 56, 57, 91, 204, 205, 216, 217, 220, 223, 224, 225, 226 234
Nucleotide-binding pocket 96, 97
Nucleotide exchange factors 24, 47, 53, 54, 55, 218, 223
  aid of 53, 54
Nucleotide-free conformation 190, 193

## O

Oligomerization 83, 84, 85, 140, 193
Oncogenic co-chaperone Bag3 10
Oncogenic proteins 2, 144
  misfolded 144
Oncoprotein 151
Organisms 1, 2, 34, 57, 74, 76, 78, 93, 96, 105, 143, 218
  cellular 1, 2
  living 74, 76, 78, 143
Organization, spatial 90, 91
Orientation 30, 36, 92, 95, 96, 114, 117, 119, 120, 122, 128, 134, 196, 197, 204, 223
  relative 92, 95, 114, 117, 128
  relative domain 120

## P

Parkinson disease (PD) 230
Pepsin 178, 181
Peptide binding domain 2
Peptide-binding site 104, 105, 132
Peptide residue Val 102
Peptides 3, 73, 74, 80, 98, 101, 102, 103, 104, 122, 130, 131, 132, 133, 134, 135, 139, 142, 153, 155, 177, 199, 221
  hybrid 153
  residue 139
  synthesized 73, 74, 98, 155
Peptide sequences, short 13, 217
Peripheral blood mononuclear cells (PBMCs) 153
Permitted protein folding 1
Peroxisomal protein transport 76
Perturbation of HSP70-Protein Interactions 14
Phage capsid protein gp23 177
PhLP1-Gβ-CCT structures 205, 206
Phosducin-like protein 190, 205
Phosphatase activity 126, 128, 133
Phosphatases 10, 130
Platforms, motif-generated protein-protein interaction 86
Polar bonds 139
Polypeptide binding 177, 217
Polypeptide chain 24, 137, 179
Polypeptides 5, 6, 79, 137, 174, 175, 176, 177, 178, 180, 181, 206, 214, 217, 226, 229
  nascent 5, 190, 194, 197, 215
  tailless 194
  target 175, 176
PolyQ aggregates 232
Positron emission tomography (PET) 149
PP5 and C-terminal EEVD residues 130
PP5 by disrupting TPR domain contacts 130
PP5 proteins 153
PP5 structures 129, 132
PP5-TPR domain 130, 131
PP5-TPR domain interactions 130
PPP family 126, 127, 128
Principal component analysis (PCA) 27, 42
Prion protein RNQ1 propagation 221

Probable structure 37, 38
Proteasomal degradation 232, 235
Proteasome 136, 144, 192, 224, 227, 231, 232
Protein acts 87
Protein aggregates 24, 228
  resolubilize 228
Protein aggregation 86, 214
Protein-binding assays 155
Protein binding specificity 76
Protein bioinformatics 106
Protein concentration 51, 53, 54, 62
Protein conformations 122
Protein Data Bank (PDB) 25, 26, 35, 61, 84, 117
Protein degradation 74, 87, 144, 152, 214, 215
  client 144
Protein degradation machinery 215
Protein denaturation 143
Protein domains 75, 191
  characteristic 75
  large 191
Protein dynamics 25, 27, 30
Protein expression 230
Protein failures 193
Protein folding 23, 73
  aberrant 73
  correct 23
Protein folding and modeling 201
Protein folding and oligomerization 193
Protein folding by type II chaperonins 190
Protein folding cavities 190
Protein-folding kinetic 85
Protein function 74, 126
Protein homeostasis 48, 154, 190, 214
Protein Hsc70 2
Protein interaction ability 114
Protein interaction module 84
Protein kinase inhibition 76
Protein kinases 10, 86, 87, 98
Protein levels 3, 229, 235
Protein ligand 80
Protein measurement 51
Protein mechanism 73
Protein misfolding 49, 214
Protein misfolding diseases 214
Protein oligomerization 85
Protein origin 142
Protein phosphatases 77, 113, 126, 129, 197
Protein-protein associations 76, 110
Protein-protein complexes 74, 80
Protein-protein interaction 174
Protein-protein interaction databases scanning 81
Protein-protein interactions 73, 74, 75, 76, 78, 102, 134
  fostered 75
Protein-protein recognition events 109
Protein provokes loss 75
Protein quality control 98, 136
Protein recognition 109
Protein-refolding 143
Protein region 57
Protein restrains 85
Proteins 1, 2, 3, 4, 5, 9, 10, 11, 12, 13, 14, 15, 23, 24, 25, 26, 27, 29, 30, 32, 33, 35, 37, 47, 48, 49, 50, 51, 52, 53, 55, 56, 57, 58, 59, 60, 61, 62, 64, 73, 74, 75, 76, 78, 79, 80, 81, 82, 83, 84, 85, 86, 87, 88, 89, 90, 91, 92, 93, 98, 99, 100, 102, 104, 106, 107, 110, 116, 125, 126, 127, 129, 133, 136, 141, 142, 143, 145, 148, 150, 152, 153, 174, 175, 178, 179, 181, 190, 191, 192, 193, 194, 197, 198, 200, 203, 204, 205, 206, 214, 215, 217, 220, 221, 223, 225, 227, 230, 231, 232, 233, 235, 236, 237
  actin polymerizing 11
  adaptor 2
  aggregation-prone 197
  α-helix-containing 125
  asymmetric 55
  β-propeller 200, 203
  cellular 174, 217
  chaperone 49
  chaperone key 1, 15
  chimeric 60, 61, 62
  co-chaperone 214
  concentrated 51
  consensus sequences 83
  conserved 86, 215
  constructed 85
  cytoskeletal 87

cytosolic 197
domain-containing 76, 84
encountering elongating 5
eukaryotic 191, 198
extra Hsp90-associated 99
folded 49, 57, 88, 191
heat-denaturated 87
heat-shock 23, 191
homologous 87
iron-sulfur 23, 24
iron-sulfur-binding 25
macrophage infectivity potentiator 116
maltose-binding 178
mitochondrial 233
modular 100
multiple client 142
mutated 192
native folded state 225
non-native 74
oncogenic client 145
potential interacting 81
protein-binding 142
ribosomal 87
transduction pathway 49
tumor suppressor 197, 206
Protein scaffolds 73, 74
Protein sequences 88, 135, 155
conserved 135
Proteins GRPEL1 223
Protein signaling 227
Protein signaling system 206
Proteins interact 33
Protein sites 78
Protein solubility 214
Protein stability 11
Protein stability Pikrl 126
Protein structure 80, 81, 93, 148
Protein structures, organized 26
Protein substrates 48, 87, 176, 190, 193
charged 176
distribute 48
unfolded 193
Protein surface 42
Protein survivin 10
Protein targets 80, 175
Protein thermo-stability 85
Protein translocation 74
Protein transport 87, 88
Protein traverses 191
Protein tyrosine phosphatases SHP1 134
Proteinuria 146
Proteostasis 48, 191, 193, 214
Protomers 75, 95, 96, 136, 137, 138, 139
Pseudogenes 2, 3, 4
Putative functional genes 89

# Q

Quaternary conformations of types 59, 60
Quaternary structure of GrpEs 56

# R

Ragged-line borders 31, 32
Rapamycin 112, 120
Reactive oxygen species (ROS) 88
Region 199, 200, 201
conserved 13, 61, 62, 88, 106, 117, 145, 195, 199, 200, 201, 205, 219, 222
flexible 61, 117
functional 13, 145
homodimerization 219, 222
proximal loop 199, 200, 201
substrate-binding 195, 205
Regulatory roles for Hsp 3, 5, 7, 9, 11, 13, 15
Release 23, 25, 33, 100, 134, 180, 190, 191, 193, 198, 201, 202, 203, 205, 206, 217, 218, 225, 226, 228, 229, 233
sequential 202, 203
Release substrate 226
Releasing substrate 226
Residue composition 81
Residues 96, 128, 131, 220
acidic 128, 131, 220
basic 128, 131
catalytic 96, 128
hydrophilic 200
Residues Asp 136, 137
Residues Lys 110, 131
Residues of Isu1 37, 39
Residues peptide 139

Residues types 78
Residues upstream 102, 116
Ribonuclease 181
Ribonucleoprotein 5
Ribosomal protein S5 domain 91
Ribosome-associated complex (RAC) 5
Ribosomes 5, 6, 191, 215
Right-handed α-helical conformation 79
RNA binding protein 11
RNA-binding proteins 87
Role of ligands in GroEL functioning 178

## S

Scattering intensity 51
Sedimentation coefficient 51, 52, 54, 56, 57
Sedimentation equilibrium 51, 52
Sedimentation velocity (SV) 51
Sequence alignment 110, 111, 155
Sequence identity 112, 113, 128, 194
Sequential release mechanism 190, 203
Sequential substrate-folding model 203
Shock protein, seventy kilodalton Heat 1
SHR ligand binding domain 99
Signaling proteins 110, 142, 147, 191
  chimeric 142
  metastable 142
Sis1, part of 223
SKCM 7, 8
Squirrel monkey FKBP51 domains 113
Stability and biological functions 83
Stabilization center 55, 56
Stabilization function 142
Stabilize oncogenic proteins 143, 144
Stable protein folding 86
Steroid hormone receptor (SHRs) 73, 75, 87, 98, 99, 100, 130, 142, 152
Sti1 domains 105
Sti1-TPR2B domain 104
Sti1 TPR domains 104
Sti1 TPR domains and peptides 104
Stomach adenocarcinoma 7, 8
Stress granules 6
Strong ATPase side 203
Structural analysis 84, 102, 109, 110, 118, 122, 130, 141, 198, 203
Structural differences 92, 106, 120, 121
Structured domain 55, 56
Structure of Isu1 33
Subdomain rotation 28, 30
Substrate affinity 229
Substrate aggregation 215
Substrate binding 12, 23, 25, 53, 63, 117, 197, 198, 199, 200, 202, 217, 218, 220, 221, 222
Substrate binding by DNAJs 221
Substrate-Binding Domain (SBD) 2, 12, 23, 24, 25, 26, 30, 31, 42, 53, 103, 216, 217, 220, 224, 225, 226
Substrate-binding site 199, 201
Substrate binding sites 53, 199, 200, 222, 228
Substrate-binding surface 193, 194
Substrate-bound complexes 203
Substrate-bound conformation 25
Substrate conformations 226
  unfolded 226
Substrate de-stabilization 12
Substrate entrance 129, 130
Substrate folding 201, 202, 203, 205, 218
Substrate polypeptide folding 180
Substrate polypeptides 174, 175, 176, 177, 180, 181, 182, 215, 218
Substrate protein dissociate 180
Substrate protein folding 179, 180, 193
Substrate protein folds 196
Substrate protein molecules 177
Substrate protein release 13
Substrate proteins 79, 130, 177, 178, 179, 180, 181, 182, 190, 191, 195, 196
  charged 178
  encapsulated 195
Substrate proteins fold 182
Substrate recognition 190, 196, 197, 198, 206
Substrate release 31, 190, 222, 228, 229
Substrate remodeling 87
Substrates 25, 154, 193, 195 202, 207, 216, 218
  bound 25, 207, 216, 218

folded 154, 193, 195, 202
phosphoprotein 129
polypeptide 217
position 204
unfold 226
Substrate sequence 229
Substrate surfaces 199
Substrate transfer 204, 218, 222, 229
Subtilisin 175, 176, 178
Subunit orientations 196, 197
Subunits 78, 79, 127, 174, 175, 197, 200
identical 174, 175
regulatory 127, 197, 200
Superhelix structure 78, 79
Suppression 9, 10
Surface plasmon resonance (SPR) 140
System proteins 230

## T

Targeting Hsp 2, 10, 12, 144
Tetragonal crystal forms 109
Tetragonal forms 107, 109
Thermosensor 55, 56
Thermosome 194, 195, 196, 198
Thr-protein phosphatase PP5 79
TIM barrel protein DapA 201
TPR-catalytic domain interface 129
TPR client proteins-Hsp90 interactions 86
TPR cochaperones 100, 123, 139
TPR-cochaperones 100
TPR cochaperones group 100
TPR consensus sequence 77, 84
TPR-containing cochaperones 124, 127, 134, 152, 154, 155
TPR-containing domains 85
TPR-containing protein distribution 78
TPR-containing protein-peptide interactions 81
TPR-containing protein rapsyn 84
TPR-containing proteins 74, 76, 78, 80, 81, 83, 84, 85, 86, 98, 135
predicted 81
TPR-containing proteins interact 124
TPR-containing proteins structure, unbound 81
TPR Domain 78, 113, 119, 128
TPR domain and αJ-helix 130
TPR domain and C-terminal αJ helix 134
TPR domain and residues upstream of EEVD 102
TPR domain associates 129
TPR domain blocks 133
TPR-domain cochaperone 99
TPR domain cochaperones 106
TPR domain co-chaperone SGT 235
TPR domain function 101
TPR domain groove 134
TPR domain-Hsp 142
TPR domain in CHIP 139
TPR domain in PP5 133
TPR domain interaction region 81
TPR domain motifs 112, 113
TPR domain of Hop 139
TPR domain of PP5 79, 101, 128
TPR domain oligomerization 84
TPR domain orientation 115
TPR domain protein architecture 81
TPR domain proteins 2, 80
TPR-domain proteins fit 78
TPR domains 73, 74, 75, 77, 79, 80, 81, 83, 84, 92, 100, 101, 102, 104, 105, 106, 107, 108, 109, 110, 112, 113, 114, 115, 116, 118, 119, 121, 122, 123, 126, 127, 128, 129, 130, 131, 132, 133, 134, 135, 136, 138, 139, 140, 142, 152, 233, 235
90-binding 130
contiguous 101
dysfunctional 77
engineered 84
first 109
isolated 133, 134
modeled 119
monoclinic CyP40 110
oligomeric 84
regulatory 126
regulatory N-terminal 127
single 139

three-unit 111
TPR domains and peptidyl-prolyl isomerase 106
TPR domains display amino acid 78
TPR domain self-association 84
TPR domain shaping 114
TPR domain stabilize 133
TPR domain structures 108, 123
TPR-interacting proteins 80
TPR-ligand protein structures 80
TPR motifs 74, 75, 76, 77, 78, 80, 82, 85, 100, 101, 102, 103, 105, 106, 108, 109, 110, 112, 113, 118, 128
TPR-pentapeptide 131, 132
TPR peptide 152, 153
TPRPRED software 112, 113, 118
TPR protein characteristics 75
TPR protein interactions 14
TPR protein ligand binding 80
TPR proteins 73, 74, 76, 77, 78, 79, 80, 81, 82, 83, 84, 85, 86, 98, 107, 133, 136, 153
consensus 84
design novel 85
glutamine-rich 74, 98
TPR proteins display 84
TPR proteins-Hsp 86
TPR proteins in life 84
TPR protein structure 73
TPR Protein Structures 79
TPR regions 125, 152
Translational diffusion coefficient 51, 52
TRPs proteins 74
Tumor cells 1, 9, 89, 144, 155
Tumorigenesis 1, 11, 15
Tumors 6, 7, 8, 9, 11, 13, 144, 146

## U

Ubiquitination 136, 140, 141, 234
Ubiquitin-interacting motifs (UIMs) 232
Ubiquitin ligases 136, 139
U-box domain 136, 137, 138

## W

Weak ATPase side 203, 206

## X

X-ray crystallography 47, 49, 84, 93, 119, 148, 174, 175, 195, 196
X-ray structures 110, 119, 148, 219

## Y

Yeast CCT 196, 197
Yeast chaperones 32
Yeast Sis1 58, 61, 62, 221
Yeast Ssq1 chaperone 40

## Z

Zinc finger-like region (ZFLR) 58, 63

Made in United States
Orlando, FL
04 November 2026

90200447R10167